Oil and Gas: Drilling and Refining Technology

Oil and Gas: Drilling and Refining Technology

Edited by Andy Margo

SYRAWOOD
PUBLISHING HOUSE

New York

Published by Syrawood Publishing House,
750 Third Avenue, 9th Floor,
New York, NY 10017, USA
www.syrawoodpublishinghouse.com

Oil and Gas: Drilling and Refining Technology
Edited by Andy Margo

© 2017 Syrawood Publishing House

International Standard Book Number: 978-1-68286-484-5 (Hardback)

Cataloging-in-Publication Data

Oil and gas : drilling and refining technology / edited by Andy Margo.
    p. cm.
Includes bibliographical references and index.
ISBN 978-1-68286-484-5
1. Petroleum. 2. Natural gas. 3. Petroleum engineering. 4. Gas engineering. 5. Oil well drilling.
6. Gas well drilling. 7. Mineral oils--Refining. I. Margo, Andy.
TN870 .O35 2017
665.5--dc23

Printed in the United States of America.

TABLE OF CONTENTS

PREFACE

This book on oil and gas technologies discusses topics related to oil and gas drilling, reservoir engineering and the use of hydraulic mechanisms for crude oil extraction. Various methods that optimize oil extraction and oil recovery are highlighted in the chapters. Profitable and effective manufacture of technology, extracting larger amounts of oil per engineering cycle and creating sustainable and eco-friendly practices are some current concerns in oil and gas engineering. This book elucidates new techniques and their applications in a multidisciplinary approach. For someone with an interest and eye for detail, this book covers the most significant topics in the field of petroleum science and engineering. With its detailed analyses and data, it will prove immensely beneficial to professionals and students involved in this area at various levels.

Significant researches are present in this book. Intensive efforts have been employed by authors to make this book an outstanding discourse. This book contains the enlightening chapters which have been written on the basis of significant researches done by the experts.

Finally, I would also like to thank all the members involved in this book for being a team and meeting all the deadlines for the submission of their respective works. I would also like to thank my friends and family for being supportive in my efforts.

Editor

Efficient desulfurization of gasoline fuel using ionic liquid extraction as a complementary process to adsorptive desulfurization

N. Farzin Nejad[1] · A. A. Miran Beigi[1]

Abstract The extractive desulfurization of a model gasoline containing several alkyl thiols and aromatic thiophenic compounds was investigated using two imidazolium-based ionic liquids (ILs), 1-butyl-3-methylimidazolium tetrachloroaluminate, and 1-octyl-3-methylimidazolium tetrafluoroborate, as extractants. A fractional factorial design of experiments was employed to evaluate the effects and possible interactions of several process variables. Analysis of variance tests indicated that the number of extraction steps and the IL/gasoline volume ratio were of statistically highly significant, but none of the interactions were significant. The results showed that the desulfurization efficiency of the model gasoline by the ILs could reach 95.2 % under the optimal conditions. The optimized conditions were applied to study the extraction of thiophenic compounds in model gasoline and several real gasoline samples; the following order was observed in their separation: benzothiophene > thiophene > 3-methylthiophene > 2-methylthiophene, with 96.1 % removal efficiency for benzothiophene. The IL extraction was successfully applied as a complementary process to the adsorptive desulfurization with activated Raney nickel and acetonitrile solvent. The results indicated that the adsorptive process combined with IL extraction could provide high efficiency and selectivity, which can be regarded as a promising energy efficient desulfurization strategy for production of low-sulfur gasoline.

✉ N. Farzin Nejad
Farzinnejadn@ripi.ir

[1] Petroleum Refining Technology Development Division, Research Institute of Petroleum Industry, 14857-33111 Tehran, Iran

Edited by Xiu-Qin Zhu

Keywords Liquid–liquid extraction · Experimental design · Adsorptive desulfurization · Gasoline · Thiophenic compounds

1 Introduction

Sulfur oxides (SO_x) resulting from the combustion of sulfur compounds in fuels have become an increasingly serious environmental problem worldwide as they are a major cause of acid rain and atmospheric pollution. Thus, in recent years, considerable attention has been paid to the deep desulfurization of gasoline and diesel fuels due to the increasingly stringent environmental regulations being imposed to reduce the S content to a very low level (Song 2003).

The removal of sulfur compounds from liquid fuels is carried out industrially via catalytic hydrodesulfurization (HDS). Although the conventional HDS has been highly effective in reduction of sulfur levels, aromatic sulfur compounds such as thiophene, benzothiophene, dibenzothiophene, and their derivatives, which are the major objectionable sulfur components present in petroleum fractions, are less reactive to this process (Song 2003). Further improvement of the HDS process for deep desulfurization is limited to increasingly severe operating conditions at high cost. Moreover, deep HDS process will require a considerable increase in the consumption of energy and hydrogen, which can substantially improve the reactivity and selectivity of the catalyst, resulting in undesirable side reactions. Such side reactions can lead to a decrease in the octane number of gasoline (Nie et al. 2006; Wang et al. 2007). Therefore, from both environmental and economic considerations, various alternative deep desulfurization processes have been extensively developed in the past few years, including adsorptive and complexation desulfurization (Mansouri et al. 2014; Sevignon et al. 2005; Shi et al. 2015), biodesulfurization (Boshagh et al. 2014; Fernandez et al. 2014), extractive desulfurization (Mokhtar et al. 2014; Domanska et al. 2014; Krolikowski et al. 2013), and oxidative desulfurization (ODS) followed by extraction (Zhang et al. 2013; Ma et al. 2014).

Among the above alternatives, extractive desulfurization deserves special attention because the extraction is a well-established and facile process that can be carried out at or around ambient temperature and pressure. However, a suitable solvent for extractive desulfurization should have a high partition coefficient for sulfur components especially aromatic sulfur compounds, negligible cross solubility, high thermal and chemical stability, nontoxicity, environmental compatibility, and low cost for commercial applications (Jiang et al. 2008). Many organic solvents, such as dimethyl sulfoxide, acetonitrile, 1-methyl-2-pyrrolidinone, dimethylformamide, and polyalkylene glycol have been used as extractants, but none of these solvents conform to all of the above requirements and their performance in removing sulfur from fuels has not been satisfactory (Li et al. 2010; Sampanthar et al. 2006).

The ionic liquids (ILs) have been recognized as promising alternatives to conventional non-desirable organic solvents and have received considerable attention as extractants for desulfurization of liquid fuels (Li et al. 2010; Mochizuki and Sugawara 2008; Kedra-Krolik et al. 2011; Hansmeier et al. 2011), or at least as a complementary technology to the HDS process (Nie et al. 2008). The ILs are environmentally friendly solvents with unique physicochemical properties, such as negligible vapor pressure, high chemical and thermal stabilities, non-flammability, and recyclability. These properties together with high affinity for sulfur-containing compounds, especially aromatic sulfur components, and immiscibility with fuels make ILs desirable extractants for desulfurization of liquid fuels (Wang et al. 2007; Kedra-Krolik et al. 2011). However, although ILs are highly effective for the extraction of some aromatic sulfur components, that is, they can lower the concentrations to desirable low levels (especially thiophenic compounds), they do not provide adequate efficiency for decreasing the total sulfur concentration to acceptable levels for many gasoline samples. Activated Raney nickel adsorbent combined with IL extraction is regarded as a promising strategy to achieve very low sulfur levels and is currently receiving increasing attention because it avoids the use of hydrogen and allows the process to be conducted at ambient conditions. This process is also appealing because the sulfur contaminants that are most resistant to HDS, such as thiophenic compounds, are the most reactive components under ambient conditions.

A survey of the literature shows that many factors, such as the type of ionic liquid, ratio of IL to gasoline, number of extraction steps, contact time (shaking time), and temperature in some instances, may significantly influence the extractive desulfurization efficiency of gasoline and diesel fuels with ILs (Wang et al. 2007; Mochizuki and Sugawara 2008; Kedra-Krolik et al. 2011). However, the influence of these factors has not been studied in detail. Experimental designs, as multivariate optimization techniques, have been widely used in various branches of chemistry such as synthesis (Zolfigol et al. 2014), chemical and biomedical analysis (Castro Sousa et al. 2008), preconcentration (Escudero et al. 2010), extraction (Chang et al. 2011), and other situations (Anunziata et al. 2008), because they are relatively fast, highly economical and effective, and allow several variables to be optimized simultaneously. This approach is applied to reduce the large amount of data so it can be easily interpreted to examine the main and interaction effects of experimental conditions on the efficiency of methods, and to obtain an estimate of the real functional relationship (response function) between the response of the system and significant factors (Morgan 1991).

In the present investigation, we attempt to optimize the extractive desulfurization of a model gasoline composed of seven sulfur compounds including alkyl thiols and aromatic compounds like thiophene (TP), 2-methylthiophene (2-MT), 3-methylthiophene (3-MT), and benzothiophene (BT) with respect to the above-mentioned parameters using experimental design. The optimized extraction conditions will be implemented for deep desulfurization of real gasoline samples following the adsorption process performed by means of activated Raney nickel as the adsorbent in acetonitrile solvent.

2 Experimental

2.1 Materials

All sulfur compounds including 1-propanethiol, 1-butanethiol, 1-pentanethiol, TP, 2-MT, 3-MT, BT, and other chemicals with the highest purity available from Merck (Darmstadt, Germany) were used without further purification. N-hexane, used as a solvent for sulfur compounds, was of analytical reagent grade from Acros (USA). The 1-butyl-3-methylimidazolium tetrachloroaluminate ([BMIM][AlCl$_4$], purity >95 %, water content = 0.3 %) and 1-octyl-3-methylimidazolium tetrafluoroborate ([OMIM][BF$_4$], purity >97 %, water content = 0.08 %) were purchased from Fluka and used as received. These ILs were used as extractants in liquid–liquid extraction for removal of sulfur compounds from model gasoline.

The stock solution of the sulfur compounds in n-hexane was prepared as a model gasoline. The solution was prepared by dissolving 1-propanethiol, 1-butanthiol, 1-pentanthiol, TP, 2-MT, 3-MT, and BT in n-hexane. The total sulfur content of the stock solution, as measured in duplicate by X-ray fluorescence (XRF) spectrometer [according to ASTMD-2622 (1998)], was found to be 520 ppm. The stock solution used for investigating the extraction efficiency of [BMIM][AlCl$_4$] at the optimized conditions was also prepared from the above-mentioned compounds, and their concentrations were determined by a gas chromatograph and a sulfur chemiluminescence detector (GC–SCD) (Table 4). The stock solutions were stored in a refrigerator at 5 °C, and used as model of gasoline samples for extractive desulfurization experiments with ILs, as well as adsorptive desulfurization and adsorptive desulfurization (ADS) + IL tests. Real samples were collected from various sources of imported and locally produced gasoline.

2.2 Apparatus and software

All the desulfurization experiments were conducted in a 40-mL three-necked jacketed glass reactor equipped with a stirrer, a condenser, and a thermometer. The reactor was connected to a thermostatic bath, the temperature of which was maintained within 1 °C. Good contact between phases was guaranteed by vigorous stirring, establishing required stirring times to achieve an equilibrium state for each one of the studied systems. The total sulfur content of the model gasoline before and after each extraction was measured in duplicate using a wavelength-dispersive XRF spectrometer (Horiba model 2800) (according to ASTM-D2622) and by the Raney nickel reduction method [according to UOP Laboratory test method (1992) pp 357–80]. The water content of the ILs was determined using a Karl-Fischer instrument according to ASTM D-1533.

The solubility of ILs in gasoline was determined by analyzing the ionic liquid-saturated gasoline using a high-performance liquid chromatography (HPLC) (Stepnowski et al. 2003) with a Waters model 510 HPLC pump, equipped with UV–Vis detector at 295 nm wavelength and a C18 column. The mobile phase was a mixture with a methanol to water volume ratio of 80:20 at a flow rate of 1 mL min^{-1}. The concentrations of the thiophenic compounds in model and real gasoline samples were determined with a gas chromatograph, Varian model CP 3800, equipped with a SCD, and CP-Sil column (30 m × 0.32 mm i.d.), operated with He carrier gas. The temperature program was from 32 to 220 °C at a ramping rate of 5 °C min^{-1}.

2.3 The fractional factorial design test

The experimental design optimization was performed using STATISTICA 6.0 software. The extractive desulfurization of the model gasoline was investigated by factorial design. For this purpose, we used five parameters including shaking time (X_1, min), temperature (X_2, °C), number of extraction steps (X_3), IL to gasoline volume ratio (X_4), and the type of IL (X_5), each variable at two levels. The [BMIM][AlCl$_4$] and [OMIM][BF$_4$] were used as the two IL levels. Imidazolium-based ILs are known to be efficient extractants in desulfurization, especially for removal of aromatic sulfur compounds (Jiang et al. 2008; Nie et al. 2007; Rogosic et al. 2014). Although a two-level full factorial design can provide sufficient information to evaluate the whole set of main effects as well as interaction effects, it will require 2^5 possible combinations, i.e., 32 experiments for five variables. The main effects and the lower-order interactions, however, are usually the most significant terms. In this work, a 2^{5-1} fractional factorial design consisting of 16 factorial runs was performed to reduce the experimental efforts. This allowed all the experiments to proceed in parallel to avoid possible impact caused by different experimental blocks. The maximum and minimum levels of each factor were chosen according to preliminary experiments (Table 1). The experiments were randomly performed in order to obtain a random distribution of unknown systematic errors.

2.4 Procedure

The extraction was carried out in the glass reactor by adding the ionic liquid to the model gasoline. The resulting mixture was stirred vigorously at 50 °C (or other specified temperatures). After a specified time, the upper phase (model gasoline) was carefully separated from the IL phase with a syringe for analysis. The total sulfur of the model gasoline was determined before and after each extraction by two methods: XRF (ASTM-D2622) and Raney nickel method (UOP 357–80) for samples with total sulfur >100 and <100 ppm, respectively.

The adsorption desulfurization experiments for real gasoline samples were performed in the same reactor described in Sect. 2.2, using a mixture of activated Raney nickel adsorbent and acetonitrile as mentioned above at 50 °C for 1 h. Activated Raney nickel was prepared by putting 0.60 ± 0.01 g of nickel–aluminum alloy in a 100-mL beaker and adding 10 mL of 2.5 m sodium hydroxide. The beaker was swirled gently in a fume hood until the vigorous evolution of hydrogen ceased. Any solids

Table 1 The value of high (+) and low (−) of the factors in 2^{5-1} fractional factorial design

Factors	Symbols	Levels	
		Low (−)	High (+)
Shaking time, min	X_1	15	60
Extraction temperature, °C	X_2	40	50
Number of extraction steps (n)	X_3	1	3
IL/gasoline volume ratio, v/v	X_4	1/5	1/1
Type of IL	X_5	[BMIM][AlCl$_4$]	[OMIM][BF$_4$]

remaining on the sides of the beaker were washed down with a minimum of water. The reaction was allowed to continue overnight in a covered vessel (such as a large desiccator minus desiccant). Then, the activated Raney nickel was stored in the beaker in the excess sodium hydroxide, in the covered vessel until ready for use. The beaker may be stored in the desiccator for a week.

Treatment of nickel–aluminum alloy with sodium hydroxide solution is one of the typical techniques for preparing activated Raney nickel. In this reaction, aluminum is oxidized to aluminate and hydrogen is evolved vigorously. This catalyst can adsorb a great amount of hydrogen through Van der Waals forces. Two reaction paths are known for the desulfurization process (Miran Beigi et al. 1999):

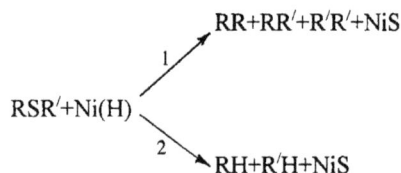

$$RSR' + Ni(H) \xrightarrow{1} RR + RR' + R'R' + NiS$$
$$RSR' + Ni(H) \xrightarrow{2} RH + R'H + NiS$$

At high temperature and when the catalyst has been stored for a long period of time, the reaction will proceed through path 1. Apparently, the adsorbed hydrogen decreases at a temperature higher than 100 °C. At low temperature and for freshly prepared catalyst, path 2 is preferred.

Activated Raney nickel and 10 mL of acetonitrile were added to 10 mL of a real gasoline sample. The mixture was stirred for 1 h and then the gasoline phase was separated and subjected to extraction by the IL ([BMIM][AlCl$_4$]), with a IL/gasoline volume ratio of 1:1, at temperature of 50 °C and contact time of 1 h.

3 Results and discussion

3.1 Optimization of the extraction method

The optimization of experimental variables in the extraction of sulfur compounds from gasoline with ILs was carried out using a two-level 2^{5-1} fractional factorial design (i.e., 16 factorial runs). The five variables considered as factors were shaking time (min), temperature of the reactor (°C), number of extraction steps (n), IL to gasoline volume ratio (v/v), and the type of IL. Table 1 presents the variables and their real values at the high (+) and low (−) levels set in the design. The experimental design matrix in which the effects of the five variables on sulfur removal were investigated, together with the total sulfur concentration and desulfurization efficiency is presented in Table 2. The runs were randomized for statistical purposes.

The significance of the effects of five variables was checked by analysis of the variance (ANOVA) (see Table 3) and by the Pareto chart in Fig. 1. It can be seen that two factors, namely number of extraction steps (X_3) and IL/gasoline volume ratio (X_4), were statistically significant with the latter variable (X_4) being the most effective. The other factors, i.e., extraction temperature (X_2), extraction time (X_1), and type of IL (X_5), were not significant in the ranges studied at the 95 % confidence level. Also, none of the interactions were significant. Although not significant, extraction time exhibited a positive effect. Bearing this in mind, we selected the high level of this factor for further studies. The same is true for the effect of X_2 factor (reactor temperature). In this case, we used the high level of this factor, because the viscosity of ILs was reduced at higher temperatures which were expected to have a positive effect on the extraction efficiency, although as reported by Wang et al. (2007), the effect of temperature on the desulfurization with ILs is very limited. Also, the type of IL (factor X_5) was not significant and showed the least effect estimate. The interaction of this factor with all the other factors was also insignificant, but due to its negative interaction effects with the other factors and considering that all the other single factors had positive effects (especially the effects of X_3, number of extraction steps, and X_4, IL/gasoline volume ratio were positive and highly significant), we decided to use the lower level of X_5 ([BMIM][AlCl$_4$]) as IL.

On the other hand, solubility of the ILs should also be considered as an important factor in assessing their applicability as extractants. Noticeable solubility of imidazolium-based IL in gasoline may contaminate the fuel and lead to NO$_x$ pollution (Jiang et al. 2008). The solubility of the

Table 2 The design matrix and analysis results from 2^{5-1} fractional factorial test. The total sulfur concentration in the model gasoline sample before performing the runs was 520 ppm

Standard run[a]	Factors					Total sulfur, after the runs, ppm	Desulfurization efficiency, %
	X_1	X_2	X_3	X_4	X_5		
1 (11)	−	−	−	−	−	388	25.4
2 (13)	+	−	−	−	+	323	37.9
3 (9)	−	+	−	−	+	337	35.2
4 (15)	+	+	−	−	−	320	38.4
5 (4)	−	−	+	−	+	228	56.1
6 (12)	+	−	+	−	−	267	48.6
7 (10)	−	+	+	−	−	210	59.6
8 (3)	+	+	+	−	+	234	55.0
9 (5)	−	−	−	+	+	207	60.2
10 (2)	+	−	−	+	−	181	65.2
11 (16)	−	+	−	+	−	309	40.6
12 (6)	+	+	−	+	+	217	58.3
13 (1)	−	−	+	+	−	143	72.5
14 (7)	+	−	+	+	+	141	72.9
15 (14)	−	+	+	+	+	144	72.3
16 (8)	+	+	+	+	−	25	95.2

[a] Values in parentheses indicate the randomized order in which the tests were run

Table 3 Results of ANOVA from the 2^{5-1} fractional factorial test for desulfurization of model gasoline

Source	Sum of squares	Degrees of freedom	Mean squares	F ratio	p value
X_1	153.76	1	153.76	64.000	0.07917
X_2	15.602	1	15.602	6.4943	0.23806
X_3^*	**1827.6**	**1**	**1827.6**	**760.69**	**0.02307**
X_4^*	**2047.5**	**1**	**2047.5**	**852.26**	**0.02180**
X_5	0.3601	1	0.3601	0.1498	0.76488
X_1X_2	51.840	1	51.840	21.578	0.13499
X_1X_3	46.240	1	46.240	19.247	0.14267
X_1X_4	112.36	1	112.36	46.768	0.09244
X_1X_5	150.06	1	150.06	62.461	0.08013
X_2X_3	145.20	1	145.20	60.438	0.08144
X_2X_4	37.823	1	37.823	15.743	0.15718
X_2X_5	50.410	1	50.410	20.982	0.13683
X_3X_5	108.16	1	108.16	45.020	0.09419
X_4X_5	30.250	1	30.250	12.5911	0.17489
Error	2.4020	1	2.4200		
Total	4779.6	15			

* Significant factors at $p < 0.05$

Number of extraction steps (X_3) and IL/gasoline volume ratio (X_4) are statistically significant with the latter variable being the most effective

ILs in gasoline was examined by analysis of the IL-saturated gasoline samples with HPLC. No IL peak was found for [BMIM][AlCl$_4$], indicating its negligible solubility in gasoline; however [OMIM][BF$_4$] showed some solubility. Higher solubility of [OMIM][BF$_4$] versus [BMIM][AlCl$_4$] may be related to the increased lipophilicity of its imidazolium moiety as the alkyl substituents in the ILs increase from butyl to octyl, although the effect of the anionic moiety should not be ignored.

The effect of the two significant variables (X_3 and X_4) was further investigated by taking several IL to gasoline volume ratios (1/5, 1/2 and 1/1) and number of extraction

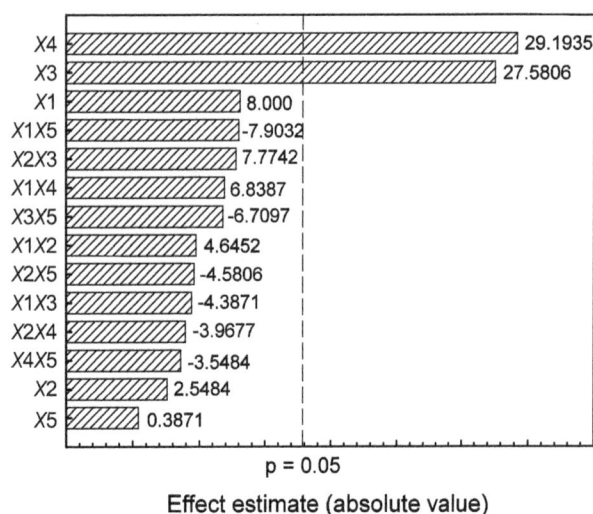

Fig. 1 Pareto chart of the standardized effects obtained for the factorial design optimization of the variables X_1 (shaking time), X_2 (reaction temperature), X_3 (number of extraction steps), X_4 (IL/gasoline volume ratio), X_5 (type of IL), and their interactions

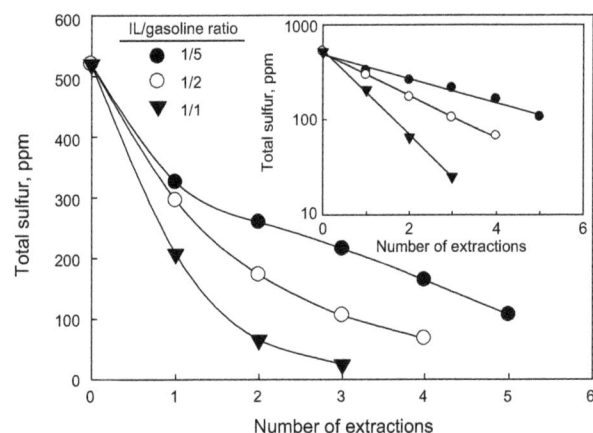

Fig. 2 Effect of number of extraction steps and IL to model gasoline volume ratio on desulfurization efficiency at 50 °C, using [BMIM][AlCl₄] and extraction time of 1 h. The *inset figure* shows the logarithm of sulfur concentration in gasoline versus the number of extraction steps

steps (1, 2, 3, 4, and 5). The results, shown in Fig. 2, clearly indicated the strong dependence of desulfurization efficiency on both of these variables. As expected, higher efficiency could be achieved at higher IL/gasoline ratio and/or larger number of extraction steps. The logarithmic plot (log of sulfur concentration in gasoline versus the number of extraction steps), shown as inset in Fig. 2, indicated that the extraction was controlled by a constant distribution coefficient according to Nernst's law. It can be seen that in this case, the degree of desulfurization increased proportionally with the increase of IL to gasoline ratio. The effect of shaking time beyond the high level used

in the design was investigated up to 6 at 1 h intervals and at different IL/gasoline ratios. No significant improvement was observed at extraction time >1 h. Based on the above results, [BMIM][AlCl₄] was used at 50 °C with IL/gasoline volume ratio of 1:1 and extraction time of 1 h for the following studies.

3.2 Extraction of thiophenic sulfur compounds from model gasoline

The results of extraction of aromatic sulfur compounds from the model gasoline with [BMIM][AlCl₄], determined using GC–SCD, are presented in Table 4. The results indicated that the extraction efficiency was in the following order: BT > TP > 3-MT > 2-MT. This behavior can be related to the π–π interaction between aromatic structures of sulfur compounds and the imidazolium ring of the IL (Krolikowski et al. 2013). BT, with its extended delocalized π system, had a strong interaction with [BMIM][AlCl₄] IL compared to TP and its alkyl substituted derivatives, therefore it was more easily extracted, with an efficiency of 96 % (Table 4). The presence of methyl substituent in 2-MT and 3-MT significantly lowered their extraction efficiency with respect to TP, possibly due to the steric hindrance effect (Nie et al. 2008).

3.3 Extraction of sulfur compounds from real gasoline samples

Table 5 gives the total sulfur and different thiophenic compound concentrations in several real gasoline samples before and after extraction with [BMIM][AlCl₄]. The samples were collected from various sources of imported and locally produced gasoline. The entry 1 shows a sample with a very high total sulfur content of 1400 ppm. While the total sulfur content of this sample decreased to 713 and 430 ppm after one and four extraction cycles, respectively, the concentration of BT decreased to < 1 ppm after the first step and those of TP, 2-MT, and 3-MT reached the same value after 4 steps. These results clearly indicated that while [BMIM][AlCl₄] exhibited a high trend to extract some aromatic constituents such as thiophenic compounds, its ability to extract many other sulfur compounds was low, this is why the total sulfur content was still high (430 ppm) even after four extraction cycles. It is known that the aromatic sulfur compounds, such as thiophenic compounds, in which a conjugation occurs between the lone-pair on S atom and the π-electrons on the aromatic ring, can preferably insert into the dynamic molecular structure of the ILs (Song 2003). On the other hand, when there is no such conjugation, as in the case of many sulfides, disulfides, and thiols, the compounds exhibit lower trends for

Table 4 Extraction efficiency of thiophenic sulfur compounds from the model gasoline with [BMIM][AlCl$_4$] as determined by GC–SCD

Compound	Initial content, ppm	After extraction, ppm	S content after extraction, ppm	Efficiency, %
TP	634	173	65.8	72.7
2-MT	224	100	32.6	55.4
3-MT	240	101	32.9	58.0
BT	280	11	2.62	96.1

Conditions: temperature 50 °C, extraction time 1 h, number of extraction steps 3

Table 5 Concentration of total sulfur and thiophenic compounds (ppm) in real gasoline samples before (B) and after (A) extraction with [BMIM][AlCl$_4$]

Gasoline sample	Total sulfur		TP		2-MT		3-MT		BT	
	B	A	B	A	B	A	B	A	B	A
1	1400	713	42	5	140	9	45	6	105	< 1
		430 (4)		<1 (4)		<1 (4)		<1 (4)		
2	124	71	20	9	21	12	26	14	29	10
		41 (2)		3 (2)		6 (2)		7 (2)		4 (2)
3	294	129	15	2	22	5	15	4	17	3
4	142	60	32	11	27	13	27	13	15	5
5	100	55	22	8	22	12	27	14	28	9
6	154	40	19	2	20	3	28	4	32	<1
7	118	91	19	7	18	10	19	6	23	<1

Conditions: temperature 50 °C; extraction time 1 h; IL to gasoline volume ratio 1

The values in parentheses indicate the number of extraction steps

extraction with ILs. A close inspection of Table 5 reveals that for most samples, the IL reduced the total sulfur to about half of their initial values (entries 1–5), but for samples 6 and 7, the remaining concentration was, respectively, 26 % and 77 %, indicating the fact that the extraction efficiency depended on the nature of the samples, which can be explained by the variety of compounds present in the gasoline samples.

3.4 Desulfurization of real gasoline samples by IL and ADS + IL extraction

The above results, lead to the conclusion that although the IL ([BMIM][AlCl$_4$]) is highly effective for the extraction of aromatic sulfur compounds and can lower their concentration to very low levels, it does not provide adequate efficiency for decreasing total sulfur concentration to acceptable levels for many gasoline samples. Therefore, in continuation of our research, we tried a combination of ADS and IL in a successive manner. At first, in the adsorption process, the acetonitrile solvent was examined alone for desulfurization. Acetonitrile solvent (10 mL) was added to 10 mL of real gasoline sample 1. The mixture was stirred at 50 °C for 1 h and then the gasoline phase was separated. After this step, the total sulfur concentration in

gasoline sample 1 decreased from 1400 to 720 ppm. In the next experiment, 10 mL of acetonitrile solvent containing 0.6 g of activated Raney nickel was added to 10 mL of gasoline sample 1 and the mixture was stirred at the same conditions. The total sulfur concentration in this sample decreased from 1400 to 267 ppm. This result clearly indicated the ability of activated Raney nickel in extracting a considerable amount of sulfur compounds. Moreover, other experiments were carried out by ionic liquid and activated Raney nickel. A 10 mL of [BMIM][AlCl$_4$] ionic liquid containing 0.6 g of activated Raney nickel was added to 10 mL of gasoline sample 1 and the mixture was stirred at 50 °C for 1 h. The total sulfur concentration reduced from 1400 to 520 ppm. This result also indicated that activated Raney nickel can be regarded as an efficient desulfurization strategy. Table 6 shows the results obtained from using IL extraction, acetonitrile solvent, acetonitrile solvent containing activated Raney nickel, IL containing activated Raney nickel, and acetonitrile solvent containing activated Raney nickel + IL (ADS + IL).

The ADS + extraction experiments were performed according to the procedure described in Sect. 2.4. The concentrations of total sulfur before and after the ADS + IL processes are presented in Table 7. Comparing the results of IL extraction alone and ADS + IL extraction

Table 6 Concentration of total sulfur (ppm) in real gasoline sample 1 before (B) and after (A) desulfurization

Gasoline sample	Total sulfur, ppm (B)	Total sulfur after treatment, ppm				
		(A) IL	(A) Acetonitrile solvent	(A) Acetonitrile solvent + activated Raney nickel	(A) IL + activated Raney nickel	(A) Activated Raney nickel + acetonitrile solvent + IL (ADS + IL)
1	1400	713	720	267	520	154

Conditions: temperature 50 °C; reaction time 1 h

Table 7 Concentration of total sulfur (ppm) in real gasoline samples before (B) and after (A) ADS + IL treatment

Gasoline sample	Total sulfur (B)	Total sulfur after treatment (A), ppm		Efficiency of ADS + IL
		IL	ADS + IL	
1	1400	713	154	89.0
2	124	71	36	71.0
3	294	129	36	87.8
4	142	60	35	75.4
5	100	55	16	84.0

Conditions: temperature 50 °C; reaction time 1 h; ADS-phase (0.60 g of activated Raney nickel in 10 mL acetonitrile solvent): gasoline volume ratio 1:1

in Table 7 shows that the ADS process is more efficient for removing total sulfur concentration in most of the samples. For example, it can be observed from these tables that the total sulfur concentration in sample 1 has reduced from 1400 to 154 ppm by the ADS + IL and 713 ppm by the IL extraction processes (efficiency of 89.0 and 49.1 %, respectively). Similarly, the sulfur removal efficiency of samples 2 and 3 by the ADS + IL processes is 71.0 and 87.8, respectively, compared to the corresponding values of 42.7 and 56.1, respectively, for the IL extraction. These results indicate that IL extraction can be used as a complementary process to the ADS. Therefore, ADS combined with IL extraction, which operates near ambient conditions and provides high efficiency and selectivity, can be regarded as a promising energy efficient desulfurization strategy for the production of low-sulfur gasoline.

3.5 Extraction of other aromatic compounds (dearomatization) from real gasoline samples by IL

[BMIM][AlCl$_4$] was used to extract other aromatic compounds from gasoline (Farzin Nejad and Karimi 2011). The dearomatization experiments were carried out with an ionic liquid to gasoline volume ratio of 1:1 at 50 °C for 1 h. The aromatic hydrocarbons removal selectivity exhibited the following order: benzene > toluene > xylene > ethylbenzene.

3.6 Regeneration of ILs

[OMIM][BF$_4$] was regenerated by direct distillation under a nitrogen atmosphere at 170 °C for 3 h. After the

regeneration was completed, the ionic liquid was mixed with fresh gasoline sample 1 and the second extraction cycle was performed. The results showed that the desulfurization efficiency was 87 % in the first regeneration cycle and 72 % in the second cycle.

In all the cases of this work, when [BMIM][AlCl$_4$] contacted with the gasoline samples, it turned black immediately. This phenomenon indicated the decomposition of [BMIM][AlCl$_4$] due to the presence of Lewis acid IL. A potential disadvantage of this particular IL is its sensitivity toward water and possible difficulty associated with its regeneration. [BMIM][AlCl$_4$] was regenerated through contacting it with an organic medium such as toluene; however, this method has not proven to be completely effective in the case of [BMIM][AlCl$_4$].

4 Conclusions

The optimization of experimental variables in the extraction of sulfur compounds from gasoline by ILs was carried out using a two-level fractional factorial design. It was found that the number of extraction steps and IL/gasoline volume ratio was statistically significant with the latter variable being the most effective. The type of IL was not significant and showed the least effect estimate. At the optimized conditions, the extraction process was applied to the desulfurization of several real gasoline samples. The results indicated that the ionic liquid provides a high tendency to extract some aromatic constituents such as thiophenic compounds in both model and real gasoline samples. Due to the inability of the ionic liquid to lower total sulfur compounds in real gasoline samples to a desirable level, an IL extraction was applied as a complementary process to the ADS, which was performed by activated Raney nickel adsorbent and acetonitrile solvent. The sulfur content of the samples after treatment with the combined desulfurization scheme presented a considerable reduction and reached to >71 % for all of the samples. The results indicated that the ADS + IL extraction provides high efficiency and selectivity and can be regarded as a promising energy efficient desulfurization strategy for the production of low-sulfur gasoline. It is expected that the desulfurization efficiency can be further improved by optimizing the ADS operating parameters, such as reaction temperature, reaction time, and the amount and proportion of reactants, to obtain ultra-low-sulfur gasoline.

Acknowledgments The authors thank National Iranian Oil Refining & Distribution Company (NIORDC) and Research & Development (R&D) center of this company for their financial support during the completion of this work.

References

Annual book of ASTM standards. ASTM D-2622: sulfur in petroleum products (X-ray spectrographic method). Philadelphia: PA; 1998.

Anunziata OA, Cussa J. Applying response surface design to the optimization of methane activation with ethane over Zn-H-ZSM-11 zeolite. Chem Eng J. 2008;138:510–6.

Boshagh F, Mokhtarani B, Mortaheb HR. Effect of electrokinetics on biodesulfurization of the model oil by Rhodococcus erythropolis PTCC1767 and Bacillus subtilis DSMZ 3256. J Hazard Mater. 2014;280:781–7.

Sousa JKC, de Sousa Dantas AN, Marques ALB, et al. Experimental design applied to the development of a copper direct determination method in gasoline samples by graphite furnace atomic absorption spectrometry. Fuel Process Technol. 2008;89:1180–5.

Chang SH, Teng TT, Ismail N. Screening of factors influencing Cu(II) extraction by soybean oil-based organic solvents using fractional factorial design. J Environ Manage. 2011;92:2580–5.

Domanska U, Walczak K, Królikowski M. Extraction desulfurization process of fuels with ionic liquids. J Chem Thermodyn. 2014;77:40–5.

Escudero LA, Cerutti S, Olsina RA, et al. Factorial design optimization of experimental variables in the on-line separation/preconcentration of copper in water samples using solid phase extraction and ICP-OES determination. J Hazard Mater. 2010;183:218–23.

Farzin Nejad N, Karimi EA. A new approach to dearomatization of gasoline by ionic liquid and liquid–liquid extraction. Pet Sci Technol. 2011;29:2372–6.

Fernandez M, Ramirez M, Gomez JM, et al. Biogas biodesulfurization in an anoxic biotrickling filter packed with open-pore polyurethane foam. J Hazard Mater. 2014;264:529–35.

Hansmeier AR, Meindersma GW, de Haan AB. Desulfurization and denitrogenation of gasoline and diesel fuels by means of ionic liquids. Green Chem. 2011;13:1907–13.

Jiang XC, Nie Y, Li CX, et al. Imidazolium-based alkylphosphate ionic liquids-a potential solvent for extractive desulfurization of fuel. Fuel. 2008;87:79–84.

Kedra-Krolik K, Fabrice M, Jaubert JN. Extraction of thiophene or pyridine from n-heptane using ionic liquids. Gasoline and diesel desulfurization. Ind Eng Chem Res. 2011;50:2296–306.

Krolikowski M, Walczak K, Domanska U. Solvent extraction of aromatic sulfur compounds from *n*-heptane using the 1-ethyl-3-methylimidazolium tricyanomethanide ionic liquid. J Chem Thermodyn. 2013;65:168–73.

Li FT, Liu Y, Sun ZM, et al. Deep extractive desulfurization of gasoline with xEt$_3$NHCl$_3$·FeCl$_3$ ionic liquids. Energy Fuels. 2010;24:4285–9.

Ma C, Dai B, Liu P, et al. Deep oxidative desulfurization of model fuel using ozone generated by dielectric barrier discharge plasma combined with ionic liquid extraction. J Ind Eng Chem. 2014;20:2769–74.

Mansouri A, Khodadadi AA, Mortazavi YA. Ultra-deep adsorptive desulfurization of a model diesel fuel on regenerable Ni–Cu/γ-Al$_2$O$_3$ at low temperatures in absence of hydrogen. J Hazard Mater. 2014;271:120–30.

Miran Beigi AA, Teymouri M, Eslami M, et al. Determination of trace sulfur in organic compounds by activated Raney nickel desulfurization method with non-dispersive gas detection system. Analyst. 1999;124:767–70.

Mochizuki Y, Sugawara K. Removal of organic sulfur from hydrocarbon resources using ionic liquids. Energy Fuels. 2008;22:3303–7.

Mokhtar WNAW, Bakar WAWA, Ali R, et al. Deep desulfurization of model diesel by extraction with N, N-dimethylformamide: optimization by Box-Behnken design. J Taiwan Inst Chem Eng. 2014;45:1542–8.

Morgan ED. Chemometrics: experimental design. ACOL. London: Wiley; 1991.

Nie Y, Li CX, Meng H, et al. N, N-Dialkylimidazolium dialkylphosphate ionic liquids: their extractive performance for thiophene series compounds from fuel oils versus the length of alkyl group. Fuel Process Technol. 2008;89:978–83.

Nie Y, Li CX, Sun A, et al. Extractive desulfurization of gasoline using imidazolium-based phosphoric ionic liquids. Energy Fuels. 2006;20:2083–7.

Nie Y, Li CX, Wang ZH. Extractive desulfurization of fuel oil using alkylimidazole and its mixture with dialkylphosphate ionic liquids. Ind Eng Chem Res. 2007;46:5108–12.

Rogosic M, Sander A, Pantaler M. Application of 1-pentyl-3-methylimidazolium bis (trifluoromethylsulfonyl) imide for desulfurization, denitrification and dearomatization of FCC gasoline. J Chem Thermodyn. 2014;76:1–15.

Sampanthar JT, Xiao H, Dou J, et al. A novel oxidative desulfurization process to remove refractory sulfur compounds from diesel fuel. Appl Catal B. 2006;63:85–93.

Sevignon M, Macaud M, Favre-Reguillon A, et al. Ultra-deep desulfurization of transportation fuels via charge-transfer complexes under ambient conditions. Green Chem. 2005;7:413–20.

Shi Y, Liu G, Wang L, et al. Efficient adsorptive removal of dibenzothiophene from model fuel over heteroatom-doped porous carbons by carbonization of an organic salt. Chem Eng J. 2015;259:771–8.

Song C. An overview of new approaches to deep desulfurization for ultra–clean gasoline, diesel fuel and jet fuel. Catal Today. 2003;86:211–63.

Stepnowski P, Muller A, Behrend P, et al. Reversed-phase liquid chromatographic method for the determination of selected room temperature ionic liquid cations. J Chromatogr A. 2003;993:173–8.

UOP Laboratory Test Method. Trace sulfur in petroleum distillate by the nickel reduction method. Des Plaines: Universal Oil Products; 1992. p. 357–80.

Wang JL, Zhao DS, Zhou EP, et al. Desulfurization of gasoline by extraction with n-alkyl-pyridinium-based ionic liquids. J Fuel Chem Technol. 2007;35:293–6.

Zhang M, Zhu W, Xun S, et al. Deep oxidative desulfurization of dibenzothiophene with POM-based hybrid materials in ionic liquids. Chem Eng J. 2013;220:328–36.

Zolfigol MA, Khazaei A, Sarmasti N, et al. Programming of microwave-assisted synthesis of new isophthalate derivatives using $ZrOCl_2$ as a catalyst under solvent-free condition by experimental design. J Mol Catal A: Chem. 2014;393:142–9.

2

A comprehensive experimental evaluation of asphaltene dispersants for injection under reservoir conditions

Hamed Firoozinia[1] · Kazem Fouladi Hossein Abad[1] · Akbar Varamesh[1]

Abstract As the efficiency of dispersants with different origins is questionable for each typical oil sample, the present study provides a reproducible and reliable method for screening asphaltene dispersants for a typical asphaltenic crude oil. Four different asphaltene dispersants (polyisobutylene succinimide, polyisobutylene succinic ester, nonylphenol-formaldehyde resin modified by polyamines, and rapeseed oil amide) were prepared and their performance on two oils from an Iranian field under laboratory and reservoir conditions was studied. A thorough analysis including ash content and SARA tests was performed on the solid asphaltene particles to characterize the nature of deposits. Then a highly efficient carrier fluid, which is crucial when injecting dispersant into the wells, was selected from a variety of chemicals by comparing their solubility. In the next step, using an optical microscope, a viscometer, and a Turbiscan, the screening of dispersants under laboratory conditions was done on a mixture of dead oil and dispersant to evaluate the onset of asphaltene precipitation and its stability when titrating by a precipitant. Finally, two different mixtures of the efficient dispersants, live oil, and carrier fluid were used with the solid detection system (SDS) and the filtration method to examine their effects on the onset pressure of asphaltene precipitation and the asphaltene content of the crude oil under reservoir conditions. The results show that the combination of experimental methods used in this work could be consistently applied to screening asphaltene dispersants. Among the four different dispersants applied here, the dispersant based on nonylphenol-formaldehyde resin modified by polyamines showed the best performance on the available live oils. This chemical modified the onset pressure of asphaltene precipitation of light oil from 4300 psi to about 3600 psi and decreased the precipitated asphaltene of heavy oil by about 30 %.

Keywords Asphaltene · Dispersant · Solvent · Onset point · Precipitation · Deposition

1 Introduction

Deposition of solids, organic (wax and asphaltene) and inorganic (scales, sand, and corrosion products), can clog production flow paths, reducing hydrocarbon production (Leontaritis 1996; Cosultchi et al. 2001; Thawer et al. 1989). Among the typical flow assurance problems, asphaltene precipitation and deposition is held to be the most complicated one. The reason behind this distinction is the intricate nature of asphaltenes with respect to other hydrocarbon compounds (Turta et al. 1997).

Asphaltenes are the heaviest and most polar constituent in crude oil. They have a wide range of molecular weights owing to various combinations of aliphatic chains and polycondensed aromatic rings in their structures (Kawanaka et al. 1991; Zendehboudi et al. 2013). Although the mechanisms of asphaltene precipitation and deposition and its characteristics are well illustrated in the literature, accurate prediction of asphaltene deposits and effective remedies have remained challenging due to the complexities (Zendehboudi et al. 2014).

✉ Hamed Firoozinia
    firooziniah@ripi.ir

[1] Research Institute of Petroleum Industry,
    Tehran 1485733111, Iran

Edited by Yan-Hua Sun

One way of preventing asphaltene precipitation and deposition is to use predictive models to design an efficient operation processes. Several models have been reported for prediction of asphaltene precipitation behavior (Hirscberg et al. 1984; Speight et al. 1985; Mansoori 1997; Junior et al. 2006). Despite the fact that the asphaltene precipitation and deposition is a nonlinear process, some of these predictive models are only valid within particular process conditions and rely on linear identification models (Zendehboudi et al. 2014). Recently, an artificial neural network has been considered as a strong technique to accurately model the highly nonlinear systems. Miri et al. (2014), for instance, used an optimization approach in order to determine the asphaltene surface deposition, entrainment, and plugging based on experimental data.

In addition to predictive tools, mechanical and biological methods can also be used to control asphaltene deposition under limited conditions. Mechanical methods are not applicable within the formation, especially near the wellbore area where maximum deposition occurs. Biological methods may need months or years to degrade noticeable amount of asphaltenes (Mansoori 2010).

Besides the mentioned methods, addition of chemicals such as dispersants is one of the most efficient techniques for mitigating asphaltene precipitation and deposition which is the subject of this paper.

Asphaltene dispersants (ADs) are a class of chemical additives which can be used to control asphaltene deposition (Oschmann 2002). They have been successfully applied to inhibiting and removing asphaltene deposits as a formation squeeze, batch, or continuous injection (Manek 1995).

ADs reduce the size of flocculated asphaltenes and keep them in suspension (Marques et al. 2004; Smith et al. 2008). In general, ADs are composed of a polar group (due to the presence of hetero-atoms like oxygen, nitrogen, phosphorous) which attach to the surface of asphaltenes and an alkyl group which prevents the adhesion of asphaltene nanoaggregates. These two groups interact with aggregated asphaltenes and with the help of a long alkyl tail they are able to change the polarity of the outer surface of aggregates. Therefore, the aggregates will have properties closer to those of crude oil and will remain dispersible in the crude oil (Barcenas et al. 2008; Ferrara 1995). An important feature of ADs is that an increase in the dosage of an AD above its optimum concentration causes the self-association of AD surfactant molecules resulting in an adverse effect on asphaltene aggregation (Barcenas et al. 2008).

ADs may have different results when using live and dead oils. That depends on the presence of resins in the solution for optimum performance of these chemicals. In fact, the tests conducted on dead oil or a solution containing asphaltenes redissolved in an aromatic solvent can be useful for primary screening of ADs (Borchardt 1989;

Bouts et al. 1995; Manek 1995; Takhar 1997; Oschmann 2002). The secondary, more reliable, screening is done under reservoir conditions to meet plausible influence of thermodynamic conditions on dispersant performance.

In the present work, the live and dead oil samples of two types of crude oil (oil A and oil B) were prepared and the effects of four dispersants on their asphaltene behavior were studied. The approach is to investigate the efficiency of these chemicals by considering the onset of asphaltene precipitation and its stability by thermodynamic conditions. The onsets of asphaltene precipitation are determined by a viscometer, an optical microscope, a solid detection system (SDS), and a filtration method. The stability of asphaltenes is measured by the Turbiscan. The onset points detected with each method differ from each other as the asphaltene aggregates should have reached a certain size in order to be recognized by these techniques. However, the aim of these experiments is to find to what extent the applied dispersant can delay the formation of asphaltene aggregates.

Since the injection of dispersant directly into the PVT cell is almost impossible due to the small quantity of dispersant needed, a set of experiments was also designed to select a proper carrier fluid. The effect of the carrier fluid on the onset of precipitation was investigated under atmospheric conditions to ensure that the correct value of onset point was obtained. Admittedly, in the field, the carrier fluid was able to remove the deposited asphaltenes from the interior tubing wall while the dispersant within the solution slowed the growth and formation of asphaltene flocculation.

Some dispersants may decompose at high temperatures, therefore, tests under reservoir conditions such as SDS and filtration methods can be considered as promising tools for choosing the best dispersant before injecting into a well.

2 Experimental

2.1 Experimental materials

The dead oil and live oil samples were obtained from two different formations of the Arvandan field, and the asphaltenes and deposited solids were sampled from the pipelines from the same field. The live oil samples were taken with a bottom-hole sampler (Ruska) to be wholly representative of the original, in-place fluids and their properties are summarized in Table 1.

The dispersants (DB, DF, DP, and DR) were synthesized by our group in the Research Institute of Petroleum Industry (RIPI). Table 2 shows the characteristics of produced dispersants. Research-grade n-heptane, toluene, and xylene were provided by Merck. Sol-100 and Sol-200 solvents were from RIPI. Gas oil and kerosene were provided by NIOPDC.

Table 1 Characteristics of Oil A and Oil B

Oil sample	Dead oil A	Reservoir oil A	Dead oil B	Reservoir oil B
Component, mol %				
H_2S	0	0.03	0	0.13
N_2	0	0.10	0	0.30
CO_2	0	2.02	0	3.33
C_1	0	43.59	0	26.50
C_2	0.07	9.47	0.1	7.37
C_3	0.45	5.49	0.49	4.51
i-C_4	0.97	1.47	0.68	0.60
n-C_4	4.63	3.06	1.47	2.03
i-C_5	2.89	1.32	1.57	2.49
n-C_5	3.39	1.53	1.65	2.96
C_6	13.81	5.19	19.69	7.56
C_7	10.5	3.89	8.38	4.44
C_8	8.06	2.96	6.52	3.77
C_9	11.76	4.27	8.08	2.82
C_{10}	7.81	2.81	5.92	3.35
C_{11}	7.86	2.83	5.33	3.29
C_{12+}	27.82	10.00	40.12	24.53
Gas oil ratio (GOR), SCF/STB	–	859.01	–	390.04
Molecular weight (MW) of the residual oil	–	216	–	296
MW of C_{12+} fraction	–	508	–	532
MW of the reservoir oil	–	93	–	174
Special gravity of C_{12+} fraction @ 60/60 °F	–	0.8958	–	0.9789
Saturates, wt%	71.84	–	49.50	–
Aromatics, wt%	20.33	–	30	–
Resins, wt%	6.69	–	9.10	–
Asphaltenes, wt%	1.14	–	11.4	–
Reservoir temperature, °F	–	270	–	205
Reservoir pressure, psi	–	9500	–	4600
Viscosity @ 20 °C, cP	–	5.2	–	8.7

Table 2 Dispersants

Chemical	Basic structure	MW, kg/mol[a]
DB	Polyisobutylene succinimide	3900
DF	Polyisobutylene succinic ester	4260
DP	Nonylphenol-formaldehyde resin modified by polyamines	8030
DR	Rapeseed oil amide	560

[a] Average molecular weight was determined by gel permission chromatography (GPC)

2.2 Experimental apparatus

The onset of asphaltene precipitation and its stability in dead oil samples were determined with a Motic BA300 POL microscope, an Anton Paar SVM-3000 viscometer, and a Turbiscan MA-2000 when titrating with n-alkane under laboratory conditions. The solid detection system (SDS) and the filtration method were applied under reservoir conditions.

2.3 Experimental methods

The concentration of ADs for the dead and live oil samples of A and B was considered being 400 and 350 ppm, respectively.

2.3.1 Visual microscopic measurements

The onset of asphaltene precipitation was determined by titration of crude oil with *n*-heptane, and particles greater than 2 μm could be observed by microscopy. Here, 500 mL of the dead oil in a beaker was stirred with a magnetic stirrer in order to have a homogeneous sample. At each step, a predetermined amount of *n*-heptane (0.2 mL) was added to 5 mL of the oil sample and the mixture was agitated vigorously for 5 min and a small drop of the mixture was observed with the microscope and the formation of asphaltene precipitates was checked. Having determined the precipitation onset for different dispersants, the efficient AD can be chosen based on the maximum ratio of *n*-heptane to oil.

2.3.2 Viscometric measurements

Similar to the previous method, the oil sample was titrated by a precipitant and the viscosity of the mixture was measured at each step. This method was only applied to oils with very low asphaltene content. Therefore, the viscosity measurements were just for oil A.

2.3.3 Turbidity measurements

The effect of a dispersant to maintain asphaltenes in a peptized state and to prevent flocculation of asphaltenes was investigated by considering its stability under different conditions. Oil with a low stability is likely to undergo flocculation of asphaltenes when stressed (e.g., the addition of a precipitant). Therefore, this method along with the two previous methods can be helpful for better understanding of the performance of dispersants.

In this technique, photons were sent into the sample. After being scattered by objects in suspension, the photons emerged from the sample and were detected by transmission and backscatter detectors. The change in the transmitted light from the oil sample (including the solvent, asphaltene, precipitant, and dispersant) due to the formation of asphaltene precipitates was measured in a fixed time interval. The less the change in transmitted light, the more the ability of the dispersant to prevent asphaltene precipitation (Fig. 1).

The Turbiscan quantitatively measures how easily the asphaltenes in the mixture of oil and a dispersant, separate upon the addition of *n*-heptane. The test method calculates a separability number (%).

When the separability number shown by the Turbiscan is greater than 10, the sample is unstable. For a separability number less than 5, the sample is stable and finally, the sample has medium stability when the measured number is between 5 and 10. The stability behavior of the mixture can

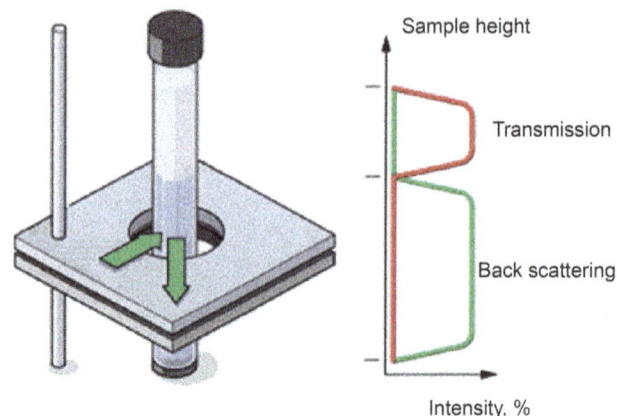

Fig. 1 Schematic of Turbiscan

also be explained by the trends of the transmitted light during the time of measurement (Fig. 2).

If an AD is effective, it will keep asphaltenes in the solution and there are no asphaltene agglomerate particles. Therefore, the mean transmittance remains constant along the Turbiscan test tube (Fig. 2a). For semi-stability mixtures, the received light varies with time (Fig. 2b). For ineffective ADs, agglomeration occurs and asphaltene particles deposit at the bottom of the test tube. Hence, the mean transmittance increases towards the top of the test tube as the upper part of the test tube becomes less dense (Fig. 2c). In this case, the change in the stability of the fluid depends mostly on the fluid characteristics.

In the first step, the separability number for different volume ratios of oil to *n*-heptane was measured. The aim of this stage is to find out the volume ratio of oil to *n*-heptane under which the mixture is unstable. Immediately after the addition of precipitant, the mixture was shaken for 5 min and 7 mL of the prepared sample was placed in the device's test tube. The whole sample was then scanned every minute for at least 15 min (the right axis in Fig. 2). For the next step, a mixture containing oil and a dispersant was prepared; 4 mL of the mixture were poured into several test tubes and different amounts of *n*-C$_7$ were added to each of them and rocked for 5 min. Then, 7 mL of the sample was placed in the Turbiscan cell and its stability was measured.

2.3.4 SDS procedure

The live oil was fed into the PVT cell which was placed in a temperature-controlled oven. The pressure and volume of the sample in the cell would be controlled by a variable-volume displacement pump. The mixer inside the PVT cell provided proper mixing of fluids and accelerated the equilibrium condition of the system. Under reservoir temperature and high pressure, more than 10,000 psi, when the sample was in equilibrium, the sample was depressurized

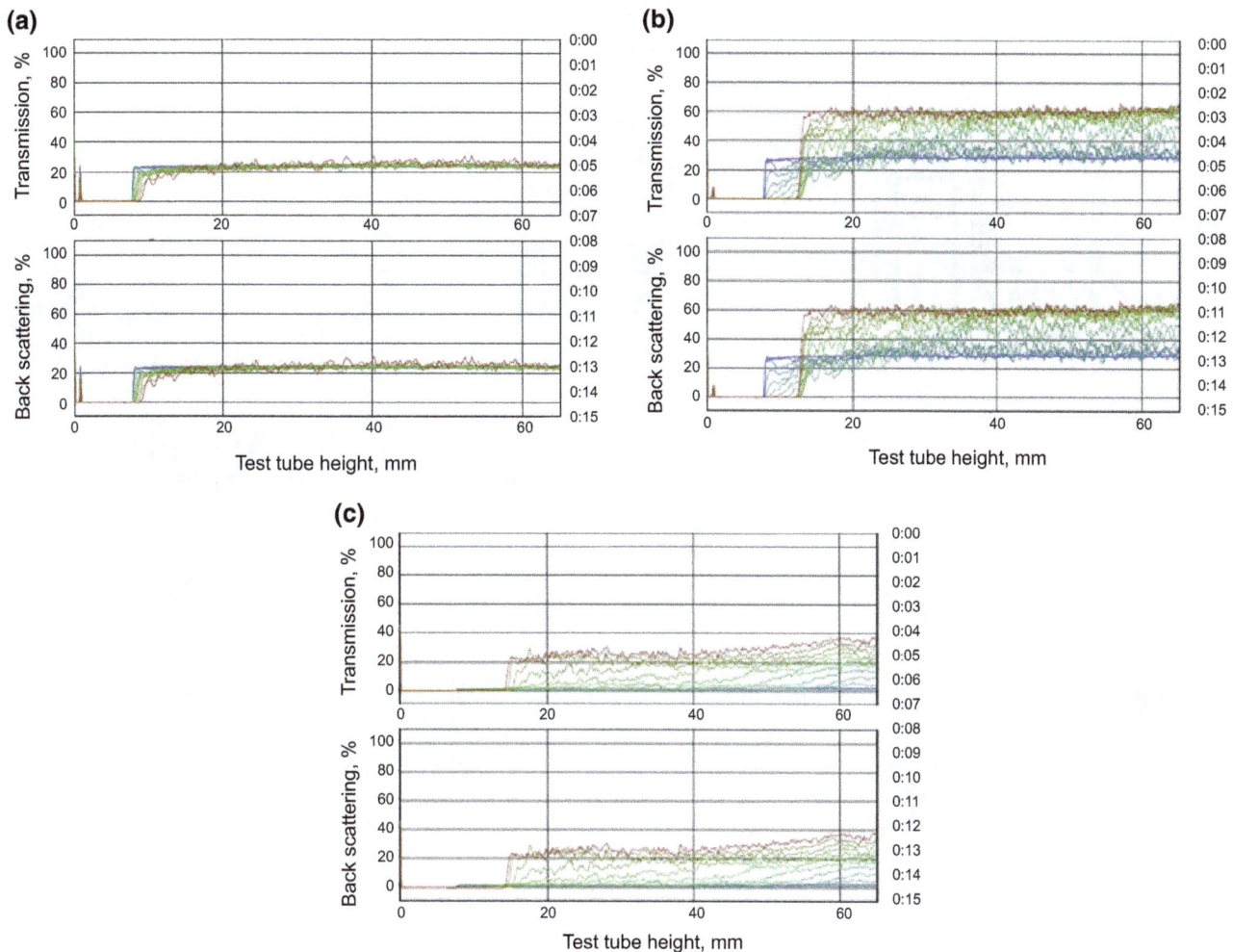

Fig. 2 Received light from stable (**a**), semi-stable (**b**), and unstable (**c**) mixtures (The horizontal axis is the height of the test tube. 0 is the tube bottom and 60 mm is its upper part)

and the data acquisition system recorded the received power. Due to depressurization, the oil expanded and there would be a reduction in the oil density. Therefore, the optical fibers received more light. At the onset pressure of asphaltene precipitation and below the asphaltene particles absorbed the transmitted light and there was a change in the slope of the graph. However, due to the density reduction it still had a rising trend. The schematic of the SDS setup is shown in Fig. 3.

For evaluation of the selected AD under reservoir conditions, it should be injected into the PVT cell. Since the amount of the AD (400 ppm) was very low compared to the live oil (50 mL), therefore one experiment was designed under laboratory conditions to determine the best carrier fluid. With 1 mL of this fluid, the AD can be added to the live oil. Before selecting the best carrier fluid, the origin of solid particles should be known. Several tests were conducted to characterize the solid particles,

including X-ray diffraction (XRD), X-ray fluorescence (XRF), ash content, IP-143, and saturates, aromatics, resins and asphaltenes (SARA) tests. These analyses can determine the presence of different elements in addition to the amount of asphaltenes in the solids.

In the first set of tests, the amount of asphaltene solids which can be dissolved in different solvents was investigated and the results were compared with dissolution power of xylene. Xylene is very useful to remove organic materials; however, it is a hazardous air pollutant and is known or suspected to cause cancer or other serious health problems. Xylene is also expensive and it is not economic to use large amounts. We tried here to decrease xylene concentration in the carrier fluid. For this purpose, the asphaltene particles were broken into small pieces and then ground with a mortar and pestle and 0.5 g of asphaltene powders were placed in a test tube. 50 mL of solvent was added to the powder. The prepared mixture was shaken

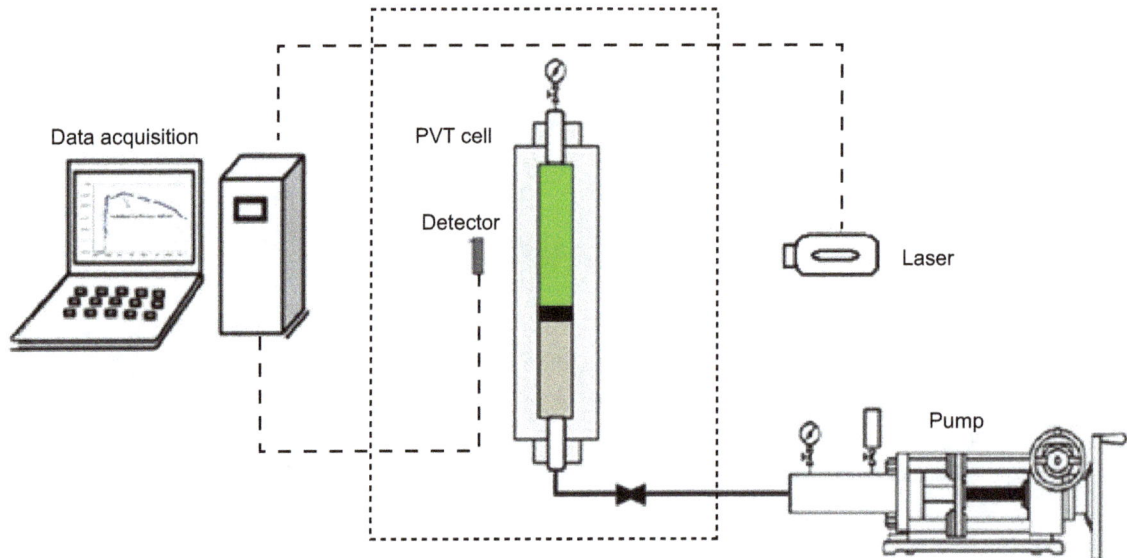

Fig. 3 The schematic of SDS setup

rigorously until the asphaltene dispersed into the solvent. After 15 min the sample was passed through a filter paper. The filter paper was then put in the oven for 2 h at 220 °C. The weights of the filter paper, before and after the experiment, can determine the amount of asphaltenes which was dissolved into the solvent. We could not perform this method for oil B as it is very dark and heavy. Therefore, all the measurements are for oil A.

2.3.5 Filtration method

The sample underwent a similar path as the SDS method; however, at each pressure step, it passed through the stainless steel filter (SWAGLOK, 0.2 μm) and the asphaltene aggregates greater than 0.2 μm remained on the filter surface. The maximum allowable differential pressure for this type of filter is 1000 psi. However, in this experiment, we tried to avoid differential pressures higher than 200 psi during sampling (Fig. 4). The change in the asphaltene content of the crude oil could determine the amount of asphaltenes precipitated and deposited in the PVT cell. For evaluating the selected ADs under reservoir conditions, the mixture of the AD and the carrier fluid was injected into the PVT cell at pressures higher than the onset pressure of asphaltene precipitation. By comparing the graphs of asphaltene weight percent versus pressure, with and without the presence of ADs, one could determine the efficiency of the AD under reservoir conditions. This method was used instead of the SDS method for heavy oils and oils with high asphaltene content. The measurements, here, are for oil B.

3 Results and discussion

3.1 Results of the tests under laboratory conditions

3.1.1 Microscopic test

The onset of asphaltene precipitation measured by titration for different oil/AD systems are shown in Tables 3 and 4. The effect of toluene and xylene on the onset of asphaltene precipitation was also investigated.

Tables 3 and 4 indicate that dispersants DB and DP show better performance than dispersants DF and DR since the precipitation onset of blank oil A changed from 0.43 to a value about 0.6. For dead oil B, the change in the onset was from 0.63 to 0.72. Dispersants DB and DP may delay the onset of asphaltene precipitation so that a higher volume ratio of n-heptane to oil is required for asphaltene precipitation. In addition, the two solvents, toluene and xylene, did not show significant effect on the onset of asphaltene precipitation. Figures 5 and 6 show the snapshots from samples under the microscope before and after the appearance of onset of asphaltene precipitation upon the addition of n-heptane on oil A. It was observed that dispersants with better performance prevent the formation of bigger aggregates, which is in agreement with the purpose of using ADs.

3.1.2 Viscometer test

In order to compare the inhibition power of ADs, the best dispersant (DP) and the worst one (DR) were selected from

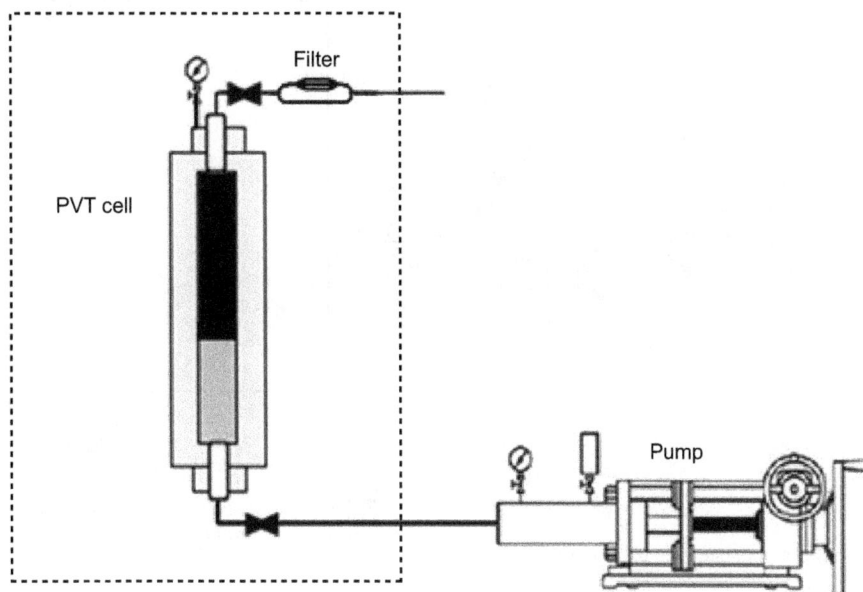

Fig. 4 The schematic of the filtration apparatus

Table 3 Onset of asphaltene precipitation in the presence of dispersants in the dead oil A

Sample no.	Oil vol. V_{oil}, mL	Dispersant	AD concentration, ppm	n-Heptane vol. $V_{n\text{-heptane}}$, mL	Onset, $V_{n\text{-heptane}}/V_{oil}$
1	40	–	–	17.0	0.43
2	40	DB	400	23.5	0.58
3	40	DF	400	20.5	0.51
4	40	DP	400	24.0	0.60
5	40	DR	400	17.5	0.44
6	40	Toluene (1 mL)	–	17.8	0.45
7	40	Xylene (1 mL)	–	18.3	0.46

Table 4 Onset of asphaltene precipitation in the presence of dispersants in the dead oil B

Sample no.	Oil vol. V_{oil}, mL	Dispersant	AD concentration, ppm	n-Heptane vol. $V_{n\text{-heptane}}$, mL	Onset, $V_{n\text{-heptane}}/V_{oil}$
8	40	–	–	25.2	0.63
9	40	DB	350	28.7	0.72
10	40	DF	350	25.9	0.65
11	40	DP	350	28.0	0.70
12	40	DR	350	25.2	0.63
13	40	Toluene (1 mL)	–	25.7	0.64
14	40	Xylene (1 mL)	–	26.0	0.65

experimental results obtained with previous methods. Figure 7 shows the changes in viscosity due to the addition of n-heptane to different samples. According to Fig. 7, the onset of precipitation can be detected by the change in the slope of the viscosity graph. Table 5 shows that the onset of asphaltene precipitation for the blank oil and the mixture of oil and DR were almost equal. Dispersant DP had better performance on delaying asphaltene onset. The results

Fig. 5 Before the formation of asphaltene precipitation (dead oil A)

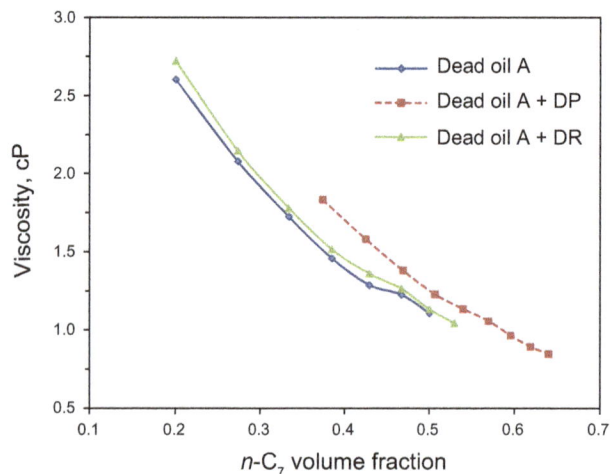

Fig. 7 Viscosity changes based on the addition of n-C_7 to dead oil A, dead oil A + DP (400 ppm), and dead oil A + DR (400 ppm)

other and such results were expected since the methods are based on distinct fundamentals (Mansur et al. 2009).

3.1.3 Turbidity test

Table 6 shows the ratio of n-heptane to oil which leads to the formation of unstable mixtures for both dead oils A and B. These ratios can be considered as the reference ratios for the evaluation of dispersants efficiency.

According to Table 7, dispersants DB, DF, and DP are promising chemicals with more positive effect on the stability of asphaltene in both dead oils A and B. It can also be observed that the dispersant DR is inefficient in both oil samples.

3.2 Results of the tests under reservoir conditions

The inhibition power of the four dispersants under laboratory conditions is as follows: DP $\approx$ DB > DF > DR. In contrast to dispersants DF and DR, it seems that ADs with nonylphenol-formaldehyde resin modified by polyamines (DP) and polyisobutylene succinimide (DB) bases show higher efficiency on these oil samples. The main focus under reservoir conditions is on dispersants DP and DB to select the best one for field applications.

3.2.1 SDS

The analyses of solid particles are presented in Tables 8 and 9. The results show that a very small proportion of solid particles (0.3 %) were made up of inorganic material and the rest were organic. As such, elemental analyses are not required, as they are not important in comparison with organic components (99.7 %). The SARA analysis

Fig. 6 Onset of asphaltene precipitation (dead oil A)

obtained here are in good agreement with the results of the two previous methods for oil A.

Before the onset of asphaltene precipitation, the addition of n-heptane dilutes the mixture that is why there is a decreasing trend in the slope of the graph. However, at the onset point, the viscosity increases owing to the formation of asphaltene aggregates. The effect of precipitating asphaltenes may not be apparent until well beyond the onset of asphaltene precipitation due to its low concentration. The change in the slope of graphs is not as sharp as it was reported by Mousavi-Dehghani et al. (2004). Any further increase in the volume of n-heptane can dilute the mixture and it will cover the effect of precipitated asphaltenes. The onset values from viscometric and microscopic methods are a little bit different from each

Table 5 Onset of asphaltene precipitation with and without a dispersant, determined by viscosity measurement

Sample	Oil vol. V_{oil}, mL	Dispersant concentration, ppm	Onset, $V_{n\text{-heptane}}/V_{oil}$
Dead oil A	1	0	0.45
Dead oil A + DP	1	400	0.56
Dead oil A + DR	1	400	0.46

Table 6 Asphaltene stability at different n-heptane/oil ratios

Oil	Oil vol., mL	n-heptane vol., mL	Separability number	Stability
Dead oil A	1	10	2.2	High
	1	20	3.7	High
	1	40	11.2	Low
Dead oil B	1	10	0	High
	1	20	1.3	High
	1	30	3.7	High
	1	50	12.7	Low

Table 7 The effect of dispersants on the asphaltene stability

Sample	Dispersant	Oil vol., mL	n-heptane vol., mL	Dispersant concentration, ppm	Separability number	Stability
Dead oil A	–	1	40	–	11.2	Low
	DB	1	40	400	0.4	High
	DF	1	40	400	0.5	High
	DP	1	40	400	0.3	High
	DR	1	40	400	11.0	Low
Dead oil B	–	1	50	–	12.7	Low
	DB	1	50	350	0.1	High
	DF	1	50	350	1.7	High
	DP	1	50	350	0	High
	DR	1	50	350	12.2	Low

Table 8 Ash content of the solid particles

Procedure	Ash content, wt%
ASTM D-482	0.3

Table 9 SARA analysis of solid particles

Asphaltenes, %	Resins, %	Aromatics, %	Saturates, %
54.0	6.6	12.9	26.5

(Table 9) indicates that the solids consisted mainly of asphaltenes.

The dissolution power of different solvents defined as the wt% of solid asphaltene particles dissolved in the solvent was measured and presented in Fig. 8. Solvents which show similar effect to xylene for dissolving solids can be considered as carrier fluids. The results show that apart from the pure xylene and toluene, a mixture of xylene and kerosene with volume ratios of 80:20 and 60:40 can give higher efficiency compared to other solvents. Here, a mixture of xylene and kerosene with a ratio of 80:20 was selected as the carrier fluid.

For the first step, the onset pressure of asphaltene precipitation for the live oil was obtained. This step is necessary for investigating the effectiveness of ADs. Since the amount of live oil was known, the prepared mixture of carrier fluid and dispersant should be such that the final mixture has the desired concentration of the AD. Here, 1 mL of the solvent accompanied by the selected AD was injected into the PVT cell containing 50 mL of live oil A. In Fig. 9, the onset of asphaltene precipitation of the live oil was determined, which is about 4300 psi. The effect of ADs on the onset pressure can be observed in Fig. 9. The figure clearly shows that dispersants DP and DB had

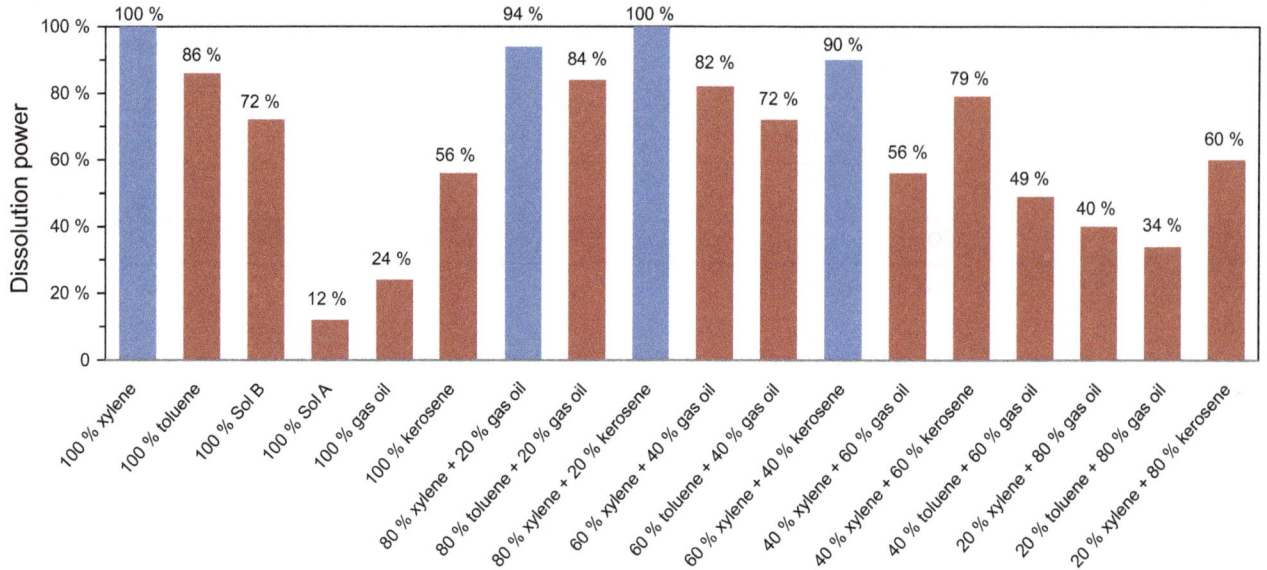

Fig. 8 The dissolution power of different solvents

altered the asphaltene onset pressure (AOP) and decreased it to 3600 and 4000 psi, respectively. Figure 9 also demonstrates the dispersant DP prevented asphaltene aggregation and kept it in the solution. Below the onset of asphaltene precipitation and above the bubble point, the pressure reduction led to an increase in the size of aggregates; however, the presence of dispersant in the solution inhibited the growth of aggregated asphaltenes.

3.2.2 Filtration

According to the literature, due to depressurization above the bubble point, the screening effect on the asphaltene nanoaggregates decreases and as a result the

interactions between them become greater; therefore precipitation occurs. Below the bubble point pressure, the solubility of asphaltene increases as the light fraction of oil separated from oil (Buenrostro-Gonzalez et al. 2004; Shokrlu et al. 2011). Figure 10 shows that the amount of asphaltenes precipitated in the presence of DP decreased to over 30 % of its initial value. The graph also shows that the addition of chemicals did not change the bubble point of the crude oil because of the addition of very small amount of chemicals in comparison to the live oil. The accuracy of this method depends considerably on the asphaltene content of the crude oil. For light oils (i.e., low asphaltene content) it is almost impossible

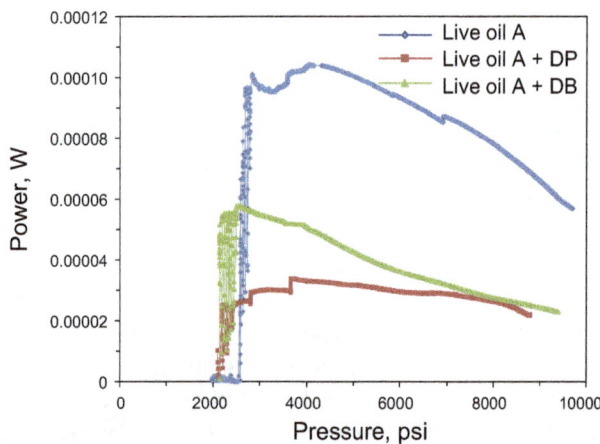

Fig. 9 Asphaltene precipitation onset for live oil A, live oil + DP (400 ppm), live oil + DB (400 ppm)

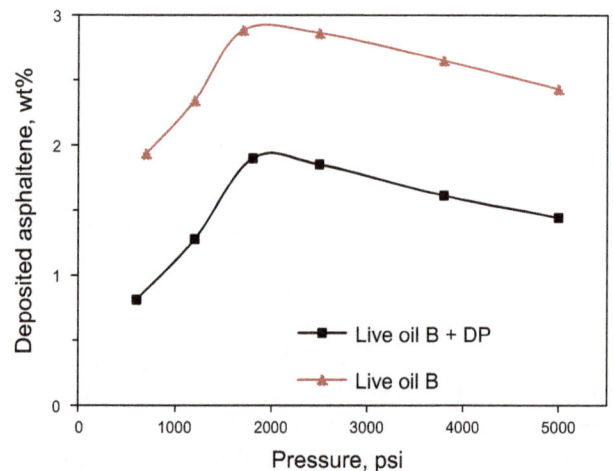

Fig. 10 Change in asphaltenes (wt%) during pressure depletion for live B and live oil B + DP (350 ppm)

Table 10 Overview of experimental results

Sample	Test condition	Oil	Experiment	Method of AD selection	DB	DF	DP	R	Best chemical
Oil A	Atmospheric	Dead	Microscopy	Onset point by high ratio of n-C$_7$/oil	$\checkmark$	$\checkmark$	$\checkmark$	$\checkmark^a$	DP, DB
		Dead	Turbiscan	Low separability number (<5.0)	$\checkmark$	$\checkmark$	$\checkmark$	$\checkmark$	DP, DB, DF
		Dead	Viscometer	Onset point by high ratio of n-C$_7$/oil	–	–	$\checkmark$	$\checkmark$	DP
	Reservoir	Live	SDS	Low onset pressure	$\checkmark$	–	$\checkmark$	–	DP
Oil B	Atmospheric	Dead	Microscopy	Onset point by high ratio of n-C$_7$/oil	$\checkmark$	$\checkmark$	$\checkmark$	$\checkmark$	DB, DP
		Dead	Turbiscan	Low separability number (<5.0)	$\checkmark$	$\checkmark$	$\checkmark$	$\checkmark$	DP, DB, DF
	Reservoir	Live	Filtration	Low asphaltene content	–	–	$\checkmark$	–	DP

[a] Test was done

to follow the effect of thermodynamic changes on the asphaltene precipitation behavior using the filtration method.

The results obtained here demonstrate that in order to select the best asphaltene dispersant for injection into wells, it is compulsory to screen the available chemicals under laboratory conditions to minimize cost and time and then choose the high efficient ones for tests under reservoir conditions. An overview of the experiments is presented in Table 10.

These experiments are valuable as they provide reliable estimation of the results that more likely happen in the field. Based on the acquired results, the injection of the selected dispersant (DP) can properly disperse the aggregating asphaltenes.

4 Conclusions

In this study, four asphaltene dispersants, polyisobutylene succinimide (DB), polyisobutylene succinic ester (DF), nonylphenol-formaldehyde resin modified by polyamines (DP), and rapeseed oil amide (DR) were evaluated under laboratory and reservoir conditions. In order to obtain reliable results, full analyses of reservoir fluids and solid asphaltene particles were done. The conditions at which the asphaltene is precipitated out of the solution, such as the addition of precipitant and changing the pressure were conducted on the oil samples with and without ADs. The following results were obtained:

(1) The solid particles contained nearly all organic material and they were mainly of asphaltenes. A mixture of 20 % kerosene and 80 % xylene could completely dissolve the asphaltene solids.

(2) Under laboratory conditions, dispersant DR, fatty oil amide based on rapeseed oil, had the lowest performance among the four different ADs on oil A and oil B.

(3) The onsets of asphaltene precipitation obtained under laboratory conditions using viscometry and microscopy methods are virtually equal. The onset of asphaltene precipitation and its stability was investigated by an optical microscope, a viscometer, and a Turbiscan under laboratory conditions. The SDS setup and filtration method were also applied under reservoir conditions to find the effect of dispersants on the onset pressure of asphaltene precipitation. Among different dispersants, the dispersant DP, based on nonylphenol-formaldehyde resin modified by polyamines, proved to have the best performance under laboratory and reservoir conditions as it moved the onset pressure from 4300 to 3600 psi. Tests on the selected dispersants under reservoir conditions show that in contrast to dispersant DB, dispersant DP in the oil phase inhibited the formation of higher aggregates below the onset pressure of asphaltene precipitation when the pressure decreased. This is exactly what the ADs are expected to do.

References

Barcenas M, Orea P, Buenrostro-Gonzalez E, et al. Study of medium effect on asphaltene agglomeration inhibitor efficiency. Energy Fuels. 2008;22(3):1917–22. doi:10.1021/ef700773m.

Borchardt JK. Chemicals used in oil-field operations. In: Oil-field chemistry, ACS Symposium Series. Washington: ACS; 1989. 396:3–54.

Bouts MN, Wiersma RJ, Muijs HM, Samuel AJ. An evaluation of new asphaltene inhibitors: laboratory study and field testing. J Pet Technol. 1995;47(9):782–7.

Buenrostro-Gonzalez E, Lira-Galeana C, Gil-Villegas A, Wu J. Asphaltene precipitation in crude oils: theory and experiments. AIChE J. 2004;50(10):2552–70. doi:10.1002/aic.10243.

Cosultchi A, Garciafigueroa E, Carcia-Borquez A, et al. Petroleum solid adherence on tubing surface. Fuel. 2001;80(13):1963–8.

Ferrara M. Hydrocarbon oil-aqueous fuel and additive compositions. WO Patent 1995:637.

Hirscberg LNJ, de Jong BA, Schipper BA, Meijer JG. Influence of temperature and pressure on asphaltene flocculation. SPE J. 1984;24(3):283–93. doi:10.2118/11202-PA.

Junior LCR, Ferreira MS, Ramos ACS. Inhibition of asphaltene precipitation in Brazilian crude oils using new oil soluble amphiphiles. J Pet Sci Eng. 2006;51(1–2):26–36. doi:10.1016/j. petrol.2005.11.006.

Kawanaka S, Park SJ, Mansoori GA. Organic deposition from reservoir fluids: a thermodynamic predictive technique. SPE Reserv Eng. 1991;6(2):185–92. doi:10.2118/17376-PA.

Leontaritis KJ. Offshore asphaltene and wax deposition: problems/solutions. World Oil. 1996;217(5):57–63.

Manek MB. Asphaltene dispersants as demulsification aids. In: SPE international symposium on oilfield chemistry, 14–17 February, San Antonio; 1995. doi:10.2118/28972-MS.

Mansoori GA. Modeling of asphaltene and other heavy organic depositions. J Pet Sci Eng. 1997;17(1–2):101–11. doi:10.1016/ S0920-4105(96)00059-9.

Mansoori GA. Remediation of asphaltene and other heavy organic deposits in oil wells and in pipelines. ELMI Sarlar J Reserv Pet Eng. 2010;SOCAR:12–23.

Mansur CRE, Guimaraes ARS, Gonzalez G, Lucas EF. Determination of the onset of asphaltene precipitation by visible ultraviolet spectrometry and spectrofluorimetry. J Anal Lett. 2009;42(16):2648–64.

Marques LCC, Gonzalez G, Monteiro JB. A chemical approach to prevent asphaltenes flocculation in light crude oils: state of the art. In: SPE annual technical conference and exhibition, 26–29 September, Houston; 2004. doi:10.2118/91019-MS.

Miri R, Zendehboudi S, Kord Sh, et al. Experimental and numerical modeling study of gravity drainage considering asphaltene deposition. Ind Eng Chem Res. 2014;53(28):11512–26. doi:10. 1021/ie404424p.

Mousavi-Dehghani SA, Riazi MR, Vafaie-Sefti M, Mansoori GA. An analysis of methods for determination of onsets of asphaltene phase separations. J Pet Sci Eng. 2004;42:145–56. doi:10.1016/j. petrol.2003.12.007.

Oschmann HJ. New methods for the selection of asphaltene inhibitors in the field. Special Publication, Royal Society of Chemistry. 2002;280:254–63.

Shokrlu YH, Kharrat R, Ghazanfaria MH, Saraji S. Modified screening criteria of potential asphaltene precipitation in oil reservoirs. Pet Sci Technol. 2011;29(13):1407–18. doi:10.1080/ 10916460903567582.

Smith DF, Klein GC, Yen AT, et al. Crude oil polar chemical composition derived from FT-ICR mass spectrometry accounts for asphaltene inhibitor specificity. Energy Fuels. 2008;22(5): 3112–7. doi:10.1021/ef800036a.

Speight JG, Wernick DL, Gould KA, et al. Molecular weight and association of asphaltenes: a critical review. Oil Gas Sci Technol: Rev. IFP. 1985;40(1):51–61. doi:10.2516/ogst:1985004.

Takhar S. A fast and effective chemical screening technique for identifying asphaltene inhibitors for field deployment. In: Proceedings of the second international conference on fluid and thermal energy conversion 1997; 83–90.

Thawer R, Nicoll DCA, Dick G. Asphaltene deposition in production facilities. SPE Prod Eng. 1989;5(4):475–80. doi:10.2118/18473-PA.

Turta A, Najman J, Fisher D, Singhal A. Viscometric determination of the onset of asphaltene flocculation. In: Annual technical meeting, 8–11, June, Calgary; 1997. doi:10.2118/97-81-MS.

Zendehboudi S, Ahmadi MA, Mohammadzadeh OR, et al. Thermodynamic investigation of asphaltene precipitation during primary oil production: laboratory and smart technique. Ind Eng Chem Res. 2013;52:6009–31. doi:10.1021/ie301949c.

Zendehboudi S, Shafiei A, Bahadori A, et al. Asphaltene precipitation and deposition in oil reservoirs–technical aspects, experimental and hybrid neural network predictive tools. Chem Eng Res Des. 2014;92(5):857–75. doi:10.1016/j.cherd.2013.08.001.

3

On the feasibility of re-stimulation of shale wells

Mohammad O. Eshkalak[1] · Umut Aybar[1] · Kamy Sepehrnoori[1]

Abstract   As a result of advances in horizontal completions and multi-stage hydraulic fracturing, the U.S. has been able to economically develop several decades of worth of natural gas. However, a considerable concern has risen on the economic viability of shale gas development for reasons associated with the fast production declines as well as recent down-turns of natural gas prices besides rises in the costs of new technologies. Therefore, an economic analysis is required to investigate the profitability of the re-fracturing treatment of unconventional gas resources. Net present value of cash flows and internal rate of return are calculated for a range of gas prices considering 20 years of natural gas production from a typical unconventional shale gas reservoir. A systematic comparison is then accomplished for three scenarios: (1) re-fracturing versus no re-fracturing, (2) combination of re-fracturing and drilling new wells, and (3) time-dependent re-fracturing treatment. Further, this paper incorporates the cost of re-fracturing treatment, the cost of drilling a new horizontal well, the water treatment cost, as well as the current and future price of natural gas in the model. The findings of this work would help the future re-stimulation development plans of the emerging unconventional shale gas plays.

Keywords   Economic analysis · Unconventional shale assets · Hydraulic re-fracturing · Net present value · Internal rate of return

✉ Mohammad O. Eshkalak
  eshkalak@utexas.edu

[1] The University of Texas at Austin – PGE, 200 E Dean Keeton, Austin, TX 78712, USA

Edited by Xiu-Qin Zhu

Abbreviations

SRR	Source rock reservoir
NPV	Net present value
IRR	Internal rate of return
LRE	Long-term re-fracturing efficiency
D&C	Drilling and completion
EUR	Estimated ultimate recovery
FC	Fixed costs
V_F	Future value of production revenue for a fracture reservoir, $
V_0	Future value of production revenue for an un-fractured reservoir, $
i	Interest rate
C_{Well}	Cost of one horizontal well, $
$C_{Fracturing}$	Cost of hydraulic fracturing, $
$C_{Re\text{-}fracturing}$	Cost of re-fracturing, $
N	Number of horizontal wells, $
MSCF	10^3 standard cubic feet, ft^3
BCF	10^6 standard cubic feet, ft^3

1 Introduction

A substantial fraction of United States natural gas production for the next decades is surmised to be supported by unconventional resources, such as shale gas plays. Large accumulations of gas shale tight formations serve as both a hydrocarbon source and a productive reservoir. Most of the gas is stored in organic-rich rock, while a lesser portion of gas in place is in pore spaces (Cipolla et al. 2010). Also, 500–1000 Tcf of potential natural gas reserve is estimated to be in place in unconventional assets (Arthur 2008). Extremely low matrix permeability as well as highly complex networks of natural fractures are unique characteristics of shale formations. Permeability of shale rocks is estimated to be between 50 nD (nano-Darcy) to 150 nD (Javadpour et al. 2007). Recent advances and innovations in hydraulic fracturing are key to the success of shale gas economic production as a viable global energy supply.

Shale gas reservoirs have some unique attributes which make hydraulic fracturing a viable option for natural gas production. Unlike conventional gas reservoirs, insufficient permeability, the ultra-low porosity of shale rock, and the limited reservoir contact, but the widespread organic matter in shale, cannot offer production in a commercial quantity without stimulation processes. Development of shale resources is still in its early stages and most wells are at the early stage of their working lifetime. Moreover, reservoir simulations and modeling of unconventional reservoirs has gained much attention in the recent years. Many studies have been conducted from the shale pore scale up to reservoir scales to improve the understanding of flow behavior in complex shale formations. Among them, researchers such as Brown et al. (2009), Cipolla et al. (2010), Moghanloo and Javadpour (2014), Omidvar Eshkalak (2013), Aybar et al. (2014a, b, c, 2015), and Eshkalak et al. (2013, 2014a, b, c, d, f) have developed and discussed numerical, quasi-static, analytical and semi-analytical reservoir models for unconventional reservoirs.

The combination of advances in hydraulic fracturing and horizontal drilling has led to the acceptance of these techniques for enhancing the production from shale strata since their first commercial implementation. Nevertheless, drilling so many horizontal wells to increase the production has not been a solution to the economic success of the shale development projects. Additionally, a recent decline in natural gas price has led to a huge shrinkage of shale gas development projects in the U.S. and operators have reduced their rig counts in these unconventional basins. Therefore, a comprehensively engineered economic model is necessary for the processes that boost natural gas production from depleted wells, which also in turn, guides decision-making processes for operators in their current development plans.

2 Re-fracturing treatment of shale gas wells

When production rates drop below economic limits, significant amounts of producible reserves still remain in the existing stimulated reservoir volume. In general, shale gas wells usually show a sharp decline at the beginning due to free gas production (existing in natural fractures or pore spaces around the wellbore) which is captured through a long transient liner flow. As a result, re-fracturing is often considered as the best option for increasing production from unconventional gas reservoirs to an economic level. Nevertheless, the re-fracturing treatment of shale wells is still in its infancy where the applicability of the technology has not yet been proven and the conditions under which it may be successful are not clearly understood for long-term profitability of shale reservoirs. Jayakumar et al. (2013) discussed that re-fracturing can be applied to shale fields because of two reasons. First, the original fracturing network has no significant contribution to the flow to the wellbore and second, the initial completion performance has degraded over time below operational or economic limits.

There are publications that address different aspects of re-fracturing treatments. For Barnett shale, Siebrits et al. (2000) reported increased production of natural gas by re-fracturing treatment. On the selection of candidate wells and time of re-fracturing treatment, researchers such as Craig and Blasingame (2005), Rousell and Sharma (2011), Moore and Ramakrishnan (2006), and Tavassoli et al. (2013) discussed and have developed criterion-based approaches. Moreover, the success of a re-fracturing treatment depends on the depleted reservoir pressure and hydraulic fracture geometry (Vincent 2010; Shekar and Hariharan 2011; Wang et al. 2013). We suggest that consideration should be given from the beginning to determine the best way to accomplish a re-fracturing treatment when the primary production has declined to a predetermined point.

Furthermore, the re-fracturing treatment is considered more beneficial because of two reasons: (1) it can be an alternative of new well drilling, and potentially can save around 1–4 million dollars (Alison and Parker 2014); and (2) the environmental impact of reusing a wellbore is dramatically less than drilling and completing a new well in a different location. However, uncertainties, associated with outcome of re-fracturing and the economic analysis of key parameters influencing its profitability, are still challenging, and need wider investigation and systematic studies. Hence, preventing a non-economic development of re-fracturing necessitates an economic analysis of the re-fracturing treatment.

There are three main challenges in re-fracturing treatment of shale gas wells, namely the selection of candidate wells, determination of optimal re-fracturing time, and the placement of new fractures. A robust procedure introduced by Tavassoli et al. (2013) is employed in this study, with which all the wells are considered to be good candidates satisfying their criteria. Further, the re-fracturing treatment is applied after 5 years of production as an optimal re-fracturing time. Also, Tavassoli et al. (2013) found that natural gas production achieved its maximum value by

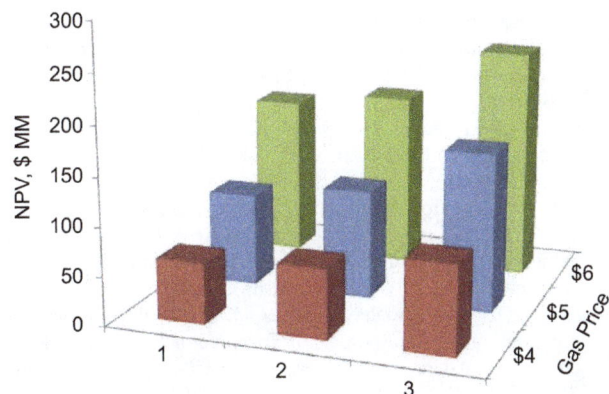

Fig. 2 Calculated NPV for 20 years for three scenarios with three different gas prices

administration costs. They represent re-occurring annual cash costs incurred throughout the economic life of a well. The value of $1 per Mscf is assumed and considered constant for the entire life of all wells ($300,000 per year assuming 0.3 BCF gas produced per year per well).

3.5 State and federal corporate income taxes

The state and federal taxes as additional expenses are accounted into the cash flow statement. A state tax of 10 % and federal income tax of 35 % are considered in this study.

The general approaches for calculating NPV and IRR are based on the theories reported by Newman (1988) and Ikoku (1985). The NPV formula is given in Eq. 1:

$$NPV = \sum_{j=1}^{n} \frac{(V_F)_j}{(1+i)^j} - \sum_{j=1}^{n} \frac{(V_0)_j}{(1+i)^j}$$
$$- \left[FC + \sum_{k=1}^{N} (C_{well} + C_{Fracturing} + C_{Re-fracturing}) \right],$$
$$(1)$$

where V_F is the future value of production revenue for a fracture reservoir, V_0 is the future value of production revenue for an un-fractured reservoir, i is the interest rate, FC is the total fixed cost, C_{well} is the cost of one horizontal well, $C_{fracture}$ and $C_{re-fracture}$ are the cost of hydraulic

fracturing and re-fracturing in a horizontal well, respectively, and N is the number of horizontal wells.

4 Results and analysis

The goal of this paper was to evaluate the profitability of the re-fracturing treatment of a typical shale gas reservoir. This economic evaluation is performed based on NPV and IRR calculations considering cash flow assumptions discussed above. Several calculations are completed in order to determine the profitability of each scenario incorporating different gas prices. The NPV is calculated for each scenario considering a discount rate of 10 %, which has been accepted as a minimum acceptable rate of return in the natural gas industry (MIT Energy Initiative, 2010). All the calculated NPVs greater than zero are considered profitable.

Figure 2 shows the calculated NPV for all the three scenarios performed with the assumptions of three different gas prices. As it demonstrates, NPV values are positive for all the three scenarios, showing the profitability of the re-fracturing treatment as a method for enhancing the gas recovery and the overall economy of a typical shale gas asset with 50 horizontal gas wells. With a slight rise in gas price, the NPV increases rapidly. This fast turnover also results in a fast development of new wells drilled in unconventional gas fields aside from re-fracturing old wells.

The calculation of IRR of the cash flow is performed after NPV is calculated. This IRR represents the interest rate that yields an NPV for the cash flow equivalent to zero. All the IRRs are considered profitable for the values above 10 %. Table 3 shows the IRR calculated for each scenario. These values are acceptable since they are greater than 10 % (MIT Energy Initiative 2010) and show that scenarios studied in this paper are profitable given the assumptions made based on the predictions for 20 years of production. It is demonstrated that with a little change in gas price, the *IRR* is raised. In this study, 3rd scenario is the most profitable one, which shows that drilling a new well must be considered in the long-term development planning of shale gas reservoirs.

Table 3 Internal rate of return (IRR)

Gas price	IRR for different scenarios, %			
	No re-fracturing	(1) Re-fracturing	(2) Re-fracturing in 5 years intervals	(3) Re-fracturing plus drilling 10 new wells at year 10
$4	15	18	22	24
$5	35	38	43	49
$6	50	54	61	65

5 Conclusions

This study provides an assessment of the importance of re-fracturing treatment of shale gas wells. This analysis allows us to predict the future pace of the re-fracturing treatment activity which also helps the large-scale economic planning of unconventional resources within the U.S. Results of the three scenarios are as follows:

(1) The calculated *NPV* for all three scenarios are positive; this demonstrates the profitability of the re-fracturing treatment considering today's gas price of $4.

(2) *NPV* is almost doubled with the increase in the gas price considering the third scenario. This shows that a slight raise in the natural gas price will make a huge jump in the development plans of unconventional shale gas.

(3) The highest *NPV* is gained when the combination of re-fracturing and new well drilling is planned. This scenario is recommended for the future development plans once the gas price rises from $4 to $6.

(4) It is also recommended that in order to have a higher level of *IRR*, more new horizontal wells must be drilled considering a constant gas price of $4.

(5) According to the economic assumptions used in this study, the re-fracturing treatment of shale gas wells is demonstrated to play an important role in the economic success of an unconventional asset.

Given these conclusions, the re-fracturing treatment of shale horizontal wells with properly identified candidates not only recoups the overall economic recovery of shale wells but also makes a profit. Moreover, pertinent information must be gathered along with an economic analysis before the treatment commences.

References

Alison D, Parker M. Re-fracturing extends lives of unconventional reservoirs. The American Oil and Gas Reporter, Exploration/Drilling/Production History, 2014

Arthur, JD. Hydraulic fracture for natural gas well of the Marcellus Shale, The ground water production council 2008 annual forum. Cincinnati; 2008.

Aybar U, Eshkalak MO, Sepehrnoori K, et al. Long term effect of natural fractures closure on gas production from unconventional reservoirs. Paper SPE 171010 presented in SPE Eastern Regional Meeting 2014 held in Charleston, West Virginia. 21–23 Oct 2014a.

Aybar U, Eshkalak MO, Sepehrnoori K, et al. The effect of natural fracture's closure on long-term gas production from unconventional resources. J Nat Gas Sci Eng. 2014;. doi:10.1016/j.jngse.2014.09.030.

Aybar U, Yu W, Eshkalak MO, et al. Evaluation of production losses from unconventional shale reservoirs. J Nat Gas Sci Eng. 2015;23:509–16.

Aybar U. Investigation of analytical models incorporating geomechanical effects on production performance of hydraulically and naturally fractured unconventional reservoirs. MSc Thesis. The University of Texas at Austin, Austin, 2014c.

Brown M, Ozkan E, Raghavan R, et al. Practical solutions for pressure transient responses of fractured horizontal wells in unconventional reservoirs. Paper SPE 125043 presented at the SPE ATCE, New Orleans, LA. 4–7 Oct, 2009.

Cipolla CL, Lolon EP, Erdle JC, et al. Reservoir modeling in shale-gas reservoirs. SPE Reserv Eval Eng. 2010;13(4):638–53 **SPE 125530-PA**.

Craig DP, Blasingame T. A new refracture candidate diagnostic test determines reservoir properties and identifies existing conductive or damaged fracture. Paper SPE 96785 presented at annual technical conference and exhibition, Dallas. 9–12 Oct 2005.

Duman RJ. Economic viability of shale gas production in the marcellus shale; indicating by production rates, cost and current natural gas price. MS thesis, Applied Natural Resources Economics, Michigan Technology Institute, 2012.

Eshkalak MO, Al-Shalabi EW, et al. Enhanced gas recovery by CO_2 sequestration versus re-fracturing treatment in unconventional shale gas reservoirs. Paper SPE 172083 presented at 2014 Abu Dhabi international petroleum and exhibition and conference held in Abu Dhabi, UAE, 10–13 Nov 2014a.

Eshkalak MO, Al-Shalabi EW, et al. Simulation study on the CO_2-driven enhanced gas recovery with sequestration versus the re-fracturing treatment of horizontal wells in the U.S. unconventional shale reservoirs. J Nat Gas Sci Eng. 2014a;. doi:10.1016/j.jngse.2014.10.013.

Eshkalak MO, Aybar U, Sepehrnoori K. An economic evaluation on the re-fracturing treatment of the U.S. shale gas resources. Paper SPE 171009 presented at the 2014 SPE eastern regional meeting held in Charleston. 21–23 Oct 2014c.

Eshkalak MO, Aybar U, Sepehrnoori K. An integrated reservoir model for unconventional resources, coupling pressure dependent phenomena. Paper SPE 171008 presented at the 2014SPE eastern regional meeting held in Charleston. 21–23 Oct 2014d.

Eshkalak MO, Mohaghegh SD, Esmaili S. Synthetic, geomechanical logs for Marcellus Shale. Paper SPE 163690 presented at the 2013 SPE digital energy conference and exhibition held in Woodlands. 5–7 Mar 2013.

Eshkalak MO, Mohaghegh SD, Esmaili S. Geomechanical properties of unconventional shale reservoirs. J Pet Eng. 2014b;. doi:10.1155/2014/961641.

Ikoku CU. Economic analysis and investment decisions. New York: Wiley; 1985.

Javadpour F, Fisher D, Unsworth M. Nanoscale gas flow in shale gas sediments. J Can Pet Technol. 2007. PETSOC-07-10-06.

Jayakumar R, Boulis A, Aura Araque-Martinez A. Systematic study for re-fracturing modeling under different scenarios in shale reservoirs. Paper SPE 165677 presented at the 2013 SPE eastern regional meeting held in Pittsburgh. 20–22 Aug 2013.

MIT Energy Initiative. The future of natural gas: an interdisciplinary MIT study interim report. Cambridge: Massachusetts Institute of Technology; 2010.

Moghanloo RG, Javadpour F. Applying method of characteristics to determine pressure distribution in 1D shale-gas samples. SPE J. 2014;19(03):361–72. doi:10.2118/168218-PA.

Moore LP, Ramakrishnan H. Restimulation: candidate selection methodologies and treatment optimization. SPE paper presented at 2006 annual technical conference and exhibition, San Antonio. 24–27 Sept 2006.

Newman DG. Engineering economic analysis. 3rd ed. Salt Lake: Engineering Press, Inc; 1988.

Omidvar Eshkalak M. Synthetic geomechanical logs and distributions for Marcellus Shale. MSc Thesis. 2013. Morgantown: West Virginia University.

Rousell N, Sharma M. Refracture reorientation enhances gas production in barnett shale tight gas well. Paper SPE 134491 presented at the 2011 SPE annual technical conference and exhibition, Denver. 30 Oct–2 Nov 2011.

Schweitzer R, Bilgesu HI. The role of economics on well and fracture design completions of Marcellus Shale wells. Paper SPE 157532 presented at 2009 eastern regional meeting held in Charleston. 23–25 Sept 2009.

Shekar S, Hariharan R. A novel screening method for selection of horizontal refracturing cadidates in shale gas reservoirs. Paper SPE 134032 presented at the 2011 North American unconventional gas conference and exhibition, The Woodlands. 14–16 June 2011.

Siebrits E, Elbel JL, et al. Refracture reorientation enhances gas production in barnett shale tight gas well. Paper SPE 63030 presented at the 2000 SPE annual technical conference and exhibition, Dallas. 1–4 Oct 2000.

Tavassoli S, Yu W, Javadpour F, et al. Well screen and optimal time of re-fracturing: a barnett shale well. J Pet Eng. 2013;. doi:10. 1155/2013/817293.

Vincent MC. Restimulation of unconventional reservoirs: when are refracs beneficial. Paper SPE 136757 presented at the 2010 Canadian unconventional resources and international petroleum conference, Calgary. 19–21 Oct 2010.

Wang SY, Luo XL, Hurt, RS. What we learned from a study of refracturing in Barnett Shale: an investigation of completion/ fracturing, and production of refractured wells. Paper IPTC 17081 presented at the 2013 international petroleum technology conference, Beijing.

4

Influence of gas transport mechanisms on the productivity of multi-stage fractured horizontal wells in shale gas reservoirs

Wei Wang[1] · Jun Yao[1] · Hai Sun[1] · Wen-Hui Song[1]

Abstract In order to investigate the influence on shale gas well productivity caused by gas transport in nanometer-size pores, a mathematical model of multi-stage fractured horizontal wells in shale gas reservoirs is built, which considers the influence of viscous flow, Knudsen diffusion, surface diffusion, and adsorption layer thickness. A discrete-fracture model is used to simplify the fracture modeling, and a finite element method is applied to solve the model. The numerical simulation results indicate that with a decrease in the intrinsic matrix permeability, Knudsen diffusion and surface diffusion contributions to production become large and cannot be ignored. The existence of an adsorption layer on the nanopore surfaces reduces the effective pore radius and the effective porosity, resulting in low production from fractured horizontal wells. With a decrease in the pore radius, considering the adsorption layer, the production reduction rate increases. When the pore radius is less than 10 nm, because of the combined impacts of Knudsen diffusion, surface diffusion, and adsorption layers, the production of multi-stage fractured horizontal wells increases with a decrease in the pore pressure. When the pore pressure is lower than 30 MPa, the rate of production increase becomes larger with a decrease in pore pressure.

Keywords Shale gas · Transport mechanisms · Numerical simulation · Fractured horizontal well · Production

✉ Jun Yao
  yaojunhdpu@126.com

[1] School of Petroleum Engineering, China University of Petroleum, Qingdao 266580, Shandong, China

Edited by Yan-Hua Sun

1 Introduction

With the increasing demand for world energy and the improvement of oil and gas exploitation technology, unconventional oil, and gas resources such as coal-bed methane, tight sandstone gas, shale oil, and gas have been paid more and more attention, especially for shale gas which has achieved commercial production (Wu et al. 2013). Shale gas resources in China are competitive with those in America and have great potential for further development (Zhang 2010). In recent years, major breakthroughs have been made in shale gas exploration and production technology in the Sichuan–Chongqing region, where the Sinopec Fuling shale gas field has already been put into commercial production (Yang 2014). The main storage space in shales is nanopores (Zou 2011), shale gas reservoirs in China are deeper, and the formation pressure is higher than those in America (Dong 2012), and gas transport mechanisms are more complicated. Therefore, it is necessary to investigate the influence of gas transport mechanisms on the productivity of fractured horizontal wells in shale gas reservoirs.

Compared with conventional reservoirs, the pore size in shale gas reservoirs is nano-scale (Javadpour et al. 2007; Loucks et al. 2009; Clarkson et al. 2013), which results in low porosity and ultra-low permeability. The matrix porosity generally is lower than 0.1, and the matrix permeability ranges from 1 nanoDarcy to 1 microDarcy (Wang et al. 2009). The shale formation acts as both the source rock and the reservoir rock, so adsorbed gas and free gas coexist in shale reservoirs (Yao et al. 2013a), where free gas is stored in the matrix pore space, and adsorbed gas can make up to 20 %–85 % of the total gas reserve (Hill and Nelson 2000). It is different pore sizes and different gas storage patterns in shales that makes the mechanisms of shale gas transport in nanopores extremely complicated. These include viscous flow, Knudsen diffusion, and molecular diffusion (which only happens when the gas is multi-component) (Bird et al. 2007). If adsorbed gas exists in porous media, surface diffusion, adsorption, and desorption should also be considered amongst the shale gas transport mechanisms (Ho and Webb 2006; Akkutlu and Fathi 2012).

Because of nano-scale porous media in shale gas reservoirs, the traditional Darcy's law cannot accurately describe the mechanism of gas transport. The existence of nanopores in shale reservoirs makes the gas slippage effect more apparent (Swami et al. 2012). However, the Klinkenberg model only applies to low-pressure gas (Klinkenberg 1941). Based on the model of micro-nano pipes, Beskok and Karniadakis put forward a volumetric flow rate formula with different flow regimes (Beskok and Karniadakis 1999). The Beskok model that describes gas flow in a single tube has been applied to research into tight gas and shale gas flow by Civan et al. in which the Knudsen number is used to describe the transport mechanism that incorporates viscous flow and Knudsen diffusion (Civan 2010; Civan et al. 2010, 2011). Javadpour (2009) considered that Darcy's law and Fick's law can describe gas transport mechanisms in micropores, and Knudsen diffusion is the main gas transport mechanism in nanopores. Thus, the Javadpour model which takes viscous flow and Knudsen diffusion into account in nanopores was established (Javadpour 2009), and used the pore radius to characterize gas transport in nanopores. Yao et al. studied matrix viscous flow, Knudsen diffusion, molecular diffusion, and adsorption and desorption based on the double porosity model and used the finite element method (FEM) for numerical simulation of vertical shale gas well production (Yao et al. 2013a, b).

Because of the nano-scale effect, the thickness of the adsorption layer greatly affects the matrix porosity and permeability (Xiong et al. 2012). Meanwhile, the thickness of the adsorption layer is a function of pressure (Sakhaee-Pour et al. 2011). Due to the existence of concentration gradients, adsorbed gas itself undergoes surface diffusion (Sheng et al. 2014), especially in nano-scale porous media, where surface diffusion is an important transport mechanism. In ultra-tight nano-scale porous media, surface diffusion even dominates the gas transport mechanisms (Sheng et al. 2014; Etminan et al. 2014; Mi et al. 2014; Ren et al. 2015).

Based on previous research, in this paper, we overall consider the shale matrix micro-nano-scale effect, which includes the influence of viscous flow, Knudsen diffusion, surface diffusion, and adsorption layer thickness and build a mathematical model of multi-stage fractured horizontal wells in shale gas reservoirs. Furthermore, the discrete-fracture model (DFM) is used to simplify the fracture description, and the FEM is applied to solve the model. Finally, the influences of different transport mechanisms on shale gas reservoir production are studied, and a parameter sensitivity analysis is used to investigate the change on production and the transport mechanism contribution to well production.

2 Shale gas transport mechanisms in the reservoir matrix

Gas transport in shale nanopores consists of several transport mechanisms, as shown in Fig. 1 (Sun et al. 2015). Molecular diffusion is caused by collision between different component gas molecules. For a single gas species, collision between molecules results in viscous flow, and Knudsen diffusion is generated from collision between molecules and the pore walls, while surface diffusion happens when adsorbed gas molecules creep along the pore surface. Considering that there only exists a single-component, methane gas, in the shale gas reservoir and free gas coexists with adsorbed gas in the shale matrix, gas transport mechanisms in the shale matrix is determined by mutual effect of viscous flow, Knudsen diffusion and adsorption layer surface diffusion.

2.1 Viscous flow

When the mean-free path of gas molecules is very small compared to the pore diameters, the probability of collisions between molecules is much higher than collisions between molecules and pore walls; thus, single-component gas transport is mainly governed by viscous flow caused by the pressure gradient. Viscous flow can be modeled by Darcy's law (Kast and Hohenthanner 2000):

$$N_v = -\frac{\rho_m k_\infty}{\mu_m}(\nabla p_m) \tag{1}$$

where N_v is the mass flux of viscous flow, kg/(m^2 s); k_∞ is the intrinsic permeability of the porous media, m^2; p_m is the matrix gas pressure, Pa; ρ_m is the matrix gas density, kg/m^3; and μ_m is the matrix gas viscosity, Pa s.

2.2 Knudsen diffusion

When the pore space is so narrow that the mean-free path of gas molecules is very close to the pore diameter, collisions between molecules and pore walls dominate. Knudsen diffusion can be expressed as (Florence et al. 2007)

$$N_k = -M_g D_k(\nabla C_m) \tag{2}$$

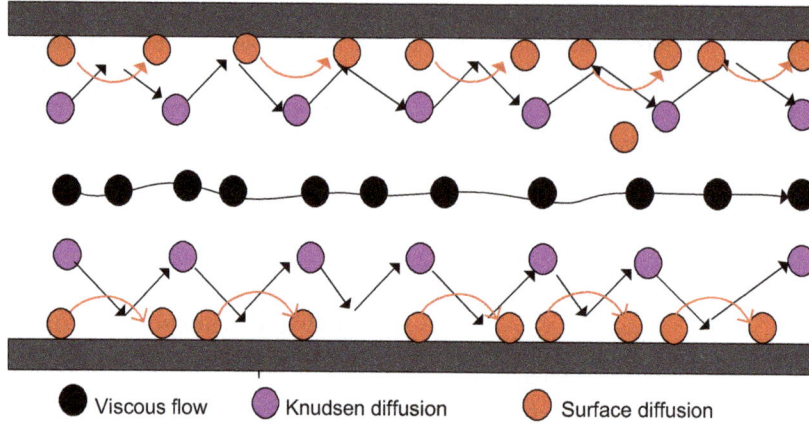

Fig. 1 Single-component gas transport in porous media

C_m can be given by $C_m = \frac{\rho_m}{M_g} = \frac{p_m}{ZRT}$, and ρ_m is obtained by $\rho_m = \frac{p_m M_g}{ZRT}$

$$N_k = -\frac{\rho_m D_k(\nabla p_m)}{p_m} \qquad (3)$$

$$D_k = \frac{4k_\infty c}{2.81708\sqrt{\frac{k_\infty}{\phi_m}}}\sqrt{\frac{\pi RT}{2M_g}}, \qquad (4)$$

where N_k is the mass flux of Knudsen diffusion, kg/(m^2 s); M_g is the molecular weight of gas, kg/mol; D_k is the Knudsen diffusivity, m^2/s; C_m is the concentration of free gas in the porous media, mol/m^3; Z is the gas compressibility factor; R is the ideal gas constant, 8.314 J/(mol K); T is the gas reservoir temperature, K; ϕ_m is the shale matrix porosity; and c is a constant close to 1. In this paper, we set $c = 1$.

2.3 Adsorption and desorption

Shale gas adsorbed on the surfaces of nanopores follows the mono-layer Langmuir isotherm adsorption equation (Civan et al. 2011):

$$q_{ads} = \frac{\rho_s M_g}{V_{std}} \cdot \frac{V_L p_m}{p_L + p_m}, \qquad (5)$$

where q_{ads} is the mass of gas adsorbed per solid volume, kg/m^3; ρ_s denotes the shale matrix density, kg/m^3; V_{std} is the molar volume of gas at standard temperature (273.15 K) and pressure (101,325 Pa), std m^3/mol; V_L is the Langmuir gas volume, std m^3/kg; p_L is the Langmuir gas pressure, Pa.

The Langmuir isotherm adsorption is based on a mono-layer model, and the adsorbed layer makes the effective diameter of the nanopore decrease. The modified effective pore radius caused by the mono-layer adsorption on the pore surface can be written as (Sun et al. 2015)

$$r_{eff} = r - d_m \frac{p_m}{p_L + p_m}, \qquad (6)$$

where r_{eff} is the effective pore radius, m; r is the pore radius, m; d_m is the diameter of a methane molecule, m.

The decrease in the pore diameter leads to a reduction in porosity. Thus, the effective porosity can be expressed as

$$\phi_{meff} = \frac{\phi_m r_{eff}^2}{r^2} \qquad (7)$$

By combining Eqs. (6) and (7), the effective intrinsic permeability that takes the adsorption layer thickness into account can be given by

$$k_{\infty eff} = \frac{\phi_{meff} r_{eff}^2}{8\tau} \qquad (8)$$

where $k_{\infty eff}$ is the effective intrinsic permeability of the shale matrix, m^2; τ is the tortuosity.

2.4 Surface diffusion

Surface diffusion only occurs in porous media where the gas is adsorbed onto the pore wall, and can be expressed as

$$N_s = -M_g D_s(\nabla C_s), \qquad (9)$$

where N_s is the mass flux of surface diffusion, kg/(m^2 s); D_s is the surface diffusivity, m^2/s; C_s is the concentration of adsorbed gas, mol/m^3. Gas adsorbed on pore surfaces follows the Langmuir isotherm adsorption and can be expressed as (Xiong et al. 2012)

$$N_s = -M_g D_s \frac{C_{s\,max} p_L}{(p_m + p_L)^2}\left(1 - \frac{r_{eff}^2}{r^2}\right)(\nabla P_m), \qquad (10)$$

where $C_{s\,max}$ is the maximum adsorbent concentration, mol/m^3.

3 Establishment and solution of the fractured horizontal well mathematical model in shale gas reservoirs

3.1 Transport equation for the matrix system

Mass transport in nano-scale porous media is the concurrent result of viscous flow, Knudsen diffusion, surface diffusion, and gas desorption. The advective–diffusive model (ADM) and dusty gas model (DGM) are generally used to incorporate the coupling mechanisms. Although DGM considers coupling effect between viscous flow and diffusion more comprehensively, the transport equation built by ADM and DGM is the same for a single-component gas (Yao et al. 2013c). Thus, ADM is applied to build the transport equation for the shale matrix:

$$N_{t,m} = N_v + N_k + N_s = -\frac{\rho_m k_{m,app}}{\mu_m}(\nabla p_m) \qquad (11)$$

with

$$k_{m,app} = k_\infty \left(1 + \frac{b_m}{p_m}\right) + M_g D_s \frac{\mu_m}{\rho_m} \frac{C_{s\,max} p_L}{(p_m + p_L)^2} \left(1 - \frac{r_{eff}^2}{r^2}\right) \qquad (12)$$

$k_{m,app}$ is the apparent permeability of the shale matrix; b_m is the Klinkenberg coefficient that considers Knudsen diffusion $b_m = \frac{D_k \mu_m}{k_\infty}$.

3.2 Continuity equation for the matrix system

We assume that there are no natural fractures in the gas reservoir and the gas is only single-component methane. Adsorbed gas and free gas coexist in the shale matrix, and

the reservoir temperature is assumed to stay constant during production. Gas adsorbed on the matrix surface conforms to the Langmuir isotherm adsorption equation. According to the mass conservation law, the continuity equation for a single-porosity matrix system can be obtained:

$$\frac{\partial}{\partial t}(\rho_m \phi_m + (1 - \phi_m)q_{ads}) + \nabla \cdot N_{t,m} = q_m \delta(M - M'), \qquad (13)$$

where $N_{t,m}$ is the mass flux, kg/(m^2 s); q_{ads} is the mass of gas adsorbed per solid volume, kg/m^3; q_m is the source sink term, kg/s; $\delta(M - M')$ is the delta function which is equal to zero at all points except point M', $\delta(M - M') = 1$.

3.3 Mathematical model for the matrix system

Assuming that the pressure is equal on the boundary of hydraulic fractures and the matrix. Γ_1 is the outer boundary of the gas reservoir, Γ_2 is the inner boundary of the production well, and Γ_3 is the boundary between hydraulic fractures and the matrix. In this study, we assume that the outer boundary Γ_1 is sealed, the inner boundary Γ_2 is under a constant pressure, and the pressure is equal across boundary Γ_3. Substituting Eqs. (5) and (11) into Eq. (13), the mathematical model for the single-porosity matrix can be derived:

$$\begin{cases} \left[\gamma \phi_m + \frac{(1 - \phi_m)M_g p_L V_L \rho_s}{V_{std}(p_L + p_m)^2}\right]\frac{\partial p_m}{\partial t} - \nabla \cdot \left[\gamma \left[\frac{p_m k_{m,app}}{\mu_m}(\nabla p_m)\right]\right] = q_m \delta(M - M') \\ p_m(x,y,z,t)|_{t=0} = p_i \\ \frac{\partial p_m}{\partial n}\Big|_{\Gamma_1} = 0 \\ p_m|_{\Gamma_2} = p_w \\ p_m(x,y,z,t)|_{\Gamma_3} = p_F(x,y,z,t)|_{\Gamma_3}, \end{cases} \qquad (14)$$

where $\gamma = \frac{M_g}{ZRT}$; p_i is the initial pressure in the gas reservoir, Pa; p_w is the wellbore pressure, Pa; and p_F is the pressure in fractures, Pa.

3.4 Mathematical model for the hydraulic fracture system

We suppose that only free gas exists in artificial fractures and that the gas obeys Darcy's law within the fracture. The pressure is equal across the boundary of hydraulic fractures and the matrix. Thus, the mathematical model for the hydraulic fracture can be represented as

$$
\begin{cases}
\dfrac{\partial}{\partial t}(\rho_F \phi_F) + \nabla \cdot \left(\dfrac{\rho_F k_F}{\mu_F} \nabla p_F \right) = q_F \delta(M - M') \\
p_F(x, y, z, t)|_{t=0} = p_i \\
p_F|_{\Gamma_2} = p_w \\
p_m(x, y, z, t)|_{\Gamma_3} = p_F(x, y, z, t)|_{\Gamma_3}
\end{cases}
\quad , \quad (15)
$$

where ρ_F is the gas density within the fracture, kg/m^3; ϕ_F is the porosity of the hydraulic fracture, which equals zero when there are no proppants within the fracture; μ_F is the gas viscosity, Pa s; k_F is the fracture permeability, m^2, which can be calculated by the following equations on condition of regular fractures:

$$
k_F = \frac{\phi_F h_F^2}{12}, \quad (16)
$$

where h_F is the fracture aperture, m.

3.5 Solution with the finite element method

The DFM can be used to simplify the description of the hydraulic fracture (Yao et al. 2010; Huang et al. 2011). Three-dimensional fractures are converted into two-dimensional surface elements (as shown in Fig. 2). In order to guarantee equal integral value, the fracture aperture should be multiplied before the surface integral. DFM greatly decreases the number of grids and improves the numerical computation efficiency.

In Eqs. (14) and (15), the DFM is applied to simplify the fracture surface. The FEM is used to solve it, in which tetrahedron elements are used in the matrix, and two-dimensional triangular elements are used in the fracture surface (Li et al. 2010; Yao et al. 2013c). The equations are nonlinear and hard to solve directly; therefore, iteration is used to solve them at an arbitrary time point. The pressure of nth time step is applied to obtain the pressure of the

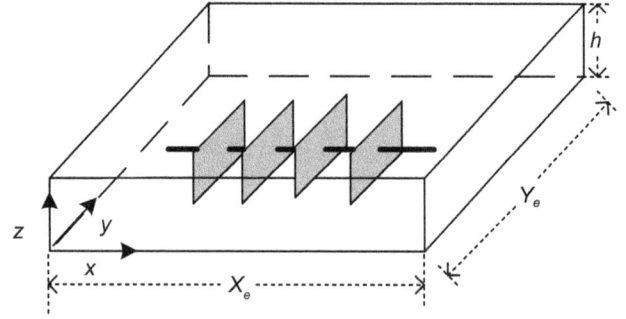

Fig. 3 3D schematic map of the fractured horizontal well

Table 1 Basic data used in the simulation of the multi-staged fractured horizontal well

Basic parameter	Value
Gas reservoir size $X_e \times Y_e \times h$, m	$1500 \times 1500 \times 50$
Initial reservoir pressure, MPa	15
Horizontal well length, m	800
Pore radius, nm	1, 5, 10, 25, 50, 100
Wellbore radius, m	0.1
Intrinsic matrix permeability, μm^2	1×10^{-7}
Matrix porosity	0.05
Number of fractures	8
Fracture half-length, m	100
Fracture spacing, m	100
Fracture aperture, m	0.005
Well bottom pressure, MPa	4
Gas reservoir temperature, °C	52
Gas component	Methane
Matrix density, kg/m^3	2600
Langmuir volume, m^3/kg	3×10^{-3}
Langmuir pressure, MPa	6

$(n + 1)$th time step. To guarantee algorithm stability, an implicit backward difference method about time is utilized.

4 Numerical example and analysis of factors affecting production

In this paper, a multi-stage fractured horizontal well in the middle of a box-shaped sealed reservoir is simulated, with its coordinate ranging from (x_1, y_0, z_0) to (x_2, y_0, z_0). Vertical hydraulic fractures are in the shape of rectangles symmetrically distributed around the horizontal well (as shown in Fig. 3). The gas reservoir is penetrated up and down by hydraulic fractures. Meanwhile, the gas reservoir is assumed to be homogeneous, and the influence of gravity is ignored. The basic parameters of the gas reservoir and fractured horizontal well are shown in Table 1.

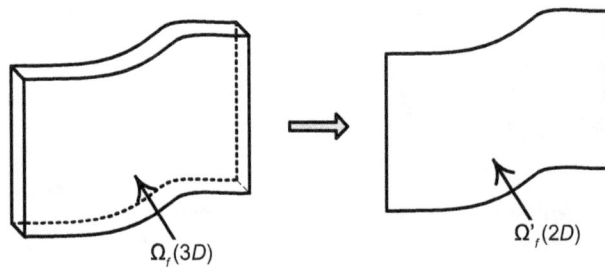

Fig. 2 Schematic diagram of the discrete-fracture model

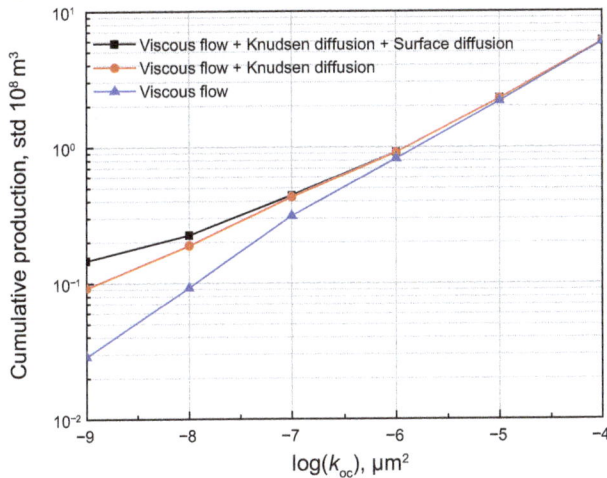

Fig. 4 Cumulative production predicted by different transport models

4.1 Influence of transport mechanisms

Because of the nanopores in shale reservoirs, the matrix porosity and permeability are extremely low. Due to different pore sizes in porous media, the intrinsic matrix permeability differs, and thus, the influence of viscous flow, Knudsen diffusion, adsorbed layer, and surface diffusion on transport mechanisms varies substantially. Figure 4 shows simulated cumulative production versus intrinsic permeability under different transport mechanisms over 20 years. As illustrated in Fig. 4, when $k_\infty > 10^{-5}$ μm^2, because of the relatively large pore size, gas transport in the porous matrix is dominated by viscous flow, Knudsen diffusion and surface diffusion have little impact on the

productivity of fractured horizontal wells and can be ignored. When 10^{-7} $\mu m^2 < k_\infty < 10^{-5}$ μm^2, with a decrease in the pore size, the influence of Knudsen diffusion becomes bigger and gradually affects the productivity of fractured horizontal wells and cannot be ignored, but the influence of the surface diffusion still can be ignored. When $k_\infty < 10^{-7}$ μm^2, because of the extremely small pore size, gas transport in a porous matrix is dominated by Knudsen diffusion, which has the biggest contribution to the production of fractured horizontal wells. Meanwhile, the impact of surface diffusion on the production of fractured horizontal wells becomes bigger and thus cannot be ignored.

4.2 Influence of intrinsic matrix permeability

For a single-porosity shale gas reservoir, the intrinsic matrix permeability is one of the main factors that affect the productivity of fractured horizontal wells. From Fig. 5, with an increase in the intrinsic matrix permeability, daily gas production and cumulative gas production increase dramatically. With a decrease in the intrinsic matrix permeability, gas transport in porous media is gradually dominated by Knudsen diffusion and surface diffusion, and the impact of viscous flow is reduced, which slows the production decline rate and guarantees long-term stable production.

4.3 Influence of an adsorption layer

As shown in Fig. 6, because of the existence of adsorbed gas, the effective pore size and effective porosity decrease. With a decrease in the pore radius, the fractured horizontal

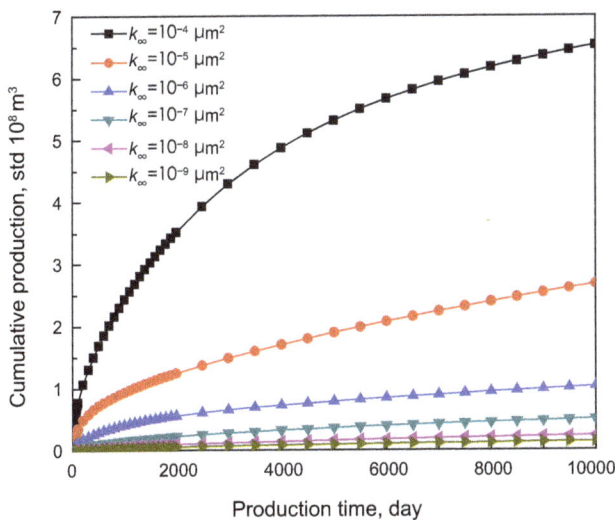

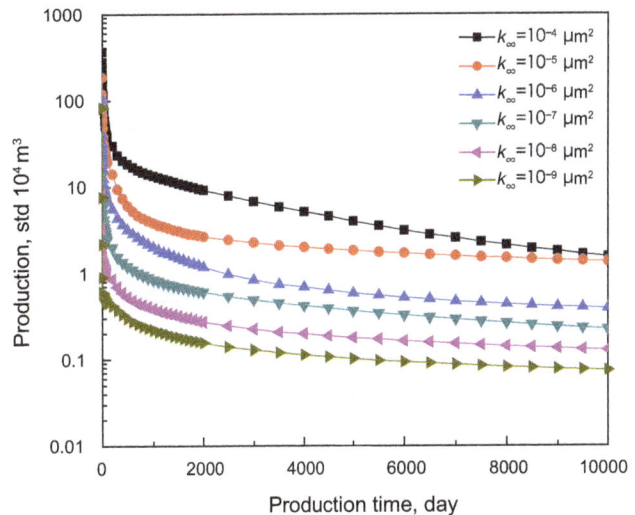

Fig. 5 Effects of intrinsic matrix permeability on production and cumulative production

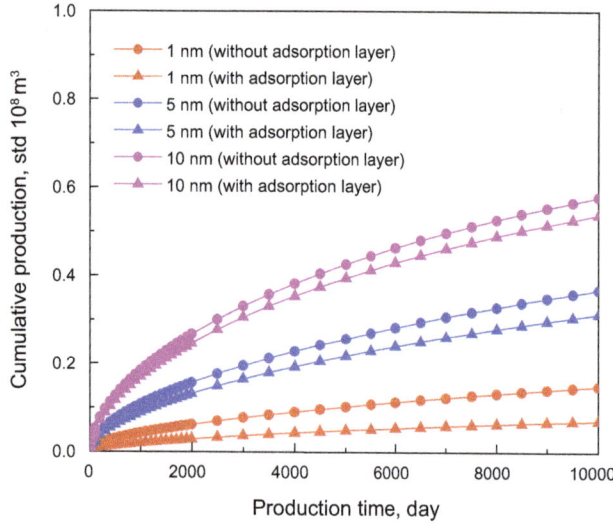

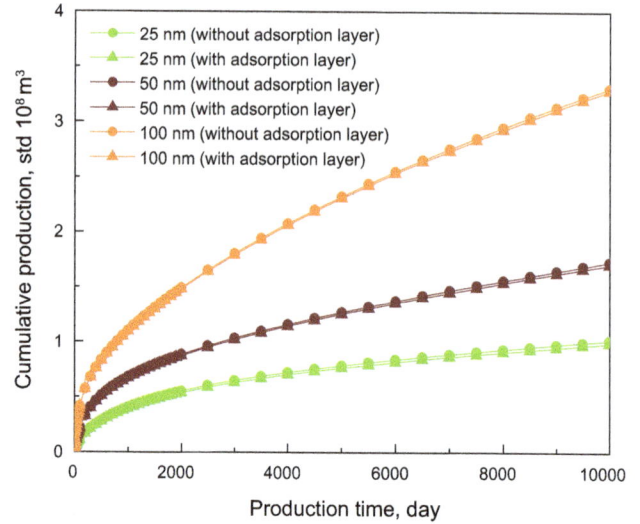

Fig. 6 Impact of the adsorption layer on fractured horizontal well productivity with different pore radii

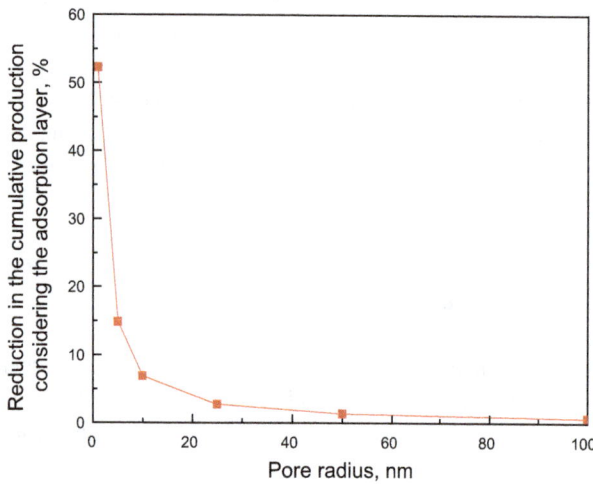

Fig. 7 Impact of adsorption layer on the reduction in the cumulative production under different pore radii

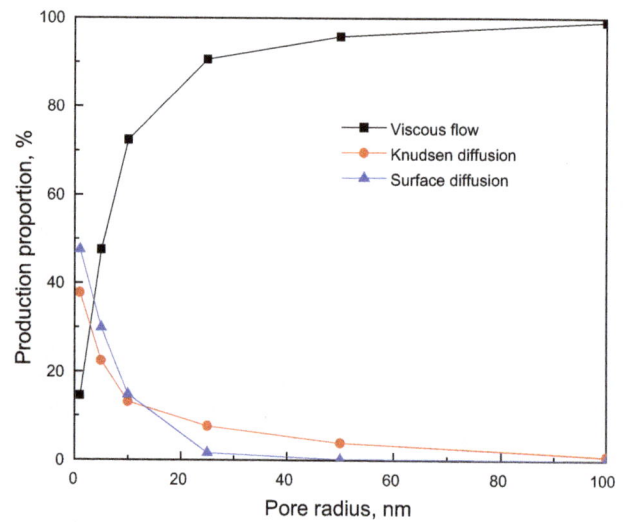

Fig. 8 Contributions to cumulative production by viscous flow, Knudsen diffusion, and surface diffusion under different pore radii

well production gradually diminishes for the same production time. The production that involves the adsorption layer is lower than production to which the adsorption layer does not contribute. Furthermore, with a decrease in the pore radius, the ratio of the adsorption layer thickness divided by the pore radius gradually increases, and the impact of adsorption layer on production becomes bigger.

The reduction in cumulative production considering the adsorption layer is a percentage that defined as the difference of the cumulative production between without and with the adsorption layer divided by the cumulative production without the adsorption layer. As shown in Fig. 7, when $r > 10$ nm, whether the adsorption layer thickness is considered or not, the production does not change much, with the reduction in cumulative production all below

10 %. When $r < 10$ nm, the influence of the adsorption layer thickness on production cannot be ignored. When r decreases to 1 nm, compared with the cumulative production of 10,000 days without the adsorption layer, the reduction in cumulative production considering the adsorption layer increased up to 52.3 %.

As can be seen in Fig. 8, when $r > 10$ nm, gas transport in a porous medium is dominated by viscous flow. Thus, with a decrease in the pore radius, considering the adsorption layer thickness, the decline rate of the fractured horizontal well production is relatively small. When $r < 10$ nm, Knudsen diffusion and surface diffusion become the main transport mechanisms. In this condition, with a decrease in the pore radius, considering the

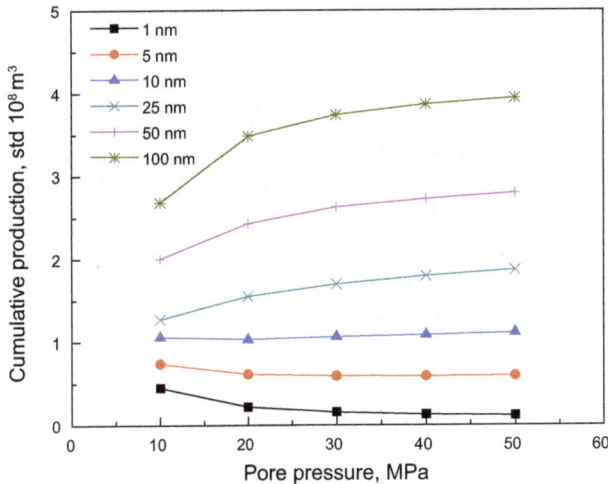

Fig. 9 Cumulative production under different pore pressures and pore radii

adsorption layer thickness, the fractured horizontal well production decrease degree gradually becomes bigger. Therefore, the influence of adsorption layer on production performance and productivity should not be ignored for nano-scale shale gas reservoirs.

4.4 Influence of gas reservoir pressure

Figure 9 shows that formation pore pressure dramatically impacts on the gas transport mechanisms in nanopores. When $r > 10$ nm, gas transport in matrix is dominated by viscous flow while the adsorption layer thickness has little impact on gas transport. Under the same drawdown pressure, with an increase in the formation pore pressure, the

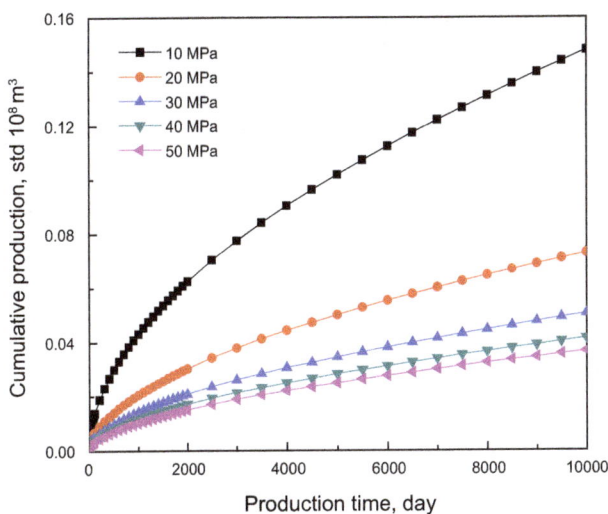

Fig. 10 Cumulative production under different pore pressures at the same drawdown pressure with $r = 1$ nm

amount of absorbed gas increases correspondingly, and the production of a fractured horizontal well gradually increases. When $r < 10$ nm, gas transport is dominated by Knudsen diffusion and surface diffusion, and the influence of the adsorption layer thickness on gas transport can no longer be ignored. In this condition, with a decrease in the formation pore pressure, the amount of absorbed gas decreases correspondingly and the thickness of adsorption layer gradually decreases. The increase in well production caused by Knudsen diffusion and surface diffusion exceeds the production loss caused by viscous flow, which makes the production of the fractured horizontal well increase with a decrease in the pore pressure. Surface diffusion varies inversely with the pore pressure, and the variation trend of its contribution to production with pressure change is the opposite to that of viscous flow.

When the pore radius is extremely small (e.g., 1 nm), with an increase in the pore pressure, the amount of absorbed gas and the thickness of the adsorption layer increase correspondingly, resulting in a decrease in the effective flow radius and effective porosity. Meanwhile, the cumulative production decreases under the same drawdown pressure (as shown in Fig. 10). When the pore pressure is relatively small, with an increase in the pore pressure, the amount of absorbed gas and the adsorption layer thickness increase significantly and the cumulative production drops dramatically. When the pore pressure is higher than 30 MPa, the increasing tendency of the amount of gas and the adsorption layer thickness slow down and tend to constant values. Meanwhile, the effective pore radius and the effective porosity gradually become stable, and the decrease in the cumulative production becomes smaller. The higher the pore pressure is, the smaller the influence of Knudsen diffusion and surface diffusion on production is. When the pore pressure is high enough, the influence of surface diffusion on production is hard to observe and thus can be ignored.

5 Conclusions

In this paper, gas transport mechanisms in shale nanopores are established, considering the influence of viscous flow, Knudsen diffusion, surface diffusion, the adsorption layer, and gas desorption. A finite element method is used to investigate the influence of shale gas transport mechanisms on the productivity of the multi-stage fractured horizontal well. The following conclusions can be made.

(1) With Knudsen diffusion and surface diffusion taken into account, the productivity of a multi-stage fractured horizontal well is higher than the production that only considers viscous flow. Furthermore,

as the intrinsic matrix permeability decreases, the increment of fractured horizontal well-cumulative production involving Knudsen diffusion and surface diffusion increases. Therefore, the influence of Knudsen diffusion and surface diffusion on the production of ultra-low permeability shale gas reservoirs should not be ignored.

(2)   The intrinsic matrix permeability dramatically influences the productivity of the multi-stage fractured horizontal well. As the intrinsic matrix permeability becomes smaller, the productivity of multi-stage fractured horizontal wells decreases and the middle- and late-period production decline rate slows down. This is because when the intrinsic matrix permeability is smaller than 10^{-5} μm^2, Knudsen diffusion and surface diffusion start to play an increasingly important role in gas transport, which guarantees the stable production of fractured horizontal wells.

(3)   The existence of an adsorption layer on the nanopore surface reduces the effective pore radius and effective porosity, which lowers the production of fractured horizontal wells. With a decrease in the pore radius, the production decline becomes bigger. When $r < 10$ nm, Knudsen diffusion and surface diffusion gradually become the main transport mechanisms. The reduction in cumulative production considering the adsorption layer increases up to 52.3 % when r decreases to 1 nm.

(4)   When $r < 10$ nm, with a decrease in the formation pore pressure, the amount of absorbed gas decreases correspondingly and the thickness of the adsorption layer gradually decreases. With Knudsen diffusion and surface diffusion taken into account, the production of multi-stage fractured horizontal wells increases with a decrease in the pore pressure. When the pore pressure is lower than 30 MPa, the increment of cumulative production becomes bigger with a decrease in the pore pressure.

The production time of fractured horizontal wells lasts longer, with lower production decline rate in the middle and late period. This is generally believed to be associated with gas adsorption and desorption. This study found that with a decrease in the formation pore pressure, gas transport in a porous matrix is dominated by Knudsen diffusion and surface diffusion, and the adsorption layer thickness decreases correspondingly, which slows the production decline rate. Therefore, the gas transport in nanopores is also the key factor that influences the production performance of shale gas wells. Gas adsorption and desorption on the matrix pore surfaces together with the gas transport mechanism in nanopores determine production behavior

and ultimate recoverable reserves of shale gas multi-stage fractured horizontal wells.

Acknowledgments   This work was supported by the National Natural Science Foundation of China (No. 51234007, No. 51490654, No. 51504276, and No. 51504277), Program for Changjiang Scholars and Innovative Research Team in University (IRT1294), the Natural Science Foundation of Shandong Province (ZR2014EL016, ZR2014EEP018), China Postdoctoral Science Foundation (No. 2014M551989 and No. 2015T80762), the Major Programs of Ministry of Education of China (No. 311009), and Introducing Talents of Discipline to Universities (B08028).

References

Akkutlu IY, Fathi E. Multiscale gas transport in shales with local Kerogen heterogeneities. SPE J. 2012;17(4):1002–11.

Beskok A, Karniadakis GE. Report: a model for flows in channels, pipes, and ducts at micro and nano scales. Microscale Thermophys Eng. 1999;3(1):43–77.

Bird RB, Stewart WE, Lightfoot EN. Transport phenomena. 2nd ed. New York: John Wiley & Sons Inc; 2007.

Civan F. Effective correlation of apparent gas permeability in tight porous media. Transp Porous Media. 2010;82(2):375–84.

Civan F, Rai CS, Sondergeld CH. Intrinsic shale permeability determined by pressure-pulse measurements using a multiple-mechanism apparent-gas-permeability non-Darcy model. In: SPE annual technical conference and exhibition, 19–22 Sept 2010, Florence, Italy. doi:10.2118/135087-MS.

Civan F, Rai CS, Sondergeld CH. Shale-gas permeability and diffusivity inferred by improved formulation of relevant retention and transport mechanisms. Transp Porous Media. 2011;86(3):925–44.

Clarkson CR, Solano N, Bustin RM, et al. Pore structure characterization of North American shale gas reservoirs using USANS/SANS, gas adsorption, and mercury intrusion. Fuel. 2013;103:606–16.

Dong DZ, Zou CN, Yang H, et al. Progress and prospects of shale gas exploration and development in China. Acta Pet Sin. 2012;33(A01):107–14 (in Chinese).

Etminan SR, Javadpour F, Maini BB, et al. Measurement of gas storage processes in shale and of the molecular diffusion coefficient in kerogen. Int J Coal Geol. 2014;123:10–9.

Florence FA, Rushing JA, Newsham KE, et al. Improved permeability prediction relations for low permeability sands. In: SPE rocky mountain oil & gas technology symposium, 16–18 April 2007, Denver, Colorado, USA. doi:10.2118/107954-MS.

Hill DG, Nelson CR. Gas productive fractured shales: an overview and update. Gas Tips. 2000;6(2):4–13.

Ho CK, Webb SW. Gas transport in porous media. The Netherlands: Springer; 2006.

Huang ZQ, Yao J, Wang YY, et al. Numerical simulation on water flooding development of fractured reservoirs in a discrete-fracture model. Chin J Comput Phys. 2011;28(1):41–9.

Javadpour F. Nanopores and apparent permeability of gas flow in mudrocks (shales and siltstone). J Can Pet Technol. 2009;48(8): 16–21.

Javadpour F, Fisher D, Unsworth M. Nanoscale gas flow in shale gas sediments. J Can Pet Technol. 2007;46(10):55–61.

Kast W, Hohenthanner CR. Mass transfer within the gas-phase of porous media. Int J Heat Mass Transf. 2000;43(5):807–23.

Klinkenberg LJ. The permeability of porous media to liquids and gases. Drill Prod Pract. 1941;1:200–13.

Li YJ, Yao J, Huang ZQ, et al. Finite element simulation of heterogeneous reservoir with full permeability tensor. Chin J Comput Phys. 2010;27(5):692–8 (in Chinese).

Loucks RG, Reed RM, Ruppel SC, et al. Morphology, genesis, and distribution of nanometer-scale pores in siliceous mudstones of the Mississippian Barnett Shale. J Sediment Res. 2009;79(12): 848–61.

Mi LD, Jiang HQ, Li JJ. Investigation of a shale gas numerical simulation method based on discrete fracture network model. Nat Gas Geosci. 2014;25(11):1795–803 (in Chinese).

Ren JJ, Guo P, Guo ZL, et al. A lattice Boltzmann model for simulating gas flow in kerogen pores. Transp Porous Media. 2015;106(2):285–301.

Sakhaee-Pour A, Bryant SL. Gas permeability of shale. In: SPE annual technical conference and exhibition, 30 Oct–2 Nov 2011, Denver, Colorado, USA. doi:10.2118/146944-MS.

Sheng M, Li GS, Huang ZW, et al. Shale gas transient flow model with effects of surface diffusion. Acta Pet Sin. 2014;35(2): 347–52 (in Chinese).

Sun H, Yao J, Fan DY, et al. Gas transport mode criteria in ultra-tight porous media. Int J Heat Mass Transf. 2015;83:192–9.

Swami V, Clarkson C, Settari A. Non-Darcy flow in shale nanopores: do we have a final answer. In: SPE Canadian unconventional resources conference, 30 Oct–1 Nov 2012, Calgary, Alberta, Canada. doi:10.2118/162665-MS.

Wang FP, Reed RM, John A, et al. Pore networks and fluid flow in gas shales. In: SPE annual technical conference and exhibition, 4–7 Oct 2009, New Orleans, Louisiana, USA. doi:10.2118/124253-MS.

Wu X, Ren ZY, Wang Y, et al. Situation of world shale gas exploration and development. Resour Ind. 2013;15(5):61–7 (in Chinese).

Xiong X, Devegowda D, Michel GG, et al. A fully-coupled free and adsorptive phase transport model for shale gas reservoirs including non-Darcy flow effects. In: SPE annual technical conference and exhibition, 8–10 Oct 2012, San Antonio, Texas, USA. doi:10.2118/159758-MS.

Yang N. Development status and prospect of shale gas at home and abroad. Pet Petrochem Today. 2014;8:16–21 (in Chinese).

Yao J, Wang ZS, Zhang Y, et al. Numerical simulation method of discrete fracture network for naturally fractured reservoirs. Acta Pet Sin. 2010;31(2):284–8 (in Chinese).

Yao J, Sun H, Fan DY, et al. Numerical simulation of gas transport mechanisms in tight shale gas reservoirs. Pet Sci. 2013a;10(4): 528–37.

Yao J, Sun H, Fan DY, et al. Transport mechanisms and numerical simulation of shale gas reservoirs. J China Univ of Pet (Edit Nat Sci). 2013b;1:91–8 (in Chinese).

Yao J, Sun H, Huang ZQ, et al. Key mechanical problems in the development of shale gas reservoirs. Sci Sin-Phys Mech Astron. 2013c;43(12):1527–47.

Zhang DW. Strategic concepts of accelerating the survey exploration and exploitation of shale gas resources in China. Oil Gas Geol. 2010;31(2):135–9 (in Chinese).

Zou CN, Dong DZ, Yang H, et al. Conditions of shale gas accumulation and exploration practices in China. Nat Gas Ind. 2011;31(12):26–39 (in Chinese).

5

Classification and characteristics of tight oil plays

Xin-Shun Zhang[1] · Hong-Jun Wang[1] · Feng Ma[1] · Xiang-Can Sun[2] · Yan Zhang[1] · Zhi-Hui Song[1]

Abstract Based on the latest conventional–unconventional oil and gas databases and relevant reports, the distribution features of global tight oil were analyzed. A classification scheme of tight oil plays is proposed based on developed tight oil fields. Effective tight oil plays are defined by considering the exploiting practices of the past few years. Currently, potential tight oil areas are mainly distributed in 137 sets of shale strata in 84 basins, especially South America, North America, Russia, and North Africa. Foreland, craton, and continental rift basins dominate. In craton basins, tight oil mainly occurs in Paleozoic strata, while in continental rift basins, tight oil occurs in Paleozoic–Cenozoic strata. Tight oil mainly accumulates in the Cretaceous, Early Jurassic, Late Devonian, and Miocene, which correspond very well to six sets of global-developed source rocks. Based on source–reservoir relationship, core data, and well-logging data, tight oil plays can be classified into eight types, above-source play, below-source play, beside-source play, in-source play, between-source play, in-source mud-dominated play, in-source mud-subordinated play, and interbedded-source play. Specifically, between-source, interbedded-source, and in-source mud-subordinated plays are major targets for global tight oil development with high production. In contrast, in-source mud-dominated and in-source plays are less satisfactory.

Keywords Tight oil · Distribution characteristics · Play · Source–reservoir relationship · Classification · Estimated ultimate recovery (EUR) · Efficiency evaluation

1 Introduction

With the rapid advances in exploration theory and technology, a majority of conventional oil resources have been discovered, leaving less and less potential oil resources in place. The great potential of unconventional oil resources has been confirmed by successive breakthroughs in exploration and development (Jarvie 2012; Jia et al. 2012; Zhao et al. 2013; Zou et al. 2014; Pang et al. 2015). As an unconventional resource that is most similar to conventional oil resources, tight oil has become a focus for global exploration and development, and a historic breakthrough has been achieved in North America (EIA 2013; BP 2015; IHS 2014a, b). In 2014, the US' tight oil production reached 3.2 MMbbl/d (165 million tons/year), which accounted for 40 % of total oil production in the country. That figure is still increasing. In the third quarter of 2014, the tight oil production in the Bakken Formation (Williston Basin) and Eagle Ford Formation (Gulf Basin) exceeded 1 MMbbl/d (50 million tons/year), respectively (Hart Energy 2014). In China, although the resources are great (Wang et al. 2015), the annual tight oil production was less than 10 million tons in 2013 and the Chang-6 and Chang-7 formations in the Ordos Basin were the main producing areas, with annual production up to 800 million tons. Breakthroughs have been made in tight oil exploration in the Qingshankou Formation of the Songliao Basin, the

✉ Xin-Shun Zhang
vvvzxs@126.com

[1] Research Institute of Petroleum Exploration & Development, PetroChina, Beijing 100083, China

[2] Oil & Gas Survey, China Geological Survey, Beijing 100029, China

Edited by Jie Hao

Permian Formation of the Junggar Basin, and the Jurassic Formation of the Sichuan Basin, but these formations are not ready for commercial development (Zou et al. 2014; Jia et al. 2014).

A lot of research has focused on pore-throat structures, development environments, and distributions of tight oil reservoirs in multiple regions (Liang et al. 2011; Kuang et al. 2012; Zhou and Yang 2012; Yang et al. 2013; Zhang et al. 2013; Pang et al. 2015). However, scholars have showed less concern about tight oil plays, since they believe that tight oil accumulated near or in source formations (Jia et al. 2012; Zou et al. 2014). This paper is a further study of "Unconventional Hydrocarbon Potential Analysis and Future Strategic Zone Selection in Global Main Areas"—a key subject under the National Oil and Gas program. This program is aimed to appraise the global unconventional resources evaluated by China National Petroleum Company (CNPC). We analyzed 28,992 wells of North American tight oil plays, including 10,653 tight oil production wells and 16,829 wells with logging data (Fig. 1). All these well data were purchased from the IHS unconventional oil and gas database, which was updated to 2014. To make the result integral and reliable, we also reviewed many other reports (C & C Reservoirs 2014), which focus on global tight oil exploration and development situations and characteristics of mature tight oil areas. Tight oil reservoirs are classified by the types of source–reservoir relationship, and the efficiencies of different tight oil reservoirs are analyzed according to the development features in North America. In this way, favorable plays are defined to provide reference for tight oil exploration.

2 Definition of tight oil

Tight oil is variously defined but mainly as follows: (1) Tight oil is one of the oil resources where the shale is the source rock and the oil also accumulates in shale or nearby. It generally refers to shale oil, similar to shale gas. Shale oil reservoirs have poor properties due to low connectivity of micro-pores in the shale. Oil in such tight shale reservoirs is explicitly defined as shale oil by IEA and some Chinese organizations. (2) Tight oil, similar to tight gas, is a petroleum resource produced from ultra-low permeability shale, siltstone, sandstone, and carbonate, which are closely related to oil-source shales. This resource is defined as "light tight oil" by IEA and "tight oil" by Statoil, EIA, and some Chinese scholars (EIA 2011; Zou et al. 2012). (3) All petroleum resources that must be produced economically from low-permeability and low-porosity reservoirs by stimulation treatments (e.g., hydraulic fracturing) are referred to as tight oil, without limitations of lithology and oil quality. This definition is similar to that of IHS and National Resources Canada (NRC 2012; IHS 2014a, b).

Although the definition of tight oil is distinct, a common understanding is that tight oil accumulates in low-porosity and low-permeability reservoirs and it can be recovered economically only by artificial stimulation

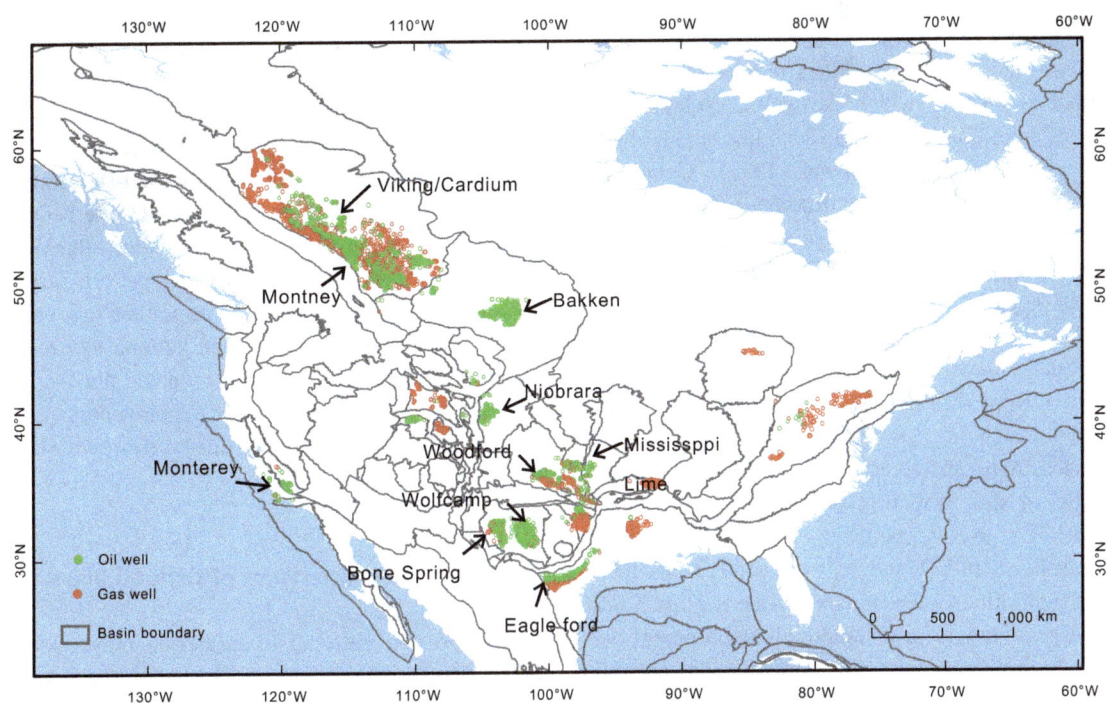

Fig. 1 Distribution map of unconventional drilling well data, North America. *Notes* data from IHS unconventional database, 2014

(Pang et al. 2014). Based on international research in recent years, tight oil is defined in this paper as the oil resource that is preserved and accumulated in low-porosity (<12 %) and low-permeability (overburden matrix permeability <0.1 mD) shale, siltstone, sandstone, carbonate, or other tight reservoirs, in or near source rocks under the control of one or more sets of high-quality source rocks (Table 1). Shale oil often accumulates continuously at large scale without trap boundaries and with almost no natural productivity.

3 Distribution of tight oil

The quality of source rocks is the most significant aspect for evaluating the unconventional resource abundance. In this study, the tight oil basins are selected by TOC higher than 1 %, vitrinite reflectance R_o of 0.7 %–1.2 %, and crude oil API higher than 38° (Ma et al. 2014; CNPC 2014). Therefore, 84 basins (137 tight oil strata series totally) are selected from 468 basins globally for evaluation (Fig. 2), and their tight oil potential is more than 240 billion barrels preliminarily estimated by volume method.

Tight oil is most prolific in North America, South America, North Africa, and Russia, but less prolific in Asia and Oceania. The hydrocarbon mainly accumulates in foreland basins, continental rift basins (Mesozoic strata), and craton basins (Paleozoic strata), and less in passive margin basins (Mesozoic strata) and back-arc basins (Cenozoic strata), as shown in Fig. 3.

Tight oil mainly accumulates in Silurian, Late Devonian, Permian, Late Jurassic, Middle Cretaceous, and Oligocene–Miocene (Fig. 4), which are well correlated with the six sets of high-quality source rocks that are globally widespread (Klemme and Ulmishek 1991). Generally, 78 % of tight oil reservoirs are marine sediments; the corresponding organic matter of source rocks are mainly Type II (48 %), Type II/III (25 %), Type I/II (18 %), Type III (5 %), and Type I (4 %); TOC mainly ranges from 2 % to 5 %, and R_o mainly from 0.9 % to 1.1 %. In view of organic matter abundance, the average TOC of tight oil reservoirs in Europe–Russia, North America, and Africa exceeds 4 %, which is significantly higher than that in South America, Asia, and Oceania. Tight oil resources are more prolific in the former regions due to the higher average TOC. The average porosity of tight oil reservoirs mainly ranges between 5 % and 7 %, or even reaches 10 % locally. The tight oil resources in North America and South America are prospective for commercial development due to relatively high average reservoir porosity. Marine sediments dominate global tight oil reservoirs, and continental sediments mainly develop in Asia.

4 Types of tight oil plays

4.1 Classification of tight oil plays

Tight oil mainly accumulates in or near source rocks under the control of one or more sets of high-quality source rocks without trap boundaries. Therefore, according to the spatial relationships between tight oil reservoirs and high-quality source rocks, tight oil plays can be classified into eight types, above-source play, below-source play, beside-source play, in-source play, between-source play, in-source mud-dominated play, in-source mud-subordinated play, and interbedded-source play (Table 1). For above-source, below-source, and beside-source plays which generally have conventional hydrocarbon features, high-quality source rocks and reservoirs are completely separated, and hydrocarbons migrate from source rocks to and accumulate in reservoirs; obvious segmentations with low gamma high resistivity of reservoirs and high gamma low resistivity of source rocks are found in well-logging curves. For in-source plays, reservoir rocks are not developed but source rocks serve as reservoirs; the reservoir space mainly consists of organic pores with high gamma in the whole section. For between-source play, hydrocarbon can be supplied from both the upper and lower source rocks of reservoirs, and the monolayer of reservoir rocks is generally very thick (usually greater than 2 m); in the well logging, reservoir rocks are often characterized by low gamma and high resistivity, which can be easily identified and can be developed as a separate reservoir.

When multiple sets of source rocks and reservoir rocks with a small monolayer thickness (less than 2 m) are interbedded vertically, the reservoir cannot be fully distinguished by well logging, and the monolayer cannot be considered separately in practice. Therefore, this reservoir can be sub-classified into in-source mud-subordinated play, in-source mud-dominated play, and interbedded-source play according to the shale-formation thickness ratio. These plays are featured by zigzag pattern in the whole-section well-logging curves with neither low gamma high resistivity of reservoirs nor high gamma low resistivity of source rocks. In-source mud-dominated plays approximate source rocks, and in-source mud-subordinated plays approximate reservoir rocks due to different shale-formation thickness ratios.

4.2 Typical characteristics of tight oil plays

Similar to conventional hydrocarbon plays, above-source plays contain major source rocks below the reservoir rocks and the tight oil reservoir closely overlying source rocks. By contrast, the above-source play has a tight reservoir,

Table 1 Types and examples of tight oil plays

Tight oil play	Above-source	Below-source	Beside-source	In-source	Between-source	In-source mud-dominated	Interbedded-source	In-source mud-subordinated
	Anadarko Basin	Gulf Basin	Anadarko Basin	Anadarko Basin	Williston Basin	Alberta Basin	Permian Basin	Denver Basin
Geological conceptual model	Cleveland	Buda	Granite Wash	Woodford	Bakken	Montney	Wolfcamp	Niobrara
Shale, sandstone, carbonate	Sandstone	Dolomite	Conglomerate, shale	Shale	Limestone, dolomite	Shale, sandstone	Shale, dolomite	Chalk, shale
GR logging response	Reservoir GR, 50–200 Source-rock GR, 200–250	Reservoir GR, 50–80 Source-rock GR, 200–250	Reservoir GR, 50–80 Source-rock GR, 100–200	Whole-section GR, >300	Reservoir GR, 20–50 Source-rock GR, >400	Whole-section GR, 100–150	Whole-section GR, 50–150	Whole-section GR, 30–150
Shale/formation thickness ratio	0.2	0	0.2	1.0	0	0.8	0.6	0.3
Analogs	Anadarko Basin Mississippi Lime	Williston Basin Three Forks, Appalachian Basin Utica	Uinta Basin Current Cr.	Alberta Basin Duvernay, Appalachian Basin Marcellus	Alberta Basin Viking and Cardium, Ordos Basin Yanchang	Piceance Basin Mesaverde, Songliao Basin Qingshankou	Gulf Basin Eagle Ford, Junggar Basin Permian	Permian Basin Bone Spring

Red thresholds are chosen for some extremely high GR; *L. CLVD* Lower Cleveland Formation; *U. CLVD* Upper Cleveland Formation; *EGFD* Eagle Ford Formation; *BKKN* Bakken Formation; *TRKS* Three Forks Formation; *WDFD* Woodford Formation; *MNTN* Montney Formation; *W.* Wolfcamp Formation; *N.* Niobrara Formation; *A.* Member A; *B.* Member B; *C.* Member C

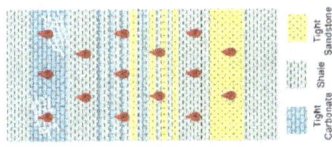

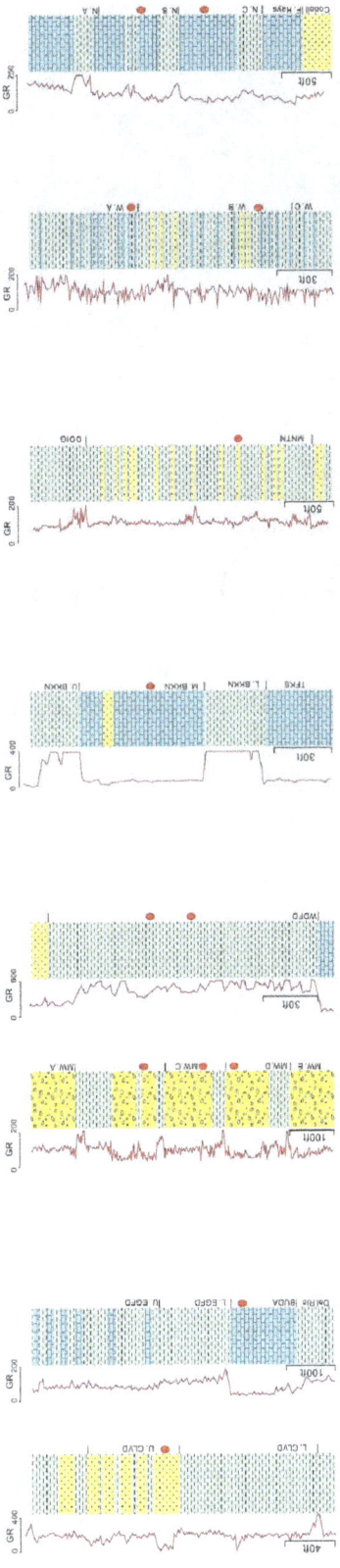

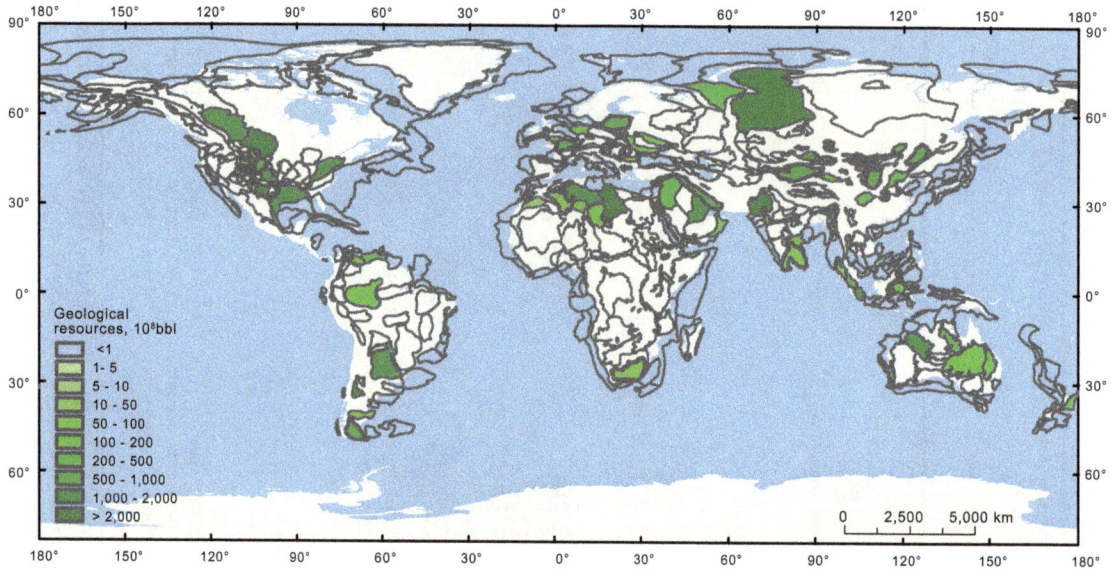

Fig. 2 Distribution of global tight oil resources (*Notes* sorted as per relevant data of USGS, EIA, and IHS)

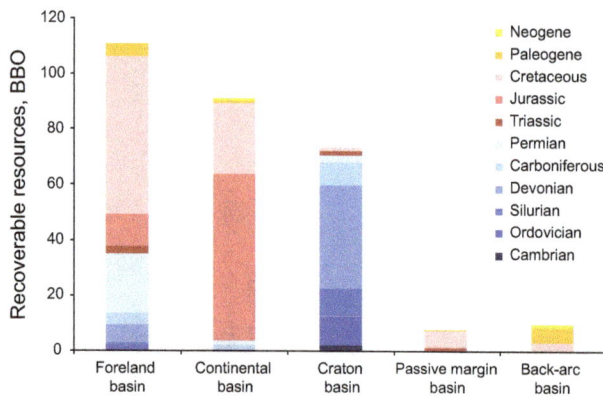

Fig. 3 Types of global basins with tight oil (*Notes* sorted as per relevant data of USGS, EIA, and IHS)

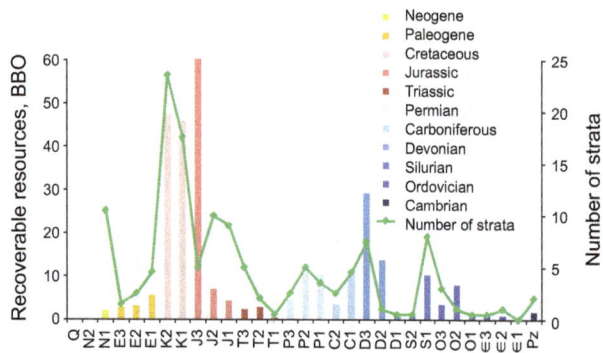

Fig. 4 Stratigraphic ages of global tight oil (*Notes* sorted as per relevant data of USGS, EIA, and IHS. From *left* to *right*, the ages become older, and *J1* Lower Jurassic; *J2* Middle Jurassic; *J3* Upper Jurassic; *Pz* Precambrian)

where oil is not controlled by buoyancy and can only migrate for a short distance. If the reservoir is not tight enough, conventional hydrocarbon rather than tight oil accumulates due to lateral hydrocarbon migration. The Cleveland tight oil play in the Anadarko Basin is a typical case, in which the major producing reservoir is a set of 15 m-thick sandstones in the Upper Cleveland Formation, and the source rocks are a set of mudstones in the Lower Cleveland Formation (Table 1) (Ambrose et al. 2011). Similarly, for the Mississippi limestone tight oil play in the Anadarko Basin, hydrocarbon is supplied by lower organic-rich marl and lower Woodford mudstone, and the high-productivity reservoir is mainly the section with relatively well-developed dolomite in the upper Mississippi Formation (Fig. 5a, b).

For below-source play, source rocks overlie tight oil reservoirs, and oil overcomes buoyancy and migrates into lower adjacent reservoirs under the action of the pressure difference between source rocks and reservoirs. In the Three Forks dolomite tight oil play in the Williston Basin (Fig. 5b), oil is mainly generated in high-quality source rocks of the Lower Bakken Formation (Nordeng and Helms 2010). The Buda dolomite tight oil play in the Gulf Basin lies below the high-quality source rocks of the Lower Eagle Ford Formation (Hentz and Ruppel 2010) (Fig. 6b). Similar plays also include the Tuscaloosa sandstone tight oil play in the Gulf Basin (Bebout et al. 1992) and Lower Qingshankou Fuyu tight oil reservoirs in the Songliao Basin.

For beside-source plays, there is obvious lateral distinction between source rocks and tight oil reservoirs, and oil from source rocks migrates laterally into tight

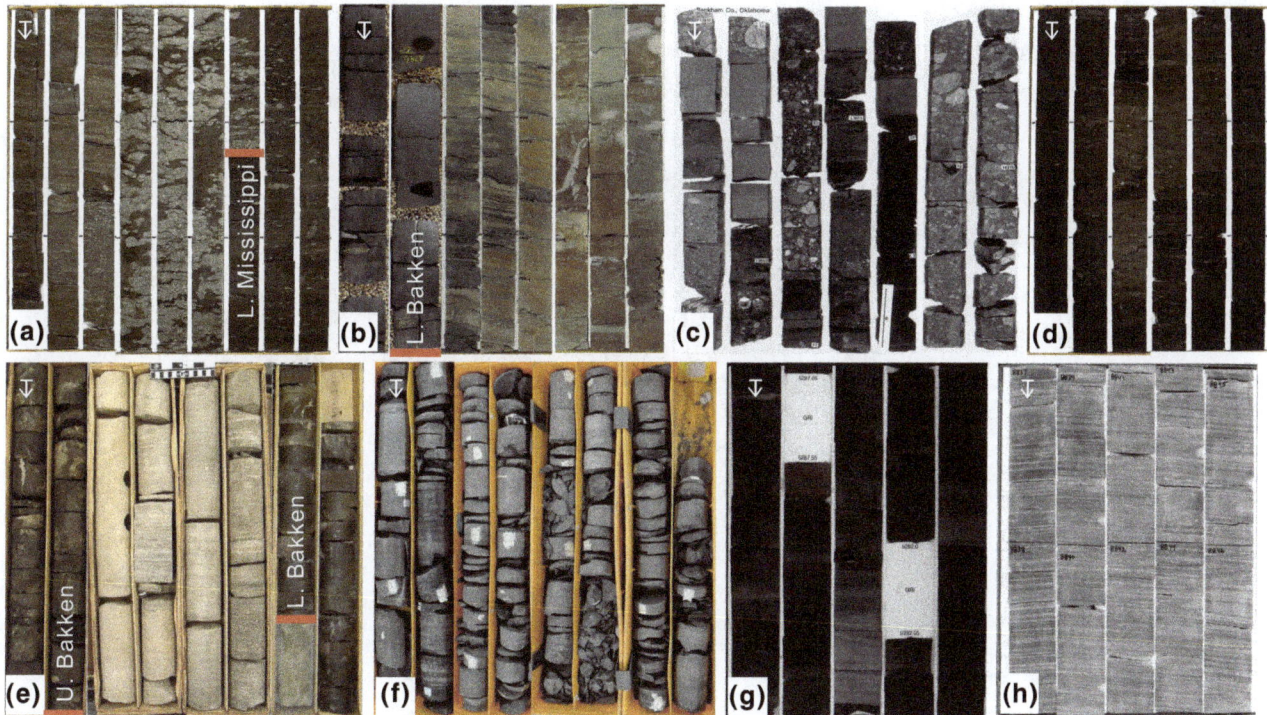

Fig. 5 Core sample photographs of different tight oil plays (*Data source* C & C Reservoir). **a** Above-source play, Mississippi limestone, *middle* and *upper* are dolomitic limestone, *lower* is organic-rich marlstone. **b** Below-source play, Three Forks Formation, mainly dolomite, *upper* is Bakken shale. **c** Beside-source play, mix of granite wash, conglomerate, sandstone, shale. **d** In-source play, Woodford Formation, mainly shale with occasional thin sandstone. **e** Between-source play, Bakken Formation, both *upper* and *lower* are organic-rich shales, *middle* is dolomitic siltstone. **f** In-source mud-dominated play, Qingshankou Formation, *middle* and *upper* are dolomitic limestone, *lower* is mudstone. **g** Interbedded-source play, Wolfcamp Formation, *upper* is mainly interbedded siltstone and shale, *lower* is interbedded dolomite and shale. **h** In-source mud-subordinated play, Bone Spring Formation, mainly sandstone and siltstone with interbedded thin shale

reservoirs. This play type often develops in steep piedmont zones, where alluvial fans exist, a large amount of detritus rapidly accumulates into the lake or sea and laterally interacts with organic-rich shale. For the Granite Wash in the Anadarko Basin, a conglomerate tight oil play, crude oil is supplied laterally by multiple sets of source rocks (Mitchell 2011); the tight oil reservoirs are very thick (usually more than 100 m), but they are not well developed and become rapidly thinner towards the basin, which are dominated by low-porosity and low-permeability conglomerate (Figs. 5c, 6c).

For between-source plays, the major reservoirs are tight formations between multiple sets of high-quality source rocks. The Bakken tight oil play in the Williston Basin is a typical case, where the middle section is a set of tight limestone reservoirs with interbedded siltstone (Fig. 5d), and both the upper and lower sections are high-quality source rocks with an average TOC of 11 % (Sonnenberg and Pramudito 2009; Angulo and Buatois 2012), which form a favorable "sandwiched" combination (Fig. 6d). Therefore, tight oil resources in this basin are the most prolific in the world. The Yanchang-6 and Yanchang-7

formations in the Ordos Basin are also attributed to this kind of play, which are major contributors of tight oil in China.

For in-source plays, tight oil generates and accumulates in source rocks, and the corresponding shale-formation thickness ratio is higher than 0.9. Well logging can hardly recognize sandstone or carbonate layers in these formations, nor reservoir rocks of massive sandstone or carbonate. Typically, in the Woodford tight oil play in the Anadarko Basin, the whole section is shale (Figs. 5e, 6e), and the GR value is 300–700 API. However, the quartz content in the vertical direction varies inside this shale formation, and the high silica section is the primary

Fig. 6 Types and examples of tight oil plays (*Data source* IHS ▶ unconventional database. Thresholds are chosen for some extremely high GR. *Red point* represents the main production layers). **a** Above-source play, Mississippi Formation, Anadarko Basin. **b** Below-source play, Buda Formation, Gulf Basin. **c** Beside-source play, Granite Wash Formation, Anadarko Basin. **d** Between-source play, Bakken Formation, Williston Basin. **e** In-source play, Woodford Formation, Anadarko Basin. **f** In-source mud-dominated play, Montney Formation, Alberta Basin. **g** Interbedded-source play, Wolfcamp Formation, Permian Basin. **h** In-source mud-subordinated play, Niobrara Formation, Denver Basin

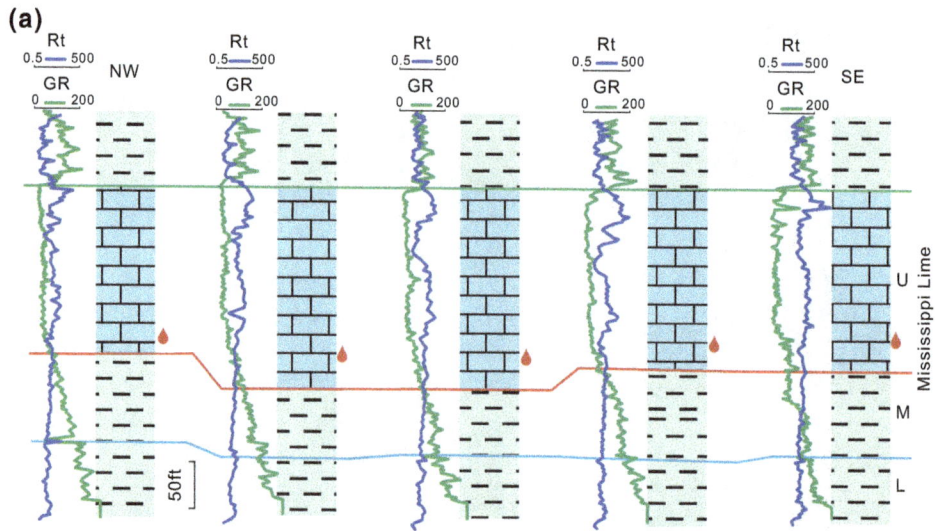

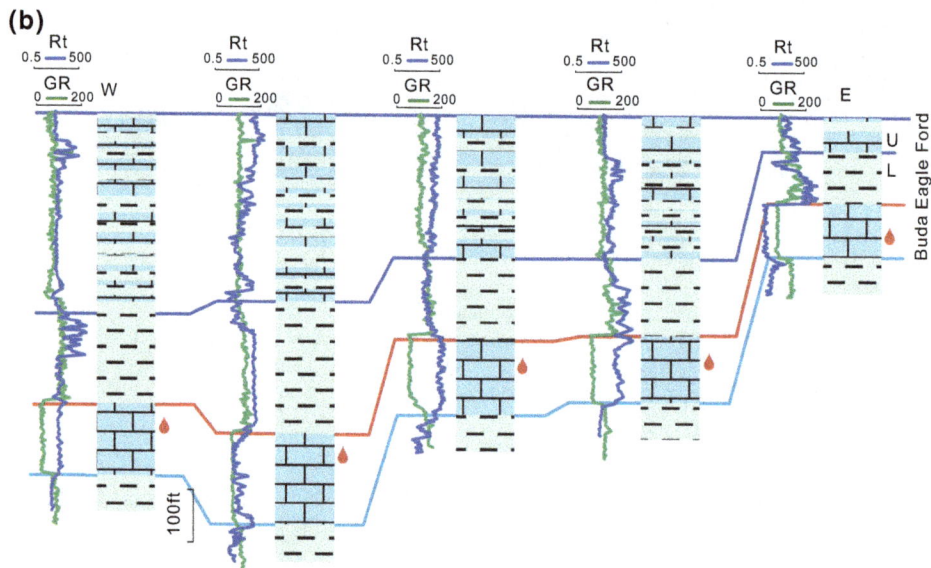

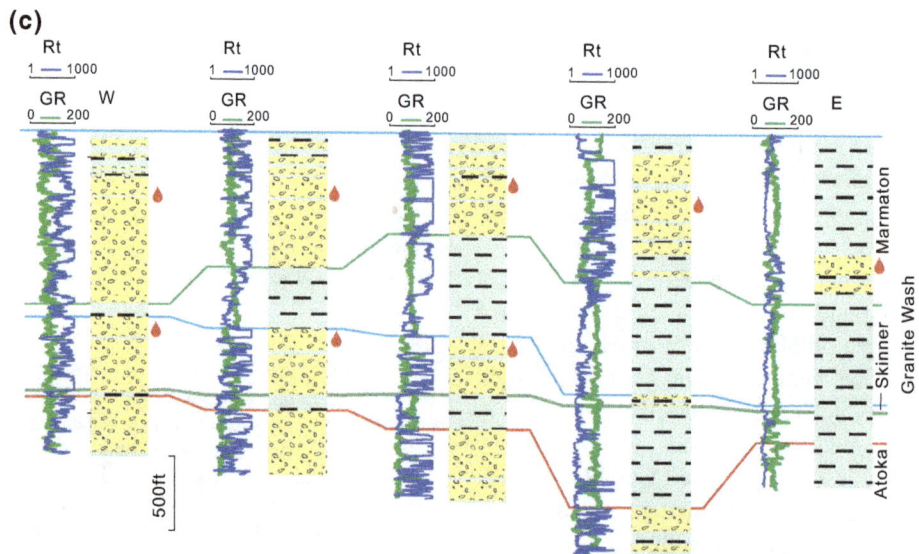

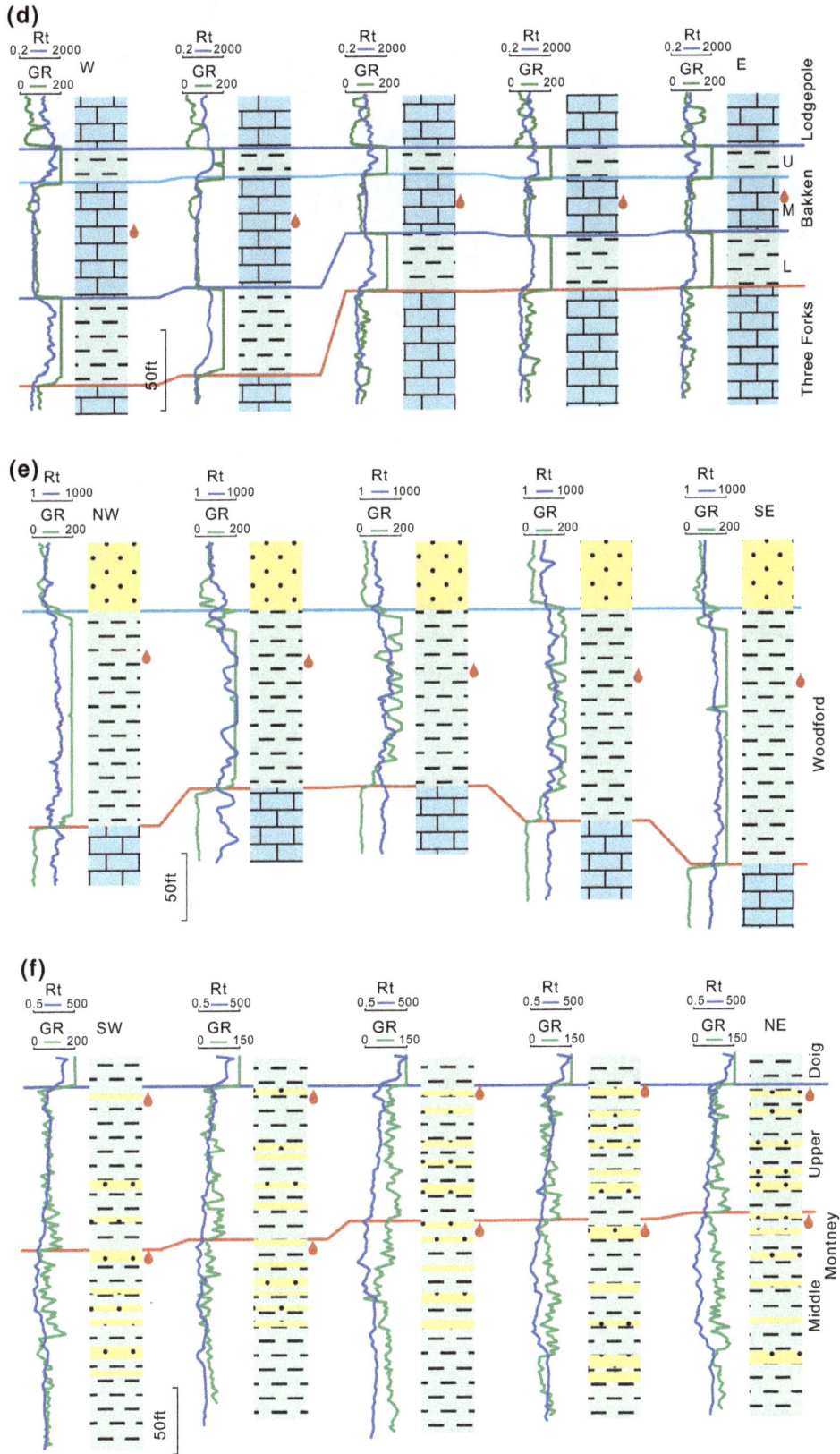

Fig. 6 continued

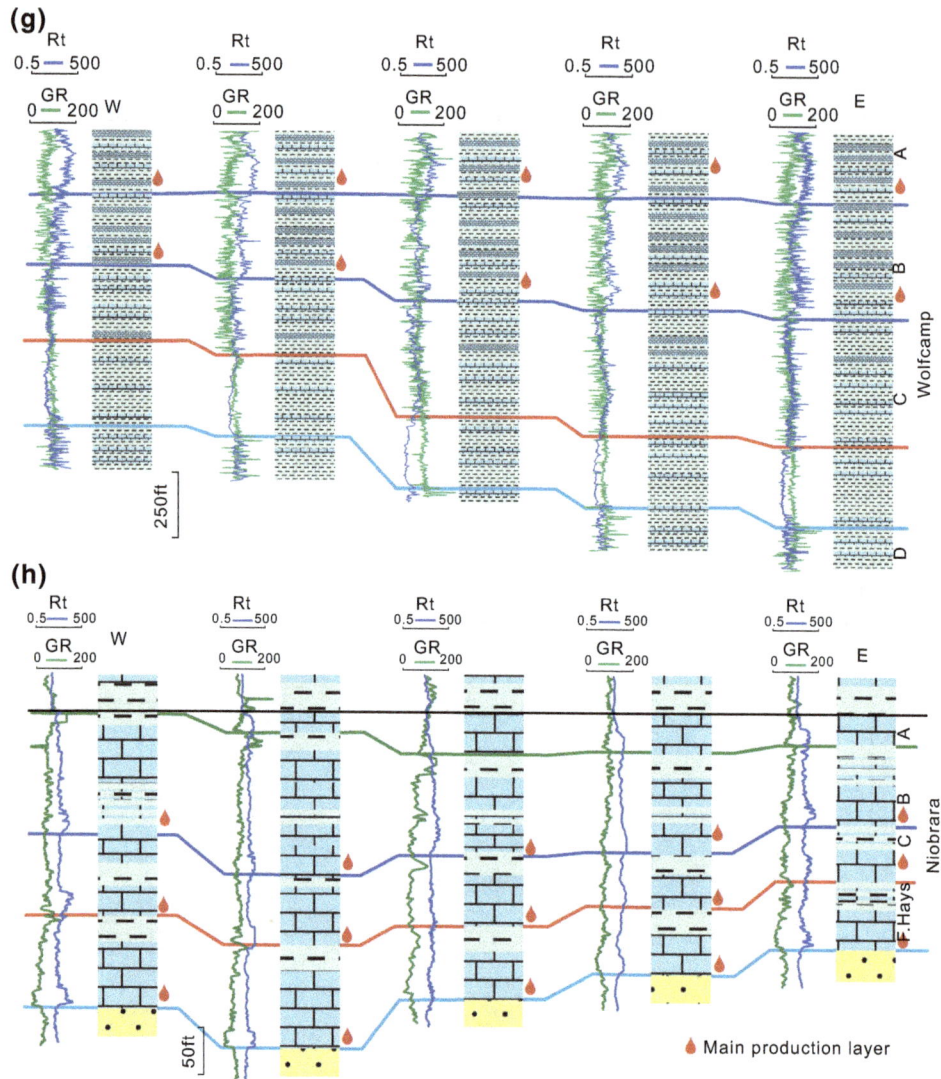

Fig. 6 continued

development target (Andrews 2010; Slatt and O'Brien 2011). In addition, the reservoir rocks in the Duvernay tight oil play in the Alberta Basin, Canada, mainly consist of shale, and the ratio of organic pore porosity to total porosity exceeds 75 % (Chow et al. 1995).

For in-source mud-dominated plays, source rocks and reservoir rocks are interbedded vertically. These mainly consist of source rocks with interbedded thin sandstone or carbonate, and the corresponding shale-formation thickness ratio ranges between 0.6 and 0.9. In the Montney tight oil play in the Alberta Basin, Canada, thick shale formations are dominant, with a small amount of thin sandstone layers and generally low porosity and permeability (Ghanizadeh et al. 2015), the average TOC is 2.5 %, and the reservoirs are featured by small monolayer thickness and relatively low gamma in the whole producing section (Utting et al. 2005). In the Cretaceous Qingshankou Formation in the

Songliao Basin, China, high-quality source rocks with tens of meters thickness are developed and interbedded with a small amount of thin sandstone (Fig. 5f). Similar characteristics are observed in Jurassic Da'anzhai tight oil play in the Sichuan Basin.

For interbedded-source plays, the shale-formation thickness ratio ranges from 0.4 to 0.6, and source rocks and reservoir rocks are interbedded in roughly equal ratios. In the Wolfcamp tight oil plays in the Permian Basin, US, source rocks and reservoir rocks are difficult to distinguish. However, intensive coring tests and analysis indicate that the whole section consists of centimeter-level dark shale, interbedded with argillaceous dolomite and argillaceous siltstone (Baumgardner et al. 2014). The average TOC of the thin shale is 5.4 %, indicating high-quality source rocks. Reservoir rocks and source rocks are difficult to distinguish from core sample photographs due to oil

Table 2 Basic characteristics of key global basins with tight oil

Basin	Formation	Country	Age	Lithology	Type of play	Sedimentary environment
Oman Basin	Athel	Oman	Cambrian	Siliceous shale	Above-source	Deep-sea anoxic sediment
Anadarko Basin	Cleveland	USA	Late Carboniferous	Sandstone, shale	Above-source	Delta–fluvial sediment
Neuquen Basin	Vaca Muerta	Argentina	Late Jurassic–Early Cretaceous	Bottom marlstone, top sandy limestone	Above-source	Lacustrine sediment
Central Sumatra Basin	Brown	Indonesia	Oligocene	Shale, carbonate	Above-source	Lacustrine sediment
Thiemann Pechora	South Devonian	Russia	Devonian	Carbonate, shale	Below-source	Shallow marine clastic zone
Northwest Basin	Posidonia	Germany, Netherlands	Jurassic	Thin black shale, marly limestone	Below-source	Shallow marine clastic zone
Appalachian Basin	Utica/Point Pleasant	USA	Late Ordovician	Calcareous shale, carbonaceous shale, limestone	Below-source	Shallow marine clastic zone
Georgina Basin	Lower Arthur Creek	Australia	Precambrian	Thick mudstone with thin interbedded sandstone	Below-source	Deep-sea anoxic sediment
Uinta Basin	Wasatch/Mesa Verde	USA	Late Paleocene–Early Eocene	Calcareous mudstone, siltstone, black shale	Beside-source	Lacustrine sediment
Anadarko Basin	Granite Wash	USA	Late Carboniferous–Permian	Conglomerate, shale	Beside-source	Lacustrine sediment
Anadarko Basin	Woodford	USA	Devonian–Early Carboniferous	Siliceous, silty shale, lenticular limestone	In-source	Shallow marine clastic zone
Alberta Basin	Duvernay	Canada	Late Devonian	Interbedded limestone and mudstone	In-source	Deep-sea anoxic sediment
West Siberian Basin	Bazhenov	Russia	Late Jurassic	Siliceous, carbonate shale	In-source	Deep-sea anoxic sediment
Ghadames Basin	Tannezuft	Libya	Early Silurian	Black shale, carbonate	In-source	Deep-sea anoxic sediment
Anglo-Dutch Basin	Limburg Group	Britain, Netherlands	Late Carboniferous	Interbedded fine sandstone and shale	In-source	Delta–fluvial sediment
Williston Basin	Bakken	USA	Late Devonian	Limestone, dolomite, a small amount of sandstone	Between-source	Shallow marine clastic zone
Cambay Basin	Tharad	India	Eocene	Calcareous shale, siltstone	Between-source	Shallow marine clastic zone
Alberta Basin	Viking	Canada	Under the Cretaceous	Sandstone, conglomerate, shale	Between-source	Shallow marine clastic zone
Central Arab Basin	Hanifa	Saudi Arabia	Jurassic	Interbedded marlstone and chalk	Between-source	Shallow marine carbonate zone
Alberta Basin	Cardium	Canada	Late Cretaceous	Mudstone, sandstone with interbedded fine-grained conglomerate	Between-source	Delta–fluvial sediment
Eromanga Basin	Merrimelia	Australia	Late Permian	Interbedded fractured sandstone, siltstone, and mudstone	Between-source	Delta–fluvial sediment
Qaidam Basin	Dameigou	China	Middle Jurassic	Black shale, gradually transitioning to fine sandstone upward	Between-source	Delta–fluvial sediment
Songliao Basin	Qingshankou	China	Late Cretaceous	Thin sandstone with interbedded shale	Between-source	Lacustrine sediment
Ordos Basin	Yanchang	China	Triassic	Frontal delta lacustrine sediment, delta, fluvial	Between-source/above-source	Lacustrine sediment

Table 2 continued

Basin	Formation	Country	Age	Lithology	Type of play	Sedimentary environment
Sichuan Basin	Lianggaoshan	China	Jurassic	Shale, limestone	Between-source/ above-source	Lacustrine sediment
Illizi Basin	Aouinet Ouenine	Algeria	Devonian	Shale, thin sandstone	In-source mud-dominated	Shallow marine clastic zone
San Joaquin Basin	Monterey	USA	Miocene	Siliceous mudstone, dolomite, chalk	In-source mud-dominated	Shallow marine clastic zone
Alberta Basin	Montney	Canada	Triassic	Shale, siltstone	In-source mud-dominated	Deep-sea anoxic sediment
Messiah Platform	Bals	Romania	Jurassic	Black shale, argillaceous siltstone	Interbedded-source	Shallow marine clastic zone
Gulf Basin	Eagle Ford	USA	Late Cretaceous	Shale, carbonate rocks, calcareous mudstone	Interbedded-source	Shallow marine carbonate area
Permian Basin	Wolfcamp	USA	Early Permian	Shale, argillaceous limestone, siltstone	Interbedded-source	Shallow marine carbonate zone
Reconcavo Basin	Candeias	Brazil	Cretaceous	Mudstone, sandstone, limestone with fractures	Interbedded-source	Lacustrine sediment
Denver Basin	Niobrara	USA	Late Cretaceous	Interbedded chalk and mudstone, siltstone, sandstone	In-source mud-subordinated	Shallow marine carbonate zone
Permian Basin	Bone Spring	USA	Early Permian	Siltstone, mudstone, carbonate rocks	In-source mud-subordinated	Deep-sea anoxic sediment

Sorted as per relevant data of USGS, EIA, and IHS

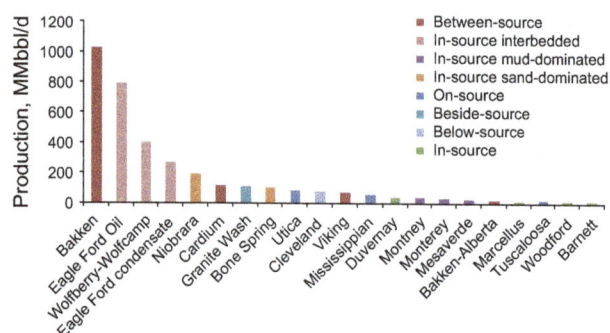

Fig. 7 Production of tight oil plays in North America (*Data source* Hart Energy 2014)

content and small monolayer thickness, but there is still a low gamma indication in well-logging curves (Figs. 5g, 6g). The Eagle Ford tight oil play in the Gulf Basin also shares these characteristics, except that the thickness ratios of lower organic-rich shale and upper carbonate are high (Treadgold et al. 2011), which can also be perceived as an on-source play to some extent. The oil content is highest in the middle of the Eagle Ford Formation, and horizontal wells are also mainly drilled along the middle of the formation. A similar case in China is the Jimusar Lucaogou tight oil play in the Junggar Basin, where tuff is interbedded with dolomite vertically.

For in-source mud-subordinated plays, source rocks and reservoir rocks are interbedded vertically, and the shale-

formation thickness ratio is below 0.4 with high thickness ratios of sandstone and carbonate. The Niobrara tight oil formation in the Denver Basin is divided into three sections of "A", "B", and "C" (Longman et al. 1998), for each of which the cumulative shale thickness is less than 10 m, while the cumulative chalk thickness is 15–20 m (Fig. 6h); the average TOC of shale is 3.8 % and the corresponding thickness ratio is less than 0.4. Both the Bone Spring Formation in the Permian Basin and the East Texas Smackover Formation in the Gulf Basin consist of sandstone with multiple interbedded thin organic-rich marine shales (Demis and Milliken 1993; Montgomery 1997).

In some tight oil plays, lithology varies greatly in the vertical direction due to complex geology, and multiple combinations usually coexist, which can be classified by sections. For example, the Eagle Ford shale in the Gulf Basin is an interbedded-source play, but the underlying Buda dolomite tight oil reservoir is a below-source play, and occasionally the overlying Austin chalk reservoir is an above-source play. In Bakken tight carbonate in the Williston Basin, between-source plays are developed, and the Three Forks dolomite below high-quality shale is also a below-source play. In the similar tight oil plays in the Buda Formation (Gulf Basin) and Three Forks Formation (Williston Basin), some companies are recovering tight oil locally. Statistics indicate that tight oil is most abundant in between-source plays (Table 2).

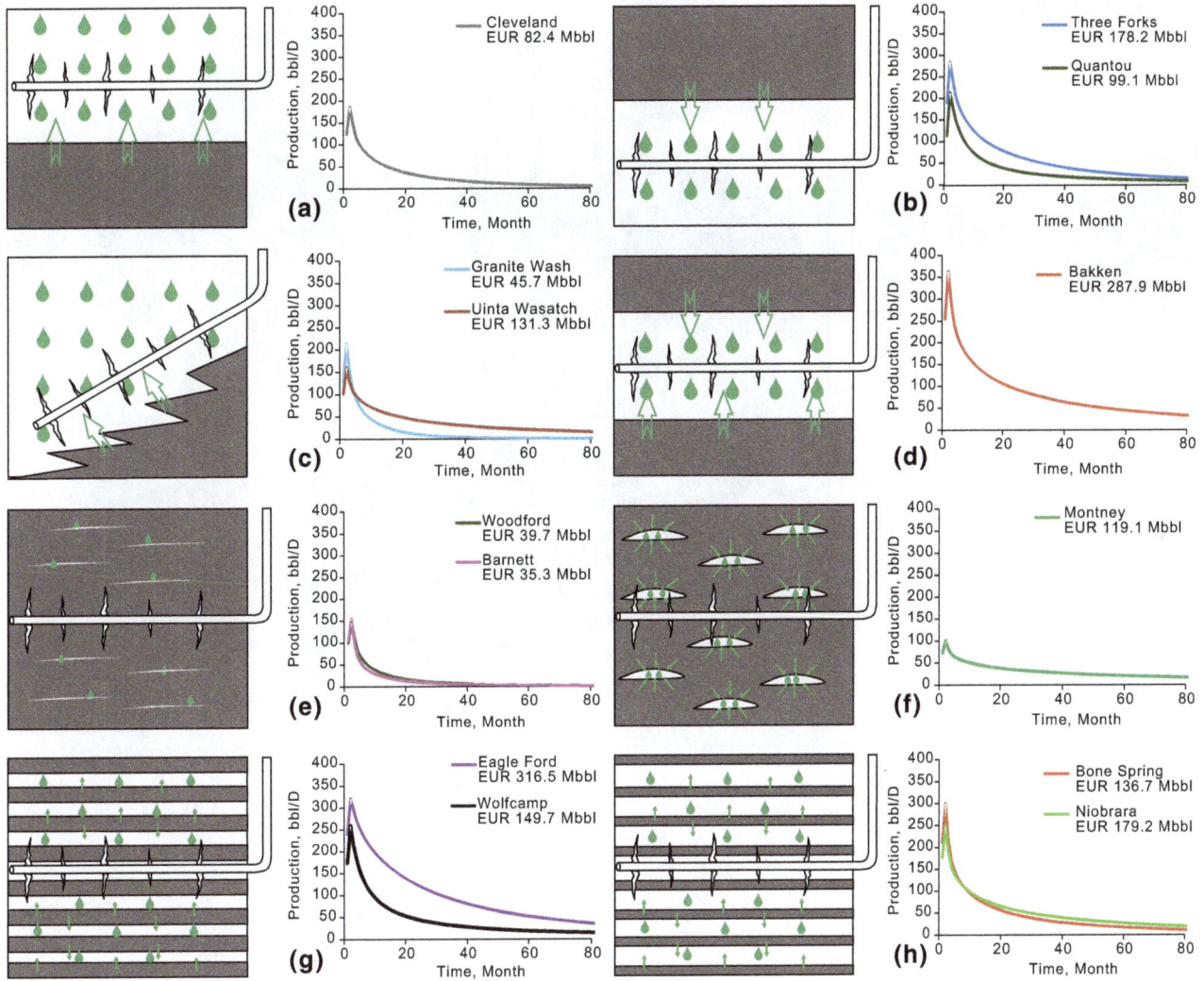

Fig. 8 Hydrocarbon supplying modes and typical production curves of tight oil plays (*Data source* IHS unconventional database; the production curves represent the average production characteristics of the tight oil plays)

5 Distribution of favorable tight oil plays

5.1 Production performances of different tight oil plays

Tight oil plays are closely related to the effective development of tight oil. Currently, more than 95 % of the global tight oil is produced in North America, where US tight oil production accounts for more than 90 % (Hart Energy 2014). According to the Hart Energy's (2014) Q3 data, more than 20 tight oil formations had been commercially developed in North America. The top ten tight oil producing formations are Eagle Ford in the Gulf Basin, Bakken in the Williston Basin, Wolfberry-Wolfcamp in the Permian Basin, Niobrara in the Denver Basin, Cardium in the Alberta Basin, Granite Wash in

the Anadarko Basin, Bone Spring in the Permian Basin, Utica in the Appalachian Basin, Cleveland in the Anadarko Basin, and Mississippi Lime in the Anadarko Basin. These tight oil plays are mainly between-source, interbedded-source, in-source mud-subordinated, and above-source plays (Fig. 7). Tight oil development in China is still in its preliminary stage. The main producing formation is the Yanchang Formation in the Ordos Basin, which is a between-source play, with annual tight oil production up to 8 million tons. Other tight oil producing formations in China include the Da'anzhai Formation in the Sichuan Basin, Qingshankou Formation in the Songliao Basin, Jimusar Sag Lucaogou Formation in the Junggar Basin, and Shulu Sag Shahejie Formation in the Bohai Bay Basin, but none of them has achieved commercial production.

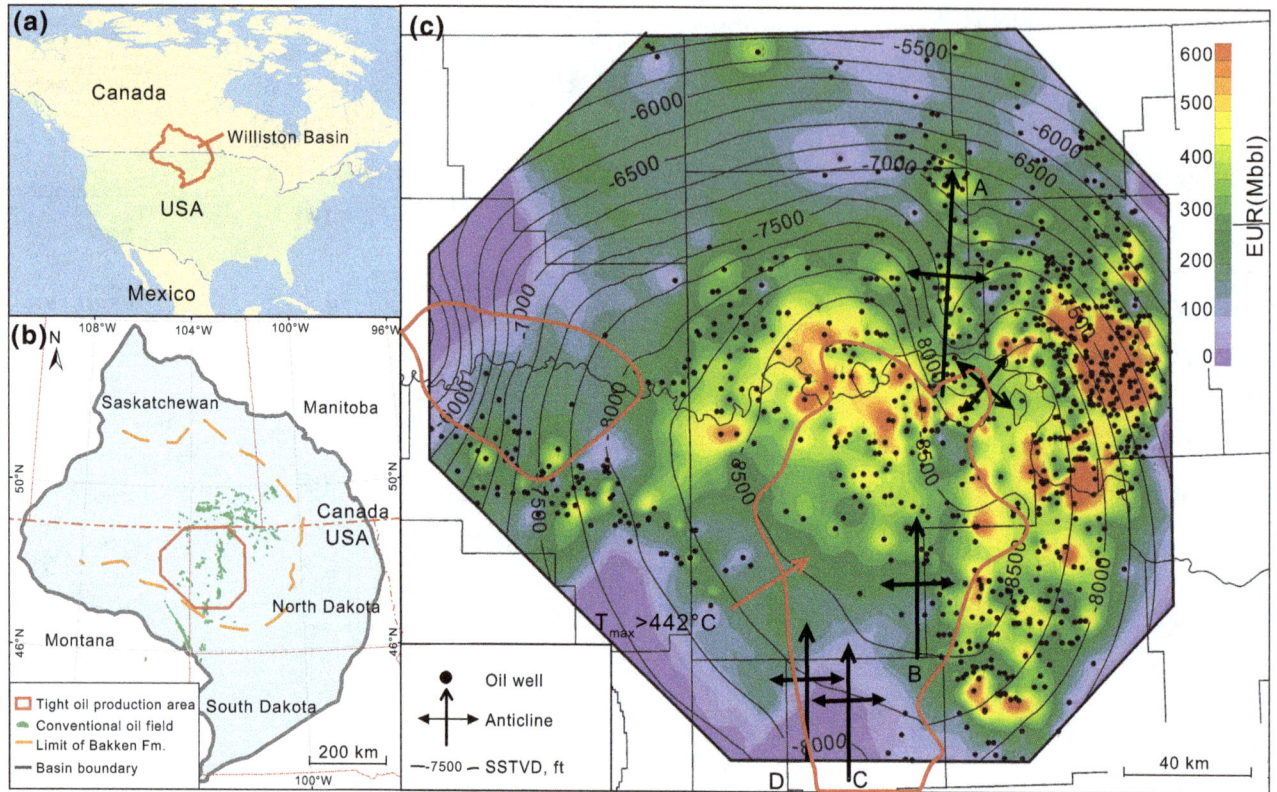

Fig. 9 Tight oil production of Bakken Formation, Williston Basin (*Note A* represents Nesson Anticline; *B* represents Little Knife Anticline; *C* represents TR Anticline; *D* represents Billings Anticline. *Red line* represents the T_{max} of 442 °C. All the wells are horizontal wells with more than 500 days production histories, and the data are from IHS unconventional database, 2014)

5.2 Analysis of favorable tight oil plays

According to the hydrocarbon-supply modes, in-source plays, in-source mud-subordinated plays, and in-source interbedded plays are all classified as bidirectional hydrocarbon-supply mode. Source rocks are developed both in the upper part and lower part of reservoir in these plays (Fig. 8) with relatively large contact area between source rocks and reservoirs, which is favorable for hydrocarbon expulsion from source rocks (Lu et al. 2012). In-source mud-dominated plays and in-source plays are classified as in-source hydrocarbon-supply mode. They are more favorable than above-source plays with unidirectional hydrocarbon-supply mode. In general, the size of pores and throats in tight oil reservoirs is higher than that in shale, and the hydrocarbon-supply is also partly controlled by fluid buoyancy (Lillis 2013). Therefore, the upward hydrocarbon-supply is more favorable than lateral hydrocarbon-supply, and the worst one is downward hydrocarbon-supply. The resource extent is related to hydrocarbon-supply mode to some extent. At present, high production is achieved in tight oil reservoirs with between-source, in-source interbedded, and in-source mud-subordinated plays.

Low clay content and high contents of quartz, feldspar, dolomite, and other brittle minerals are favorable for the implementation of hydraulic fracturing and other reservoir stimulation treatments (Cipolla et al. 2012). In above-source, below-source, between-source, and beside-source plays, the hydrocarbon mainly accumulates in the tight formations that are adjacent to reservoirs. These tight formations featured a high brittle mineral content and better reservoir quality, which is favorable for development. In-source plays and in-source mud-dominated plays mainly consist of source rock with a relatively high clay content, which is difficult to develop (Miller et al. 2013). Fracture or fracture-pore is the dominant reservoir space with strong heterogeneity in these two plays, and fractures rarely develop on a large scale, which results in difficult reservoir prediction. The in-source mud-subordinated plays and in-source interbedded plays fall in between the two above-mentioned classifications. However, a tight lithologic-stratigraphic reservoir usually develops due to the seal of upper and lower source rocks, which is similar to a conventional reservoir and is easy to develop.

Tight oil reservoirs are adjacent to high-quality source rocks, which leads to relatively little difference in hydrocarbon-supply efficiency (Jia et al. 2014). Therefore, comparing with the hydrocarbon-supply mode, reservoir quality is more crucial for the development of tight oil reservoirs. The single-well production performances indicate that the single-well initial production rate (IP) and estimated ultimate recovery (EUR) are significantly higher in the tight oil reservoirs with between-source plays, in-source interbedded plays, and in-source mud-subordinated plays, and the corresponding IP and EUR are 200–400 bbl/d and 150–300 Mbbl, respectively. The IP and EUR in the tight reservoirs with above-source, below-source, and beside-source plays are 150–250 bbl/d and 50–150 Mbbl, respectively. The IP and EUR in the tight reservoirs with in-source mud-dominated plays and in-source plays are 100–150 bbl/d and 30–100 Mbbl, respectively.

In addition, the in-source plays of the Anadarko Basin Woodford Shale, Appalachian Basin Marcellus Shale, and Fort Worth Basin Barnett Shale are the major producing areas with huge shale gas production (IHS 2014a, b; Hart Energy 2014). In comparison, there is a great deal of difference in favorable plays between tight oil reservoirs and shale gas reservoirs, which cannot be equally treated. This mainly results from the big difference in physical properties between oil and natural gas. In shale gas reservoirs, the produced gas includes not only the free gas stored in reservoir space but also the adsorbed gas stored in shale. However, the adsorbed oil in shale is mainly heavy oil which is barely produced, with high contents of asphaltene and non-hydrocarbons. In addition, the oil wettability of shale will affect the tight oil recovery factor (Mwangi et al. 2013), and the reservoirs are difficult to fracture due to low brittle mineral content. Therefore, an in-source play cannot be classified as favorable in spite of rich tight oil resources. Similarly, desired development cannot be achieved in in-source mud-dominated plays.

In summary, in-source interbedded, in-source mud-subordinated, and between-source plays are the most favorable tight oil plays, followed by above-source, below-source, and beside-source plays. In-source plays and in-source mud-dominated plays are the worst plays.

5.3 Distribution of favorable zones for development

Although the type of tight oil play has a significant influence on the production, favorable plays for the development are still controlled by many other factors (Pang et al. 2014). Previous studies believed that tight oil plays are largely free of buoyancy, and structural aspects are neglected in the demonstration of favorable areas (Zou et al. 2012). In fact, statistics show that the present tight oil exploration and development are mainly concentrated in structural slope areas.

The Bakken tight oil play in the Williston Basin is taken as an example, and 884 tight oil production wells are selected for the research in this paper. These wells are completed in 2010–2012, the production times all exceed 500 days and all have entered the stable production stage. All these wells with lateral lengths of more than 3300 ft generally cover the tight oil production areas in the Bakken Basin. The Arps hyperbolic decline model (Robertson 1988) is used to calculate EUR of every well, and a planar EUR distribution (Fig. 9) is established through interpolation. High-production wells are mainly distributed in the basin slope with gentle gradient. There are a few high-production wells in the basin center and high positions such as anticlines.

The structural slope area adjoins high-quality source rocks. The organic-rich shale in the Upper and Lower Bakken Formation is widely developed (Nordeng and Helms 2010), and maturity is the key factor controlling its hydrocarbon generation capacity. Thus, the high-maturity area in the basin center is considered as hydrocarbon generation center. The high-production tight oil area of the Bakken Formation is precisely located at the slope which is the edge of the hydrocarbon generation center. Reservoir space is relatively developed in the structural slope area. In general, the slope area is closer to provenience and is featured by higher granularity and better reservoir capacity. The sediment of Bakken Formation mainly comes from the north east. The sedimentary thickness and the siltstone proportion in the middle Bakken Formation increase to the north east. The middle Bakken Formation in the south west mainly consists of dolomite (Sonnenberg and Pramudito 2009). Small-scale "lithologic-stratigraphic traps" often develop in the structural slope area. These "traps" are sweet spots for tight oil development, which can significantly increase tight oil production. In addition, a number of low-amplitude structural traps are developed in the slope area, which is favorable for fracture development (Sonnenberg et al. 2011). Of course, there are also other possible factors that need to be further researched.

In the Gulf Basin Eagle Ford tight oil play, Permian Basin Wolfcamp tight oil play, and some other plays, the drilled wells are all distributed in the basin slope area (Hart Energy 2014).

6 Conclusions

(1) Tight oil is most prolific in North America, South America, Africa, and Russia, mainly in foreland basins, craton basins (Paleozoic strata), and continental rift basins (Mesozoic strata). Tight oil resources mainly accumulate in Upper Silurian–Middle Ordovician, Upper Devonian–Lower Carboniferous, Permian, Lower Jurassic, Cretaceous, and Oligocene–

Miocene, which are well correlated with the six sets of high-quality source rocks globally. Tight reservoirs adjacent to or in the high-quality source rocks are favorable targets for tight oil exploration.

(2) According to the spatial relationships between reservoir rocks and high-quality source rocks, the tight oil plays can be classified into eight types. These are above-source plays, below-source plays, between-source plays, beside-source plays, in-source plays, in-source mud-dominated plays, in-source mud-subordinated plays, and interbedded-source plays. Between-source, above-source, and in-source mud-subordinated plays are the most favorable types, which are the dominant plays in existing major producing areas and the favorable exploration areas. In contrast, in-source mud-dominated, in-source, and below-source plays are less prospective for development.

(3) The structural gentle-slope areas with favorable play types are the favorable zones for tight oil development because the structural slope areas are generally characterized by proximity to the mature source rocks, relatively better reservoir space, weak structural activities, and more "sweet spots".

References

Ambrose WA, Hentz TF, Carr DL, et al. Oil- and gas-production plays and trends in the Pennsylvanian Marmaton Group and Cleveland Formation, Anadarko Basin, north Texas and western Oklahoma. Report of Investigations—University of Texas at Austin. Bureau of Economic Geology. 2011, p. 67–98.

Andrews RD. Production decline curves and payout thresholds of horizontal Woodford wells in the Arkoma Basin, Oklahoma (Part 2). Shale Shak. 2010;60(4):147–56.

Angulo S, Buatois LA. Integrating depositional models, ichnology, and sequence stratigraphy in reservoir characterization: The middle member of the Devonian-Carboniferous Bakken Formation of subsurface southeastern Saskatchewan revisited. AAPG Bull. 2012;96(6):1017–43.

Baumgardner RW, Hamlin HS, Rowel HD. High-resolution core studies of Wolfcamp/Leonard basinal facies, southern Midland Basin, Texas. AAPG Southwest Section Annual Convention, Midland. 2014;5:11–4. http://www.searchanddiscovery.com/pdfz/documents/2014/10607baumgardner/ndx_baumgardner.pdf.html.

Bebout DG, White WA, Garrett CM, et al. Atlas of major central and eastern Gulf Coast gas reservoirs. Austin: Bureau of Economic Geology, University of Texas at Austin; 1992. p. 88.

BP. Statistical review of world energy. 2015. http://www.bp.com/statistics.

C & C Reservoirs. Field analogs, North America. Houston, 2014.

Chow N, Wendte J, Stasiuk LD. Productivity versus preservation controls on two organic-rich carbonate facies in the Devonian of Alberta: sedimentological and organic petrological evidence. Bull Can Pet Geol. 1995;43(4):433–60.

Cipolla C, Lewis R, Maxwell S, et al. Appraising unconventional resource plays: separating reservoir quality from completion effectiveness. International Petroleum Technology Conference 2012, IPTC. 2012;2:1524–50.

CNPC. Unconventional hydrocarbon potential analysis and future strategic zone selection in global main areas. Beijing, 2014. (unpublished).

Demis WD, Milliken JV. Shongaloo field: a recent Smackover (Jurassic) discovery in the Arkansas-Louisiana State Line Graben. Gulf Coast Assoc Geol Soc Trans. 1993;43:109–19.

EIA. Review of emerging resources: US shale gas and shale oil plays. 2011. http://www.eia.gov/analysis/studies/usshalegas/.

EIA. Technically recoverable shale oil and shale gas resources: an assessment of 137 Shale Formations in 41 Countries Outside the United States. 2013. http://www.eia.gov/analysis/studies/worldshalegas/.

Ghanizadeh A, Clarkson CR, Aquino S, et al. Petrophysical and geomechanical characteristics of Canadian tight oil and liquid-rich gas reservoirs: I. Pore network and permeability characterization. Fuel. 2015;153:664–81.

Hart Energy. North American Shale Quarterly. 2014. http://nasq.hartenergy.com/.

Hentz TF, Ruppel SC. Regional lithostratigraphy of the Eagle Ford Shale: Maverick Basin to East Texas Basin. Gulf Coast Assoc Geol Soc Trans. 2010;60:325–38.

IHS. Going global: predicting the next tight oil revolution. 2014. http://www.ihs.com/products/cera/.

IHS. Unconventional Frontier: prospects for tight oil in North America. 2014. http://www.ihs.com/products/cera/.

Jarvie DM. Shale resource systems for oil and gas: part 2. Shale-oil resource systems. AAPG Mem. 2012;97:89–119.

Jia CZ, Zheng M, Zhang YF. Four important theoretical issues of unconventional petroleum geology. Acta Pet Sin. 2014;35(1):1–10 (in Chinese).

Jia CZ, Zou CN, Li JZ, et al. Assessment criteria, main types, basic features and resource prospects of the tight oil in China. Acta Pet Sin. 2012;33(3):343–9 (in Chinese).

Klemme HD, Ulmishek GF. Effective petroleum source rocks of the world: stratigraphic distribution and controlling depositional factors. AAPG Bull. 1991;75(12):1809–51.

Kuang LC, Tang Y, Lei DW, et al. Formation conditions and exploration potential of tight oil in the Permian saline lacustrine dolomitic rock, Junggar Basin, NW China. Pet Explor Dev. 2012;39(6):657–67 (in Chinese).

Liang DG, Ran LH, Dai DS, et al. A re-recognition of the prospecting potential of Jurassic large-area and non-conventional oils in the central-northern Sichuan Basin. Acta Pet Sin. 2011;32(1):8–17 (in Chinese).

Lillis PG. Review of oil families and their petroleum systems of the Williston Basin. Mt Geol. 2013;50(1):5–31.

Longman MW, Luneau BA, Landon SM. Nature and distribution of Niobrara lithologies in the Cretaceous Western Interior Seaway of the Rocky Mountain Region. Mt Geol Rocky Mt Assoc Geol. 1998;35(4):137–70.

Lu SF, Huang WB, Chen FW, et al. Classification and evaluation criteria of shale oil and gas resources: discussion and application. Pet Explor Dev. 2012;39(2):249–56 (in Chinese).

Ma F, Wang HJ, Zhang GY, et al. Tight oil accumulation characteristics and selection criteria for potential basins. Xinjiang Pet Geo. 2014;35(2):243–7 (in Chinese).

Miller C, Hamilton D, Sturm S, et al. Evaluating the impact of mineralogy, natural fractures and in situ stresses on hydraulically induced fracture system geometry in horizontal shale wells. Society of Petroleum Engineers—SPE Hydraulic Fracturing Technology Conference. 2013, p. 695–710.

Mitchell J. Horizontal drilling of deep granite wash reservoirs, Anadarko Basin, Oklahoma and Texas. Shale Shak. 2011;62(2):118–67.

Montgomery SL. Permian Bone Spring Formation: sandstone play in the Delaware Basin, Part I—slope. AAPG Bull. 1997;81(8):1239–58.

Mwangi P, Thyne G, Rao D. Extensive experimental wettability study in sandstone and carbonate-oil-brine systems: part 1—screening tool development. International Symposium of the Society of Core Analysts held in Napa Valley, California, USA. 2013. http://www.scaweb.org/assets/papers/2013_papers/SCA2013-084.pdf.

National Resources Canada. North American tight oil. 2012. http://www.nrcan.gc.ca/energy/sources/crude/2114#oil1.

Nordeng SH, Helms LD. Bakken source system—Three Forks Formation assessment. North Dakota Department of Mineral Resources. 2010. p. 22.

Pang XQ, Jia CZ, Wang WY. Petroleum geology features and research developments of hydrocarbon accumulation in deep petroliferous basins. Pet Sci. 2015;1:1–53.

Pang XQ, Jiang ZX, Huang HD, et al. Formation mechanisms, distribution models, and prediction of superimposed, continuous hydrocarbon reservoirs. Acta Pet Sin. 2014;35(5):795–828 (in Chinese).

Robertson S. Generalized hyperbolic equation. 1988, SPE-18731-MS.

Slatt RM, O'Brien NR. Pore types in the Barnett and Woodford gas shales: contribution to understanding gas storage and migration pathways in fine-grained rocks. AAPG Bull. 2011;95(12):2017–30.

Sonnenberg SA, Pramudito A. Petroleum geology of the giant Elm Coulee field, Williston Basin. AAPG Bull. 2009;93:1127–53.

Sonnenberg SA, LeFever JA, Hill R. Fracturing in the Bakken Petroleum System, Williston Basin. In: Robinson JW, LeFever JA, Gaswirth SB, editors. The Bakken—Three Forks Petroleum System in the Williston Basin. Denver: Rocky Mountain Association of Geologists; 2011. p. 393–417.

The unconventional oil subgroup of the resources & supply task group. Potential of North American unconventional oil resource. Working Document of the NPC North American Resource Development Study. 2011. p. 8–11.

Treadgold G, Campbell B, McLain W, et al. Eagle Ford prospecting with 3-D seismic data within a tectonic and depositional system framework. Lead Edge. 2011;30(1):48–53.

Utting J, Zonneveld JP, MacNaughton RB, et al. Palynostratigraphy, lithostratigraphy and thermal maturity of the Lower Triassic Toad and Grayling, and Montney Formations of western Canada, and comparisons with coeval rocks of the Sverdrup Basin, Nunavut. Bull Can Pet Geol. 2005;53(1):5–24.

Wang JL, Feng LY, Mohr S, et al. China's unconventional oil: a review of its resources and outlook for long-term production. Energy. 2015;82:31–42.

Yang H, Li SX, Liu XY. Characteristics and resource prospects of tight oil and shale oil in Ordos Basin. Acta Pet Sin. 2013;34(1):1–11 (in Chinese).

Zhang K, Zhang KY, Zhang LL. On tight and shale oil and gas. Nat Gas Ind. 2013;33(9):17–22 (in Chinese).

Zhao JZ, Li J, Cao Q, et al. Hydrocarbon accumulation patterns of large tight oil and gas fields. Oil Gas Geol. 2013;34(5):573–83 (in Chinese).

Zhou QF, Yang GF. Definition and application of tight oil and shale oil terms. Oil Gas Geol. 2012;33(4):541–4 (in Chinese).

Zou CN, Yang Z, Zhang GS, et al. Conventional and unconventional petroleum "orderly accumulation" theoretical recognition and practical significance. Pet Explor Dev. 2014;41(1):14–27 (in Chinese).

Zou CN, Zhu RK, Wu ST, et al. Types, characteristics, genesis and prospects of conventional and unconventional hydrocarbon accumulations: taking tight oil and tight gas in China as an instance. Acta Pet Sin. 2012;33(2):173–87 (in Chinese).

6

Diffusion coefficients of natural gas in foamy oil systems under high pressures

Yan-Yu Zhang[1] · Xiao-Fei Sun[1,2,3] · Xue-Wei Duan[1] · Xing-Min Li[4]

Abstract The diffusion coefficient of natural gas in foamy oil is one of the key parameters to evaluate the feasibility of gas injection for enhanced oil recovery in foamy oil reservoirs. In this paper, a PVT cell was used to measure diffusion coefficients of natural gas in Venezuela foamy oil at high pressures, and a new method for determining the diffusion coefficient in the foamy oil was developed on the basis of experimental data. The effects of pressure and the types of the liquid phase on the diffusion coefficient of the natural gas were discussed. The results indicate that the diffusion coefficients of natural gas in foamy oil, saturated oil, and dead oil increase linearly with increasing pressure. The diffusion coefficient of natural gas in the foamy oil at 20 MPa was 2.93 times larger than that at 8.65 MPa. The diffusion coefficient of the natural gas in dead oil was 3.02 and 4.02 times than that of the natural gas in saturated oil and foamy oil when the pressure was 20 MPa. However, the gas content of foamy oil was 16.9 times higher than that of dead oil when the dissolution time and pressure were 20 MPa and 35.22 h, respectively.

Keywords Foamy oil · Diffusion coefficient · Heavy oil · Gas injection · High pressure

1 Introduction

Field trials in China, Venezuela, and Canada have shown that the primary depletion production of several heavy oil reservoirs is anomalous compared to conventional oil reservoirs under solution gas drive. Once below the equilibrium bubble point pressure, the producing gas–oil ratio (GOR) increases slowly, and a higher primary recovery factor (5 %–25 %) than expected has been reported from some of those reservoirs (Guan et al. 2008; Mu et al. 2009; Mu 2010; Li et al. 2012; Liu et al. 2013b). The most plausible explanation of this anomalous behavior appears to be the foamy oil phenomenon. Such phenomenon occurs when the solution gas is released from oil to form small dispersed, trapped gas bubbles inside the oil which flow with the oil because of the high viscosity of heavy oil. As a result, the oil is expanded and its viscosity is reduced due to gas bubbles (Yu and Shen 2008; Peng et al. 2009; Wang et al. 2009; Torabi et al. 2012). However, when the reservoir pressure is below the pseudo-bubble point pressure, the gas–oil ratio of the reservoir increases quickly, and the oil production rate decreases sharply because the gas bubbles trapped in the oil begin to coalesce together to form free gas phase (Liu et al. 2011; Sun et al. 2013).

Natural gas injection is regarded as one of the most effective methods for enhancing heavy oil recovery after primary production (Garcia 1983; Xu et al. 2009; Zhu et al. 2010; Guo et al. 2010a; Dong et al. 2013). That is because natural gas can take advantage of residual solution gas in the reservoir after primary production, as well as many mechanisms involved in the gas injection process, such as, oil viscosity reduction, oil swelling, and foamy oil

✉ Xiao-Fei Sun
sunxiaofei540361@163.com

[1] College of Petroleum Engineering, China University of Petroleum, Qingdao 266580, Shandong, China

[2] School of Geosciences, China University of Petroleum, Qingdao 266580, Shandong, China

[3] The Department of Chemical and Petroleum Engineering, University of Calgary, Calgary, AB T2N 1N4, Canada

[4] Research Institute of Petroleum Exploration and Development, PetroChina, Beijing 100083, China

Edited by Yan-Hua Sun

formation. However, doubts arose about whether the injected gas is able to dissolve into the oil by molecular diffusion and enhance oil recovery by the above-mentioned mechanisms. To overcome these doubts, the emphasis should be made on the accurate and convenient prediction of diffusion coefficients of natural gas in foamy oil, which is one of the most important parameter to evaluate the potential of gas injection process for foamy oil reservoirs after primary production (Zhang et al. 2000).

In the literature, there are many methods for predicting diffusion coefficients of gases in hydrocarbon systems. These methods can be roughly categorized into direct and indirect methods. In the first category, first measurements of diffusion coefficients were performed by Hill and Lacey (1934) for the methane–decane system at low pressures. Later, Woessner et al. (1969) reported some experimental data for gases in heavy oil and bitumen at reservoir pressure. The above-mentioned direct methods involve compositional analysis of liquid samples extracted at different times during the diffusion process, which is tedious and expensive. To eliminate this requirement, Riazi (1996) developed a simple and indirect way to predict diffusion coefficients of gases—the pressure decay method. It is an experimental method for predicting diffusion coefficients of gases in liquids using a PVT cell. The step changes in the pressure within the PVT cell in combination with a developed mathematical model were used to predict the diffusion coefficients. Later, Jamialahmadi et al. (2006) and Etminan et al. (2010) reported a new method for predicting diffusion coefficients of gases in heavy oil from experimental volume–time profiles. In this method, the volume changes of the PVT cell were recorded and used to calculate diffusion coefficients of gases in liquid hydrocarbon systems instead of the pressure changes.

The present investigations systematically estimate the molecular diffusion of gases, such as methane and CO_2, in dead oils (those with very little dissolved gas) at low pressures. Experimental data on molecular diffusion at high pressures are scarce, at least in part, because conducting these experiments is difficult, expensive, and time consuming, and none of the available experimental data on molecular diffusion and methods is suitable for the prediction of diffusion coefficients of natural gas in foamy oil at high pressures. In this paper, a high pressure PVT cell was used to measure diffusion coefficients of natural gas in typical Venezuela foamy oil at high pressures (5, 8.65, 12, 16, 20 MPa). Furthermore, a new method for determining the diffusion coefficients of natural gas in foamy oil, saturated oil, and dead oil was developed on the basis of experimental data. This paper provides a better understanding of the diffusion process of natural gas in foamy oil and the effects of types of the liquid phase and pressure on the diffusion coefficient of natural gas, which is a critical factor for feasibility evaluation of the gas injection process in foamy oil reservoirs after primary production.

2 Experimental

The oil state in porous media is related to reservoir pressure of foamy reservoirs, as shown in Fig. 1. When the reservoir pressure is above the bubble point pressure (P_b), the gas exists as solution gas in the oil phase. Thus, the oil is in the saturated state. Once the reservoir pressure is below the bubble-point pressure, the released solution gas is trapped and dispersed in oil. The oil is foamy and is in a pseudo-undersaturated state. As the reservoir pressure reaches the pseudo-bubble-point pressure (P_{pb}), the dispersed gas in the heavy oil disengages from oil completely and becomes a movable phase. The oil in some part of the reservoir becomes dead oil after the gas is produced by the wells. In order to understand the natural gas diffusion process in foamy oil which is different from that in saturated oil and dead oil, and to determine the injection timing (reservoir pressure) for gas injection, diffusion coefficients of natural gas in saturated oil, foamy oil, and dead oil at high pressures were measured with a high-pressure PVT apparatus.

2.1 Experimental materials and setup

Crude oil from the MPE3 reservoir in the Orinoco Belt, Venezuela was supplied by China National Petroleum

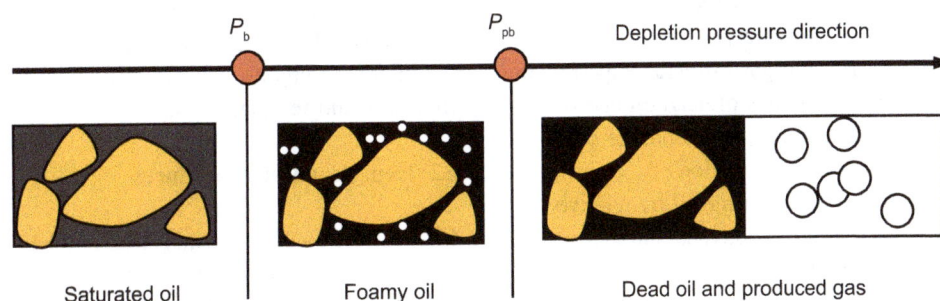

Fig. 1 Three oil states during the production process of foamy oil reservoirs

Table 1 Summary of fluid samples from the MPE3 block in Venezuela

Flash data	GOR, m^3/m^3	15
	Dead oil density, g/cm^3	1.013
Viscosity, mPa s	@ 50 °C	24,715
	@ 65 °C	5559
	@ 80 °C	1620
	@ 95 °C	644
Recombined oil properties	FVF, m^3/m^3	1.173
	Density, g/cm^3	0.957
	Bubble pressure, MPa	4.95
	Pseudo bubble pressure, MPa	
	@ 60 min for each depletion step	3.44
	@ 12 h for each depletion step	2.74
	@ 1 days for each depletion step	1.89

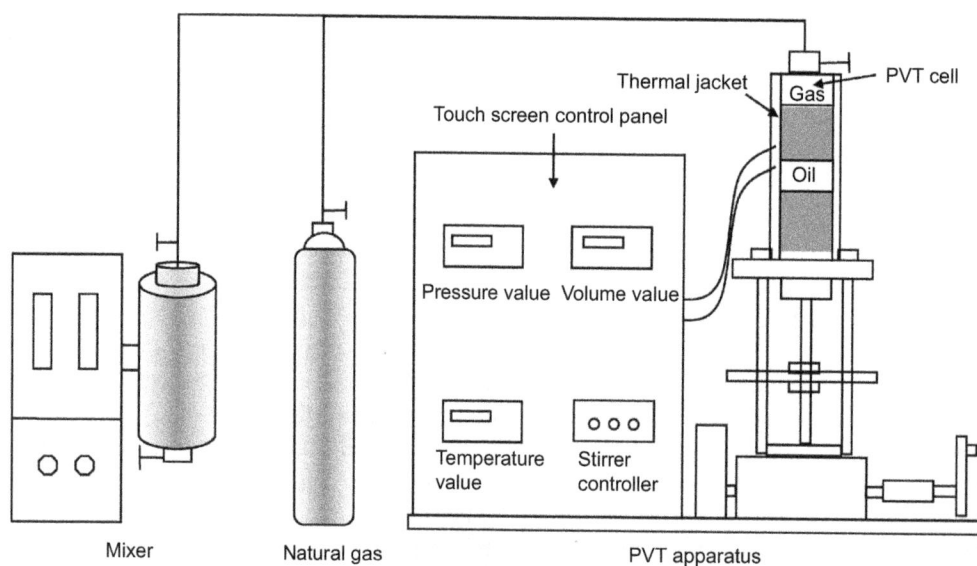

Fig. 2 Diagram of the experimental apparatus for gas diffusion experiments

Corporation, which can easily form foamy oil during production (Bondino et al. 2009; Guo et al. 2010b). According to flashed gas component analysis, methane and carbon dioxide account for 86.7 % and 10.8 %, respectively. Therefore, the natural gas used in experiments consisted of methane and carbon dioxide with a volume ratio of 8:1. The crude oil was recombined with natural gas at the reservoir temperature and pressure (54.2 °C and 8.65 MPa, respectively) to yield recombined reservoir oil for use in gas diffusion experiments. Table 1 lists the fluid characteristics of dead oil and recombined oil. The bubble pressure and pseudo-bubble pressures are important parameters for determination of the foamy oil state in the reservoir which were estimated by relative volume versus pressure curves in the conventional and unconventional PVT tests (Sun et al. 2013).

The experimental apparatus used for this study is shown schematically in Fig. 2. It is a traditional PVT apparatus

(Haian HWGX-60) designed for operation at high temperatures and high pressures. The internal cross-sectional area of the PVT cell is 8.33 cm^2 and the cell height is 30.0 cm. The apparatus provides the pressure, total volume, and temperature values on a touch-screen control panel. The volume of the PVT cell and its pressure can be adjusted by a pump which is controlled by means of a hand-actuated piston or electrically. The fluids in the cell can be mixed vigorously using a magnetic stirrer at desired pressures and temperatures.

2.2 Experimental procedures

After the PVT cell was evacuated for 24 h, the reconstituted oil was introduced into the cell from the mixer to obtain an oil column of a desired height, and the cell pressure and temperature were maintained at initial reservoir conditions

(8.65 MPa and 54.2 °C, respectively). Then, the cell pressure decreased to current reservoir pressure (4.0 MPa) which was between the bubble pressure and pseudo-bubble pressure (Table 1). The unconventional method was used to simulate the foamy oil behavior, that is the PVT cell was not rocked, avoiding a rapid artificial nucleation of the gas micro-bubbles and hence forming a separated gas phase. In this way, the oil would be a pseudo-phase that contains oil and gas bubbles trapped within the oil (foamy oil). A day later, natural gas was injected slowly into the PVT cell from the top until the cell pressure rose to 5.0 MPa. Because the injection gas dissolved into the oil, the cell pressure decreased correspondingly. The volume of the PVT cell was adjusted and recorded to maintain the cell pressure constant. When the PVT cell pressure stayed constant over a period of time, the equilibrium between the gas phase and the aqueous phase was considered to be reached. The natural gas dissolution tests were started from a low pressure. After the first measurement, the above natural gas diffusion process was carried out at 4 higher cell pressures (8.65, 12, 16, and 20 MPa).

Natural gas diffusion experiments in saturated oil and dead oil at the same pressures were studied in a similar process except the method to simulate saturated oil and dead oil before the diffusion process. The saturated oil was simulated by decreasing the cell pressure from the reservoir pressure to 4.0 MPa, and disengaging the released solution gas from oil by strong agitation in the PVT cell. The dead oil was made by decreasing the cell pressure to atmospheric pressure to release all the solution gas in the recombined oil, and then the cell pressure increased to 4.0 MPa.

3 Experimental results and discussion

The gas contents and cumulative diffusion volumes of the foamy oil, saturated oil, and dead oil with time measured in experiments at different pressures are presented in Tables 2, 3 and 4 and Figs. 3, 4 and 5.

In all cases, the cumulative diffusion volumes of natural gas in three types of oils at different pressures show a similar tendency with time, which can be roughly divided into two stages (Figs. 3, 4, 5). The cumulative diffusion volumes of natural gas in these three types of oils rose sharply at the initial stage. Subsequently, the cumulative diffusion volumes kept increasing, but increased slowly compared with the initial stage.

Pressure had an important effect on gas diffusion in foamy oil, saturated oil, and dead oil. The cumulative diffusion volumes and gas contents of the three types of oil increased with an increase in pressure. For gas diffusion in foamy oil, the cumulative diffusion volume at 20 MPa was 4.8 times of that at 8.65 MPa when the diffusion time was 35.22 h. The cumulative diffusion volumes of gas in dead

oil were larger than that in saturated oil and foamy oil at the same pressure and time. When the dissolution time was 35.22 h, the cumulative dissolution volume of gas in dead oil was 1.43 times larger than that in foamy oil, indicating that natural gas has a better diffusion capacity in dead oil than in foamy oil. However, the gas content in foamy oil was higher than that in saturated oil and dead oil. For example, when the dissolution time and pressure were 35.22 h and 20 MPa, the gas content in foamy oil was 1.11 and 16.93 times higher than that in saturated oil and dead oil. This increase is because of the foamy oil system which is a non-equilibrium system and has an amount of trapped gas and solution gas. For this reason, it is harder for natural gas to diffuse into foamy oil.

4 Mathematical model

Descriptions for gas diffusion in foamy oil, saturated oil, and dead oil are schematically shown in Fig. 6. The major assumptions are as follows:

1) Swelling of foamy oil, saturated oil, and dead oil caused by gas solution is negligible.
2) The concentration at the interface is the equilibrium concentration.
3) The PVT cell temperature is constant during gas diffusion.
4) The diffusion coefficient does not change significantly with concentration over the range of concentrations encountered in the test.
5) The foamy oil, saturated oil, and dead oil are non-volatile, and the natural gas is assumed to be a single component gas.
6) When the natural gas is injected into the PVT cell, the dispersed gas in foamy oil dissolves instantaneously into the liquid phase.

Molecular diffusion of gases in oil plays a role in heavy oil recovery processes, such as solution gas drive and gas flooding. In order to corroborate experimental results and predicate diffusion coefficients of natural gas in foamy oil, saturated oil, and dead oil, a model for a one-dimensional diffusion PVT cell without chemical reaction was used as follows:

$$\frac{\partial C}{\partial t} = D \frac{\partial^2 C}{\partial x^2} \tag{1}$$

where C is the mass concentration of solute, kg/m^3; D is the diffusion coefficient, m^2/s; x is the coordinate direction, m; t is time, s.

Before diffusion of gas in oil, the concentration of gas in foamy oil (C_i) is assumed to have two parts according to the characteristics of foamy oil: (1) $C_{solution}$ is the

Table 2 Gas content (GC) in foamy oil versus time at different pressures

5 MPa		8.65 MPa		12 MPa		16 MPa		20 MPa	
Time, h	GC, cm³	Time, h	GC, cm³	Time, h	GC, cm³	Time, h	GC, cm³	Time, h	GC, cm³
0.17	41.14	0.17	41.19	0.08	41.29	0.08	41.46	0.08	41.43
0.33	41.18	0.33	41.23	0.09	41.3	0.42	41.74	0.22	41.90
0.83	41.19	0.83	41.24	0.14	41.32	0.75	41.94	0.38	42.16
1.17	41.21	1.00	41.25	0.17	41.35	0.92	42.1	0.55	42.19
1.33	41.21	1.17	41.26	0.25	41.37	1.08	42.12	0.72	42.27
12.33	41.25	1.33	41.26	0.42	41.4	1.25	42.13	1.05	42.34
18.83	41.27	2.33	41.26	0.58	41.42	1.42	42.1	1.22	42.41
23.83	41.28	12.33	41.3	1.08	41.45	1.58	42.16	1.55	42.42
26.83	41.28	18.83	41.33	2.47	41.45	2.75	42.22	1.72	42.52
37.33	41.30	23.83	41.35	15.47	41.56	3.75	42.24	8.00	42.59
42.83	41.31	26.83	41.42	17.98	41.62	4.75	42.26	11.72	42.64
50.00	41.33	37.33	41.44	29.47	41.74	28.75	42.46	17.72	42.72
55.00	41.35	42.83	41.53	46.47	41.95	46.75	42.54	35.22	42.84

Table 3 Gas content (GC) in saturated oil versus time at different pressures

5 MPa		8.65 MPa		12 MPa		16 MPa		20 MPa	
Time, h	GC, cm³	Time, h	GC, cm³	Time, h	GC, cm³	Time, h	GC, cm³	Time, h	GC, cm³
0.10	36.50	0.12	36.52	0.17	36.82	0.08	37.17	0.12	37.51
0.28	36.54	0.28	36.54	0.33	36.93	0.17	37.32	0.20	37.76
0.50	36.56	0.50	36.55	0.50	36.97	0.33	37.45	0.37	37.87
0.98	36.67	0.98	36.67	1.00	37.00	0.50	37.56	0.53	37.97
2.70	36.75	1.50	36.79	1.75	37.04	0.83	37.63	1.00	38.03
6.20	36.77	2.70	36.85	3.00	37.05	1.83	37.71	1.58	38.09
16.60	36.80	6.20	36.9	5.75	37.1	2.83	37.76	2.08	38.09
23.38	36.81	6.60	36.9	10.75	37.14	3.83	37.78	3.00	38.18
50.00	36.87	23.38	36.96	15.00	37.19	4.83	37.8	3.40	38.2
68.00	36.92	50.00	37.11	23.08	37.22	16.16	37.95	13.00	38.32
73.00	36.94	78.00	37.2	40.00	37.34	25.53	37.97	26.50	38.44
80.00	36.96	83.00	37.22	47.53	37.38	39.80	38.12	36.17	38.54
85.00	36.97	90.00	37.25	51.00	37.42	45.00	38.16	60.00	38.76

concentration of the solution gas, and (2) $C_{dispersed}$ is the concentration of the dispersed gas which is trapped in the oil phase. The concentration of gas in saturated oil (C_i) is the concentration of the solution gas, $C_{solution}$. The concentration of gas in dead oil is negligible in that the solution gas was released from oil when the pressure was decreased to atmospheric pressure. Thus, the initial condition of the diffusion PVT cell is

Foamy oil:

$$C_i = C_{solution} + C_{dispersed} \qquad (2)$$

Saturated oil:

$$C_i = C_{solution} \quad \text{for } t = 0 \quad 0 \leq x \leq Z_x \qquad (3)$$

Dead oil:

$$C_i = 0. \qquad (4)$$

According to Whitman's theory (1923), the gas and liquid phases at the interface are thermodynamically in equilibrium. Thus, the concentration of gas at the interface, C_{eq}, remains constant. The first boundary condition of Eq. (1) is

$$C = C_{eq} \quad \text{for } t > 0 \quad x = Z_0. \qquad (5)$$

In these experiments, the natural gas cannot diffuse to reach the bottom of the PVT cell because diffusion coefficients of gas are sufficiently low and the experimental time is relatively short. Hence, the semi-infinite system assumption is valid. Hence, the second boundary condition of Eq. (1) can be defined as

$$C = C_i \quad \text{for } x = 0 \quad t \geq 0. \qquad (6)$$

The Laplace transform was applied to reducing the partial differential equation [Eq. (1)] using the initial

Table 4 Gas content (GC) in dead oil versus time at different pressures

5 MPa		8.65 MPa		12 MPa		16 MPa		20 MPa	
Time, h	GC, cm³	Time, h	GC, cm³	Time, h	GC, cm³	Time, h	GC, cm³	Time, h	GC, cm³
0.08	0.04	0.08	0.09	0.50	0.62	0.50	1.20	0.15	1.30
1.77	0.19	0.43	0.15	1.00	0.78	1.00	1.41	0.65	1.61
2.27	0.28	0.77	0.19	2.00	0.83	2.00	1.51	2.15	1.80
3.77	0.31	1.27	0.21	2.50	0.83	6.50	1.67	12.15	2.17
8.47	0.52	1.77	0.24	4.50	0.93	11.50	1.79	24.65	2.36
13.27	0.58	2.27	0.27	6.50	1.12	24.00	1.89	29.15	2.53
14.27	0.59	3.77	0.51	11.50	1.21	33.50	1.99	35.65	2.56
37.27	0.79	12.27	0.73	23.00	1.31	35.00	2.01	48.65	2.59
51.00	0.90	13.27	0.77	33.00	1.40	54.00	2.24	54.00	2.71
64.00	0.99	14.27	0.81	34.50	1.42	62.00	2.42	58.00	2.80
70.00	1.05	37.27	0.91	54.00	1.68	65.00	2.45	60.00	2.86
75.00	1.09	51.00	1.20	60.00	1.75	68.00	2.59	63.00	2.96
80.00	1.13	54.00	1.25	65.00	1.90	70.00	2.63	67.00	3.05

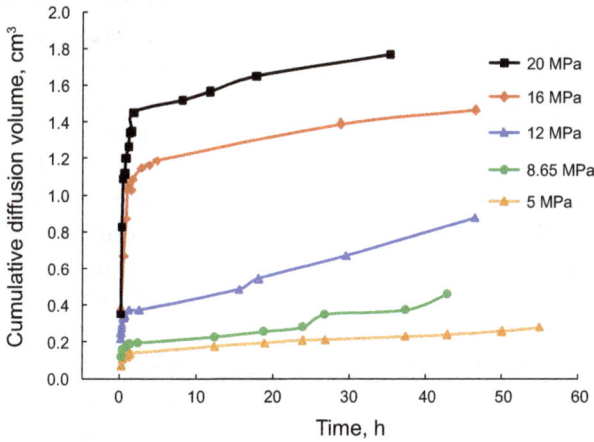

Fig. 3 Measured cumulative diffusion volumes of natural gas versus time for foamy oil

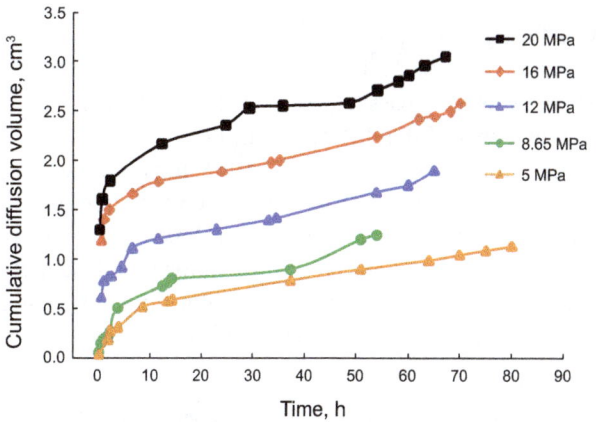

Fig. 5 Measured cumulative diffusion volumes of natural gas versus time for dead oil

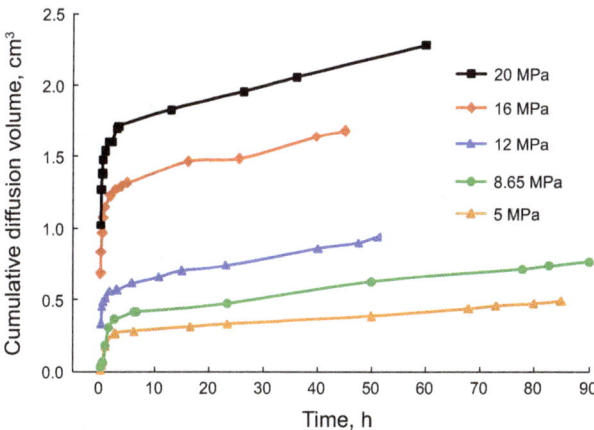

Fig. 4 Measured cumulative diffusion volumes of natural gas versus time for saturated oil

conditions and an analytical closed form of the Laplace inverse is available as follows:

$$C(x, t) = (C_{eq} - C_i) erfc\left(\frac{x}{2\sqrt{Dt}}\right) + C_i, \quad (7)$$

where $C(x, t)$ is the solute concentration at position x at time t, kg/m³.

The mass of the natural gas transferred into the oil phase after time t can be calculated from the integration of Eq. (7) over the volume of the PVT cell:

$$m = \int_0^{Z_x} C(x, t)\, dV = A \int_0^{Z_x} C(x, t)\, dx$$

$$= 2S(C_{eq} - C_i)\sqrt{\frac{Dt}{\pi}} + m_i = K\sqrt{t} + m_i, \quad (8)$$

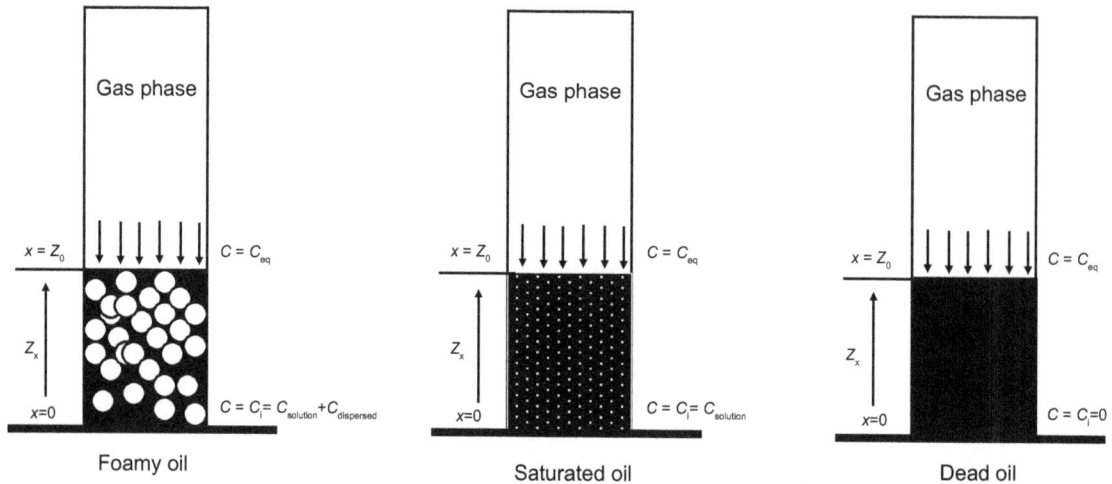

Fig. 6 Schematic and dimensions of diffusion process models

with

$$K = 2A(C_{eq} - C_i)\sqrt{\frac{D}{\pi}}, \qquad (9)$$

where m is the mass of the solute, kg; A is the cross-sectional area of the diffusion cell, m^2.

According to Eq. (9), a plot of the mass of natural gas transferred into the oil phase versus the square root of time should provide a straight line with a slope of K. Thus, the diffusion coefficients of natural gas in foamy oil, saturated oil, and dead oil can be predicted from Eq. (9) by calculating the slope of K from the experimental data.

5 Determination of diffusion coefficients

For the application of the model for determination of diffusion coefficients, gas compressibility factors under different pressures must be calculated based on the volume and the cross-sectional area of the cell, and the temperature and initial pressure of the system. Later, the interfacial concentration of gas (C_{eq}) and the initial concentration of gas in the liquid phase (C_i) should be determined by the characteristics of the three oil types. It is noted that the initial concentration of gas in foamy oil can be obtained from the trapped free gas coefficient (α) which represents the capability of the heavy oil to trap the released gas after depressurization. The factor is the volume of gas entrained in the foamy oil ($x - y$) divided by volume of gas (x) released after depressurization which can be obtained by comparing the GOR behavior with pressure in both conventional and unconventional differential liberation PVT tests (Fig. 7).

The volume of gas (x) released after depressurization can be measured from the GOR obtained from a conventional differential liberation PVT test shown in Fig. 7. During this test, the oil and gas are under equilibrium

conditions. Thus, the gas released is the maximum quantity at every pressure. For the unconventional differential liberation PVT test, the volume of the released gas existing as free gas bubbles (y) can be measured. Therefore, the volume of gas entrained in foamy oil can be estimated by subtracting the released gas (y) during an unconventional differential liberation PVT test from the volume of gas (x) released in a conventional differential liberation PVT test. The initial concentration of gas in saturated oil shown in Eq. (3) can be determined by the conventional differential liberation PVT test (Wang et al. 2012; Mohammad et al. 2012; Liu et al. 2013a).

After determination of all the parameters mentioned above in Eq. (9), the mass m of foamy oil, saturated oil, and dead oil were plotted against the square root of time $\sqrt{t}$ as shown in Figs. 8, 9 and 10.

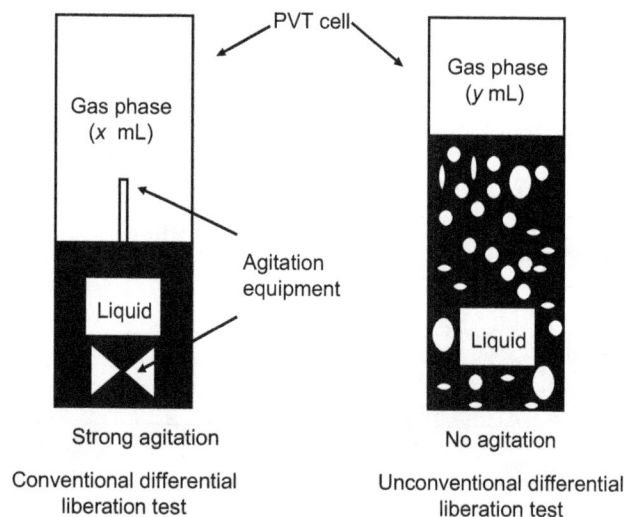

Fig. 7 Schematic comparison of conventional and unconventional PVT tests showing the method to determine the initial concentration of gas in the foamy oil

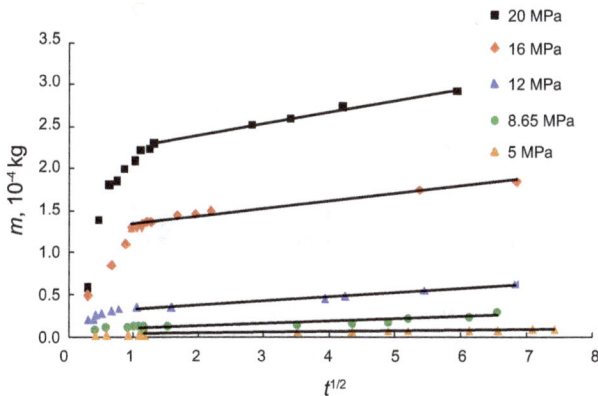

Fig. 8 Estimation of diffusion coefficients of natural gas in foamy oil at different pressures using Eq. (9)

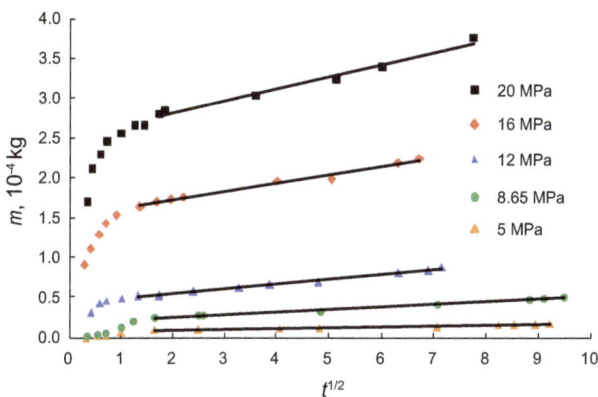

Fig. 9 Estimation of diffusion coefficients of natural gas in saturated oil at different pressures using Eq. (9)

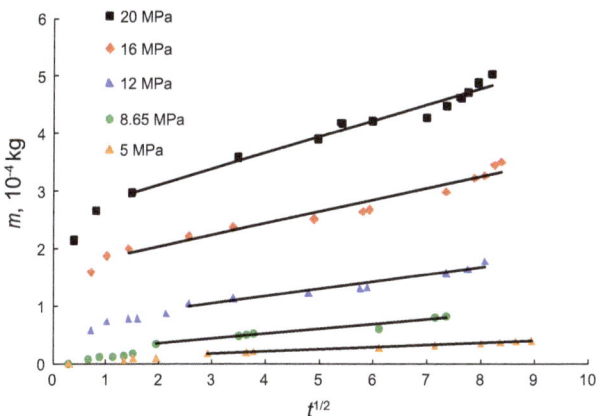

Fig. 10 Estimation of diffusion coefficients of natural gas in dead oil at different pressures using Eq. (9)

It is seen that experimental plots m versus $\sqrt{t}$ can be divided into two stages in Figs. 8, 9 and 10. The early stages of diffusion are often affected by convective mixing arising from initially high mass transfer rates and surface tension-driven instabilities. This initial period or "incubation period" (Renner 1988) increases as the operating pressure increases. Since the effects of the incubation period decay with increasing contact time, middle and late time data are more reliable for estimation of diffusion coefficients from the experimental data. The diffusion coefficients were determined from the slopes of these straight lines by Eq. (9). The results of these calculations are given in Table 5 and Fig. 11 as a function of pressure for foamy oil, saturated oil, and dead oil.

The following information can be obtained from Fig. 11 and Table 5:

(1) It can be seen from R^2 shown in Table 5 that all experimental data in the plots m versus $\sqrt{t}$ conform well to the straight lines after the incubation period, which corroborates that the proposed model is suitable to determine diffusion coefficients of gas in foamy oil, saturated oil, and dead oil.

(2) Diffusion coefficients of natural gas in foamy oil, saturated oil, and dead oil increase steadily with increasing operating pressure. The diffusion coefficients of natural gas in foamy oil at 20 and 8.65 MPa are 5.53×10^{-9} and 1.89×10^{-9} m²/s, respectively, indicating that diffusion coefficient of natural gas in foamy oil at 20 MPa is 2.93 times larger than that at 8.65 MPa.

(3) A comparison of diffusion coefficients of natural gas in foamy oil, saturated oil, and dead oil at the same pressure indicates that diffusion coefficients of natural gas in foamy oil are lower than that of gas in saturated oil and dead oil. For example, the diffusion coefficients of natural gas in dead oil are 3.02 and 4.02 times than that of gas in saturated oil and foamy oil when the pressure is 20 MPa. This is the reason why the cumulative diffusion volume of gas in dead oil is larger than that of gas in saturated oil and foamy oil.

(4) From Fig. 11, it is observed that diffusion coefficients of natural gas in foamy oil, saturated oil and dead oil increase linearly with increasing operating pressure. The slope of the straight line for dead oil is larger than that for saturated oil and foamy oil. This means that the growth rate of the diffusion coefficient with increasing pressure in foamy oil is less than that in saturated oil and dead oil.

Several investigators have reported the diffusivity of methane and carbon dioxide in bitumens and heavy oils. However, these results are molecular diffusion of gases in dead oils at low pressures. Thus, in order to prove the validity of experimental results, correlation of gas diffusion coefficient in dead oil (Fig. 11) determined using the

Table 5 Diffusion coefficients of natural gas in foamy oil, saturated oil, and dead oil

Pressure, MPa	Oil type	Slope	R^2	Diffusion coefficient, 10^{-9} m^2/s
20	Foamy oil	0.0000130	0.9956	5.53
	Saturated oil	0.0000151	0.9797	7.36
	Dead oil	0.0000278	0.9199	22.41
16	Foamy oil	0.0000090	0.9704	4.87
	Saturated oil	0.0000105	0.9854	5.58
	Dead oil	0.0000202	0.9451	17.92
12	Foamy oil	0.0000049	0.8483	2.66
	Saturated oil	0.0000060	0.9902	3.88
	Dead oil	0.0000125	0.9381	13.73
8.65	Foamy oil	0.0000028	0.9864	1.89
	Saturated oil	0.0000033	0.9866	2.52
	Dead oil	0.0000082	0.9365	11.63
5	Foamy oil	0.0000008	0.9711	0.73
	Saturated oil	0.0000010	0.9729	1.05
	Dead oil	0.0000036	0.9931	7.66

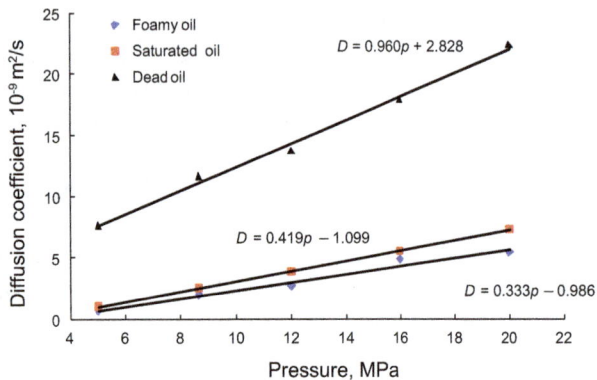

Fig. 11 Diffusion coefficients of natural gas in foamy oil, saturated oil, and dead oil as a function of pressure

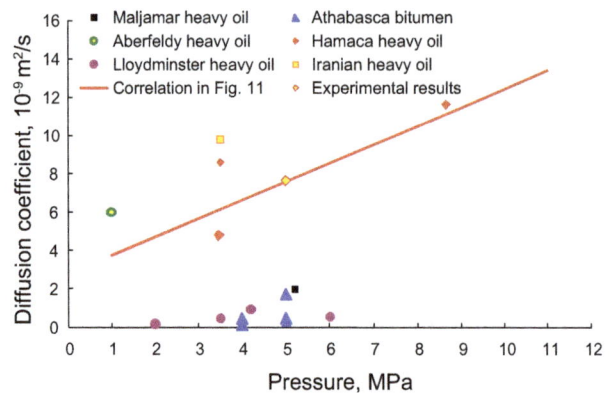

Fig. 12 Comparison of measured diffusion coefficients in different oils

equation of the straight line regression of experimental data is compared with the reported values for similar systems, shown in Table 6. A comparison (Fig. 12) shows that the correlation of natural gas diffusivities in Orinoco heavy oil obtained in this study lies in the same range to those in the literature data for Hamaca heavy oil in Venezuela. They are much larger than the gas diffusivities in Lloydminster heavy oil, Athabasca heavy oil, and Maljamar heavy oil

Table 6 Available diffusivity data of gases in bitumens and heavy oils

Reference	Pressure, MPa	Temperature, °C	Gas–liquid phase	Diffusivity 10^{-9} m^2/s
Grogan et al. (1988)	5.2	25	CO_2–Maljamar heavy oil	2
Schmidt (1989)	5	20–200	Methane–Athabasca bitumen	0.28–1.75
	5	50	CO_2–Athabasca bitumen	0.5
Nguyen and Farouq Ali (1998)	1	23	CO_2–Aberfeldy heavy oil	6
Zhang et al. (2000)	3.51	21	Methane–Hamaca heavy oil	8.6
	3.47	21	CO_2–Hamaca heavy oil	4.8
Upreti and Mehrotra (2000)	4	25–90	CO_2–Athabasca bitumen	0.16–0.47
Tharanivasan et al. (2004)	3.5–4.2	23.9	CO_2–Lloydminster heavy oil	0.46–0.94
Yang and Gu (2005)	2–6	29.3	CO_2–Lloydminster heavy oil	0.199–0.551
Jamialahmadi et al. (2006)	3.5	50	Methane–Iranian heavy oil	9.8

probably because of the different oils and test conditions, such as experimental temperature and method.

6 Conclusions

(1) A new model for determining the diffusion coefficient of natural gas in foamy oil, saturated oil, and dead oil was developed based on experimental data. The diffusion coefficient of natural gas in these three types of oil can be predicted accurately and conveniently by determining the slope of the plot of the mass of the oil as it absorbs gas against the square root of time.

(2) During the determination of diffusion coefficient from experimental data, the initial concentration of gas in foamy oil can be obtained by the trapped free gas coefficient which can be obtained by comparing the GOR behavior with pressure in both conventional and unconventional differential liberation PVT tests.

(3) The diffusion coefficients of natural gas in foamy oil are lower than those of gas in saturated oil and dead oil. The diffusion coefficient of natural gas in dead oil is 3.02 and 4.02 times than that of gas in saturated oil and foamy oil when the pressure is 20 MPa. However, the gas content of foamy oil is higher than that of saturated oil and dead oil. The gas content of foamy oil is 1.11 and 16.9 times higher than that of saturated oil and dead oil when the dissolution time and pressure are 20 MPa and 35.22 h, respectively.

(4) The diffusion coefficients of natural gas in foamy oil, saturated oil, and dead oil increase lineally with increasing operating pressure. The diffusion coefficient of natural gas in foamy oil at 20 MPa is 2.93 times larger than that at 8.65 MPa.

(5) The growth rate of diffusion coefficient with increasing pressure in foamy oil is less than that in saturated oil and dead oil.

Acknowledgments The authors acknowledge the financial support from the Major Subject of National Science and Technology (2011ZX05032-001) and the Fundamental Research Funds for the Central Universities (NO.11CX06022A). The authors wish to express their thanks to Research Institute of Petroleum Exploration and Development, PetroChina for providing oil samples.

References

Bondino I, McDougall SR, Hamon G. A pore-scale modeling approach to the interpretation of heavy oil pressure depletion experiments. J Petrol Sci Eng. 2009;65:14–22.

Dong ZL, Li Y, Liu M, et al. A study of the mechanism of enhancing oil recovery using supercritical carbon dioxide microemulsions. Petrol Sci. 2013;10(1):91–6.

Etminan SR, Maini BB, Chen ZX, et al. Constant-pressure technique for gas diffusivity and solubility measurements in heavy oil and bitumen. Energy Fuels. 2010;24(1):533–49.

Garcia FM. A successful gas-injection project in a heavy oil reservoir. In: 58th annual technical conference and exhibition, San Francisco, California; 1983 (SPE 11988).

Grogan AT, Pinczewski VW, Ruskauff GJ, et al. Diffusion of CO_2 at reservoir conditions: models and measurements. SPE Reserv Eng. 1988;2(1):93–102 (SPE 14897).

Guan WL, Wu SH, Zhao J. Utilizing natural gas huff and puff to enhance production in heavy oil reservoir. 2008 SPE international thermal operation and heavy oil symposium, Calgary, Alberta; 2008 (SPE 117335).

Guo P, Deng L, Yang XF, et al. Dry gas huff and puff effect evaluation of gas wells in low-permeable condensate gas reservoirs with rich condensate oil. Petrol Explor Dev. 2010a;37(3):354–7 (in Chinese).

Guo JX, Wang HY, Chen CG, et al. Synthesis and evaluation of an oil-soluble viscosity reducer for heavy oil. Petrol Sci. 2010b;7(4):536–40.

Hill ES, Lacey WN. Rate of solution of propane in quiescent liquid hydrocarbons. Ind Eng Chem. 1934;26(2):1324–7.

Jamialahmadi M, Emadi M, Müller SH. Diffusion coefficients of methane in liquid hydrocarbons at high pressure and temperature. J Petrol Sci Eng. 2006;53:47–60.

Li SY, Li ZM, Lu T, et al. Experimental study on foamy oil flow in porous media with Orinoco Belt Heavy Oil. Energy Fuels. 2012;26:6332–42.

Liu HB, Xiao LZ, Guo BX, et al. Heavy oil component characterization with multidimensional unilateral NMR. Petrol Sci. 2013a;10(3):402–7.

Liu PC, Wu YB, Li XL. Experimental study of the stability of the foamy oil in developing heavy oil reservoirs. Fuel. 2013b;111:12–9.

Liu SQ, Sun XM, Li SL. Foamy oil recovery mechanism in cold production processes of super heavy oil in Venezuela MPE-3 block. Spec Oil Gas Reserv. 2011;18(4):102–4 (in Chinese).

Mu LX, Han GQ, Xu BJ. Geology and reserve of the Orinoco heavy oil belt, Venezuela. Petrol Explor Dev. 2009;36(6):784–9 (in Chinese).

Mu LX. Development actualities and characteristics of the Orinoco heavy oil belt, Venezuela. Petrol Explor Dev. 2010;37(3):338–43 (in Chinese).

Mohammad TS, Behnam SS, Fariborz R. Genetic algorithm application for matching ordinary black oil PVT data. Petrol Sci. 2012;9(2):199–211.

Nguyen TA, Farouq Ali S. Effect of nitrogen on the solubility and diffusivity of carbon dioxide into oil and oil recovery by the immiscible WAG process. J Can Petrol Technol. 1998;37(2):24–31 (SPE 95–64).

Peng J, Guo QT, Kovscek AR. Oil chemistry and its impact on heavy oil solution gas drive. J Petrol Sci Eng. 2009;66:47–59.

Riazi MR. A new method for experimental measurement of diffusion coefficients in reservoir fluids. J Petrol Sci Eng. 1996;14:235–50.

Renner TA. Measurement and correlation of diffusion coefficients for CO_2 and rich-gas applications. SPE Reserv Eng. 1988;5(1):517–23 (SPE 15391).

Schmidt T. Mass transfer by diffusion. In: AOSTRA technical handbook on oil sands, bitumens and heavy oils. Alberta oil sands technology and research authority. Edmonton, Alberta, Canada; 1989.

Sun XF, Zhang YY, Li XM, et al. A case study on foamy oil characteristics of the Orinoco Belt, Venezuela. Adv Petrol Explor Dev. 2013;5(1):37–41.

Tharanivasan AK, Yang C, Gu Y. Comparison of three different interface mass transfer models used in the experimental measurement of solvent diffusivity in heavy oil. J Petrol Sci Eng. 2004;44:269–82.

Torabi F, Qazvini FA, Kavousi A, et al. Comparative evaluation of immiscible, near miscible and miscible CO_2 huff-n-puff to enhance oil recovery from a single matrix–fracture system (experimental and simulation studies). Fuel. 2012;93(3):443–53.

Upreti SR, Mehrotra AK. Experimental measurement of gas diffusivity in bitumen: results for carbon dioxide. Ind Eng Chem. 2000;39(4):1080–7.

Whitman WG. The two-film theory of absorption. Chem Metall Eng. 1923;29:147–52.

Woessner E, Snowden BS, George RA, et al. Dense gas diffusion coefficients for the methane-propane system. Ind Eng Chem. 1969;5(4):780–7.

Wang BJ, Wu YB, Jiang YW, et al. Physical simulation experiments on PVT properties of foamy oil. Acta Petrol Sin. 2012;33(1): 76–9 (in Chinese).

Wang J, Yuan YZ, Zhang LH, et al. The influence of viscosity on stability of foamy oil in the process of heavy oil solution gas drive. J Petrol Sci Eng. 2009;66:69–74.

Xu ZY, Yue DL, Wu SH, et al. An analysis of the types and distribution characteristics of natural gas reservoirs in China. Petrol Sci. 2009;6(1):38–42.

Yang CD, Gu YA. New experimental method for measuring gas diffusivity in heavy oil by the dynamic pendant drop volume analysis (DPDVA). Ind Eng Chem. 2005;44(12):4474–83.

Yu CC, Shen SK. Progress in studies of natural gas conversion in China. Petrol Sci. 2008;5(1):67–72.

Zhang YP, Hyndman CL, Maini BB. Measurement of gas diffusivity in heavy oils. J Petrol Sci Eng. 2000;25:37–47.

Zhu GY, Zhang SC, Liu QC, et al. Distribution and treatment of harmful gas from heavy oil production in the Liaohe Oilfield, Northeast China. Petrol Sci. 2010;7(3):422–7.

7

Three-dimensional physical simulation and optimization of water injection of a multi-well fractured-vuggy unit

Ji-Rui Hou[1,2,3] · Ze-Yu Zheng[1,2,3] · Zhao-Jie Song[1,2,3] · Min Luo[1,2,3] ·
Hai-Bo Li[1,2,3] · Li Zhang[1,2,3] · Deng-Yu Yuan[1,2,3]

Abstract With complex fractured-vuggy heterogeneous structures, water has to be injected to facilitate oil production. However, the effect of different water injection modes on oil recovery varies. The limitation of existing numerical simulation methods in representing fractured-vuggy carbonate reservoirs makes numerical simulation difficult to characterize the fluid flow in these reservoirs. In this paper, based on a geological example unit in the Tahe Oilfield, a three-dimensional physical model was designed and constructed to simulate fluid flow in a fractured-vuggy reservoir according to similarity criteria. The model was validated by simulating a bottom water drive reservoir, and then subsequent water injection modes were optimized. These were continuous (constant rate), intermittent, and pulsed injection of water. Experimental results reveal that due to the unbalanced formation pressure caused by pulsed water injection, the swept volume was expanded and consequently the highest oil recovery increment was achieved. Similar to continuous water injection, intermittent injection was influenced by factors including the connectivity of the fractured-vuggy reservoir, well depth, and the injection–production relationship, which led to a relative low oil recovery. This study may provide a
constructive guide to field production and for the development of the commercial numerical models specialized for fractured-vuggy carbonate reservoirs.

Keywords Multi-well fractured-vuggy unit ·
Three-dimensional physical model · Similarity criteria ·
Bottom water drive · Optimization of water injection mode

1 Introduction

Carbonate reservoirs are the important sources of hydrocarbons, accounting for half of world's oil and gas reserves (Li and Chen 2013; Yousef et al. 2014). The Tahe Oilfield is the largest oilfield found in the Paleozoic marine carbonates in China (Li et al. 2014a). The successful exploitation of this oilfield can significantly reduce the crisis of the hydrocarbon shortage in China.

The Tahe Oilfield is located in the middle of the South Tarim Uplift in the Tarim Basin, China, with an area of 2400 km[2] (Li et al. 2014b). The Ordovician reservoir in the Tahe Oilfield is an example of fractured-vuggy carbonate reservoirs (Li 2013). The reservoir space is complicated because of the co-existence of vugs, fractures, and large caves in these reservoirs (Xu et al. 2012). Large caves serve as the main storage space, while fractures act as the primary flow paths with only a little oil accumulated in them (Lv et al. 2011; Zhang and Wang 2004). On the other hand, the carbonate matrix exhibits ultra-low permeability and almost no storage ability. Therefore, a fractured-vuggy unit is considered as the basic production unit during the exploitation of these carbonate reservoirs (Yi et al. 2011; Rong et al. 2013). This type of reservoir is characterized by serious heterogeneity, random distribution, and complex co-location of fractures and vugs, and various filling types

---

✉ Ze-Yu Zheng
zhengzeyu890808@outlook.com

1 Research Institute of Enhanced Oil Recovery, China University of Petroleum, Beijing 102249, China

2 Key Laboratory of Marine Facies, Sinopec, Beijing 102249, China

3 Key Laboratory of Petroleum Engineering, Ministry of Education, Beijing 102249, China

Edited by Yan-Hua Sun

and filling percentages of storage space. Therefore, fractured-vuggy carbonate reservoirs are very complex with multiple oil–water systems and the existence of bottom water, which makes hydrocarbon production difficult (Popov et al. 2009; Lucia et al. 2003).

And so far, in most of the numerical simulations of fractured-vuggy carbonate reservoirs, this type of reservoir is taken as equivalent to sandstone reservoirs, e.g., CMG, Eclipse (Presho et al. 2011; Wu et al. 2011; Guo et al. 2012). A commercial numerical simulator specialized for fractured-vuggy carbonate reservoirs has not yet been developed. Consequently, fundamental experimental research is required to provide a theoretical foundation for numerical simulations.

Previous research has been conducted on fractured-vuggy reservoirs through physical simulations (Cruz et al. 2001; Zheng et al. 2010; Zhang et al. 2011; Jia et al. 2013; Xu et al. 2013). Cruz et al. (2001) performed water-displacing oil experiments using a two-dimensional vuggy fractured porous cell, took photographs of the water front, measured the flow area corresponding to oil and water, and finally developed a theoretical model based on the experimental results. Zheng et al. (2010) designed and constructed a cylindrical core model, according to actual conditions of a fractured-vuggy carbonate reservoir. Water-displacing oil experiments were conducted and it is found that oil recovery, water cut, and water breakthrough were significantly impacted by fractured-vuggy structures. Zhang et al. (2011)

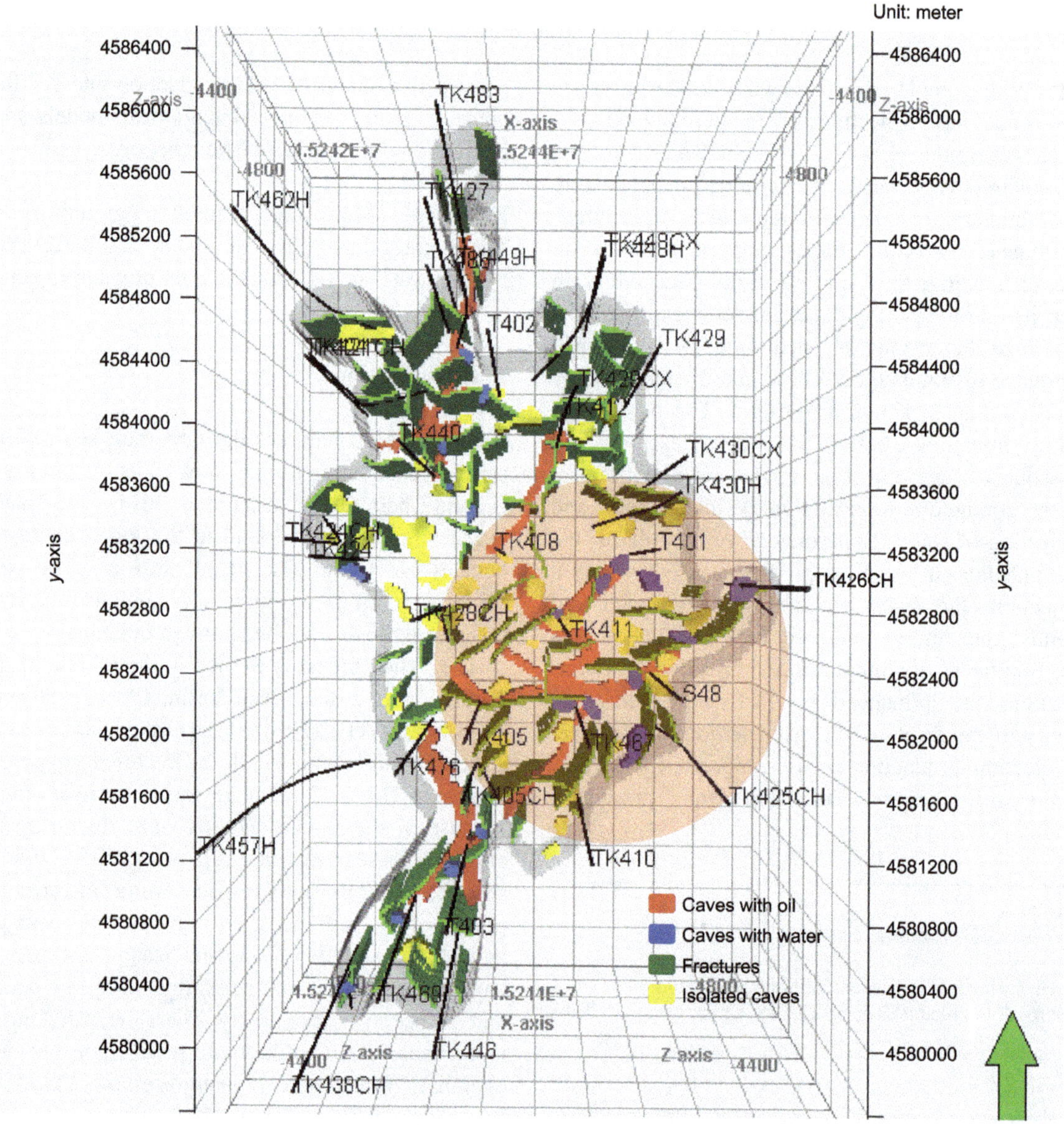

Fig. 1 Geological conditions of the example reservoir

built a single-phase fluid flow pattern in fractured-vuggy media with different vug densities, and concluded that with an increase in vug density, the pressure gradient at the same velocity declined and the critical velocity increased. However, due to the limitations of dimensionality, similarity design, and other factors, three-dimensional fluid flow in porous media is difficult to simulate reliably in these models. In this paper, a three-dimensional physical model is designed and built based on a geological example unit in the Tahe Oilfield and similarity criteria. Bottom water drive is conducted to validate the 3D physical model according to well history and then the optimization of water injection

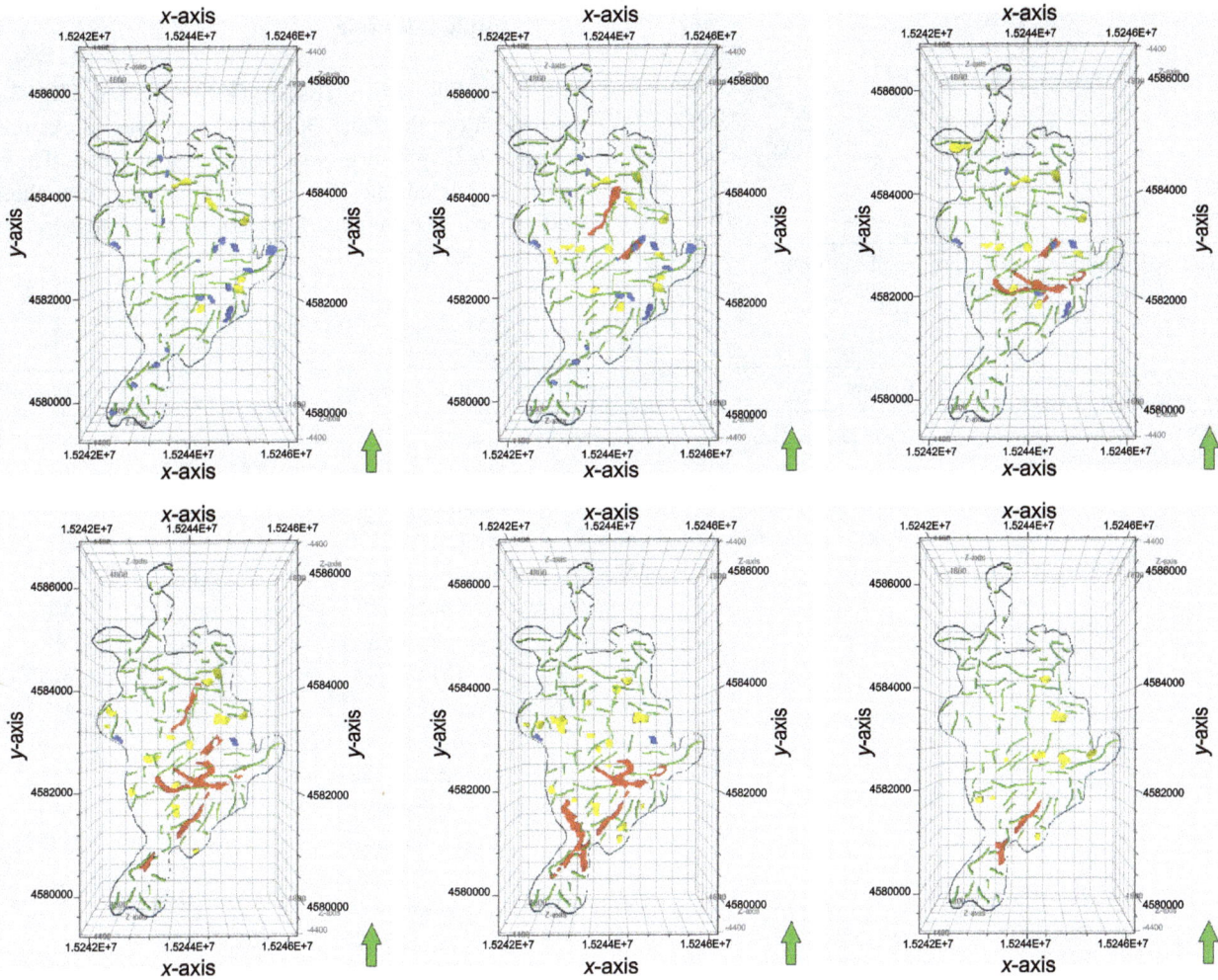

Fig. 2 Longitudinal sections of the reservoir

Table 1 Production histories of 5 wells with bottom water drive in the S48 unit

Well	Production date	Production duration, y	Crude output, 10^4 m^3	Liquid output, 10^4 m^3	Annual liquid output, 10^4 m^3/y
S48	1997	12	80.04	94.02	7.83
T401	1998	11	50.17	63.24	5.75
TK411	1999	10	12.87	18.61	1.86
Tk426CH	2000	9	2.650	10.73	1.19
TK467	2002	7	9.714	14.21	2.03

Table 2 Annual data of the S48 unit with bottom water drive

Year	Annual liquid output, 10^4 m^3	Daily liquid output, m^3/d
1997*	2.889	481.4
1998	17.36	475.7
1999	31.13	852.9
2000	41.50	1137
2001	29.33	803.5
2002	23.40	373.2
2003	12.20	334.2
2004	9.340	255.9
2005	9.328	255.6
2006	10.04	275.0
2007	11.91	326.4
2008	14.78	404.9

* The wells only produced for two months in 1997

modes is performed by means of continuous, intermittent, and pulsed injection of water.

The experimental results may provide an insight into the later production with water injection and numerical simulations of fractured-vuggy carbonate reservoirs.

2 Geological example and well history

2.1 Example reservoir

Five wells, S48, T401, TK411, TK426CH, and TK467, in the S48 unit in the Tahe Oilfield were chosen as examples to design a 3D physical model, as shown in Fig. 1. The red part in Fig. 1 indicates the area of influence of the chosen well groups. The longitudinal sections of the reservoir are presented in Fig. 2.

Table 3 Physical parameters and dimensions related to oil and water two-phase flow in a multi-well fractured-vuggy unit

Category	Number	Physical parameter	Symbol	Dimension
Basic physical parameter	1	Quality	m	M
	2	Time	t	T
	3	Length	l	L
Dimensional parameter	4	Oil flow rate	u_o	LT^{-1}
	5	Water flow rate	u_w	LT^{-1}
	6	Oil density	ρ_o	ML^{-3}
	7	Water density	ρ_w	ML^{-3}
	8	Oil viscosity	μ_o	$ML^{-1}T^{-1}$
	9	Water viscosity	μ_w	$ML^{-1}T^{-1}$
	10	Volume flow rate of oil	Q_o	L^3T^{-1}
	11	Volume flow rate of water	Q_w	L^3T^{-1}
	12	Injection pressure	P	$ML^{-1}T^{-2}$
	13	Acceleration of gravity	g	LT^{-2}
	14	Fracture permeability	K_f	L^2
	15	Cave height	H	L
	16	Fracture aperture	B	L
	17	Cave diameter	d	L
	18	Fracture density	n_f	L^{-1}
	19	Cave density	n_v	L^{-2}
	20	Fluid compressibility	C_L	LT^2M^{-1}
	21	Total compressibility	C_t	LT^2M^{-1}
Dimensionless parameter	22	Fracture porosity	Φ_f	
	23	Cavernous porosity	Φ_v	
	24	Oil relative permeability	K_{ro}	
	25	Water relative permeability	K_{rw}	
	26	Oil saturation	S_o	
	27	Water saturation	S_w	
	28	Pseudo coordination number	ξ	
	29	Filling percentage	η	

2.2 Well history

A 3D physical model was constructed based on the example of the geology of the S48 unit. Bottom water drive was employed in the S48 unit in the early stage of production. The physical model was required to be similar to actual reservoir conditions so that experimental results are as reliable as possible. Hence, it is imperative to understand and analyze the practical production situations during bottom water drive. Tables 1 and 2 depict well history information of the S48 unit. According to well history and other information reported earlier (Zhu et al. 2009; Hu et al. 2013), an oil recovery of 20 %–23 % was obtained at the end of bottom water drive in this unit.

3 Physical model design and fabrication

3.1 Similarity design

According to previous studies of physical simulation (Elkaddifi et al. 2008; Zhang et al. 2008; Bünger and Herwig 2009; Zuta and Fjelde 2011; Wang et al. 2014), some similarity criteria, such as geometric similarity, kinematic similarity, dynamic similarity, and characteristic parameter similarity, were adopted to insure the similarity between the physical model and the geological example.

However, it was impossible to achieve similarity of all parameters. Thus, a dimensional analysis method was introduced to simplify the similarity criterions.

The characteristics of oil and water two-phase flow in a multi-well fractured-vuggy unit should be taken into account and relevant physical parameters were considered as illustrated in Table 3.

According to Buckingham's pi-theorem (Haftkhani and Arabi 2013), ρ, u, and l were chosen as basic parameters, and the physical parameters with the same dimensions were classified as a group. The similarity criterion group is obtained as shown in Table 4.

For a specific flow model, the flow field in the physical model is generally similar to that in the actual reservoir when some main similarity criteria are achieved (Kang and Tang. 2014; Nam et al. 2014). The ratio of the cave diameter to the diameter of the area of influence was chosen as a criterion for geometric similarity. Since the bottom water drive is required to be simulated, the liquid production rate and production duration should be similar to actual situations, so $\pi 6$ and $\pi 7$ in Table 4 were chosen as kinematic similarity criteria. The Reynolds number should be considered when dynamic similarity is discussed. Meanwhile, the ratio of pressure to gravity that influences the fluid distribution in displacement processes needed to be considered. In addition, the physical model should comply with the similarity theorem on some characteristic parameters, such as pseudo coordination number and

Table 4 Similarity criterion group of the multi-well fractured-vuggy unit

Geometric similarity	Kinematic similarity	Dynamic similarity	Characteristic parameter similarity
$\pi 1 = d/l$	$\pi 6 = Q/(ul^2)$	$\pi 8 = \mu/(\rho ul)$	$\pi 11 = \xi$
$\pi 2 = B/l$	$\pi 7 = tu/l$	$\pi 9 = P/(\rho u^2)$	$\pi 12 = \eta$
$\pi 3 = K_f/l^2$	–	$\pi 10 = gl/u^2$	$\pi 13 = n_f l$
$\pi 4 = \Phi_f$	–	–	$\pi 14 = n_v l^2$
$\pi 5 = \Phi_v$	–	–	–

Table 5 Comparison of parameters of the example reservoir and the physical model

Parameter	Example reservoir	Physical model	Similarity coefficient
Cave diameter	5–25 m	3–12 cm	166.7–200.0
Differential pressure	2–14 MPa	10–60 kPa	200.0–233.3
Oil viscosity	19.7–28.5 mPa s	65 mPa s	0.3038–0.4377
Oil density	0.92 g/cm³	0.8 g/cm³	1.150
Acceleration of gravity	9.8 m/s²	9.8 m/s²	1.000
Fracture density	3–50/m	50/m	0.05880–1.000
Fracture aperture	0.5–5 mm	2–4.5 mm	0.2500–1.110
Fluid flow velocity	30–135 m/d	0.022–0.152 m/s	0.001585–0.001903
Liquid production rate	255.6–1137 m³/d	4–10 mL/min	44.37–78.97
Production duration	1 year	5 min	1.051×10^5
Filling percentage	0 %–100 %	0 %–100 %	1
Coordination number	1–5	1–5	1

filling percentage. Since it is difficult for a physical model to satisfy multiple similarity criteria simultaneously, the Reynolds number should always be given priority.

The index of the similarity criteria was calculated according to similarity coefficients. A similarity coefficient is the ratio of field parameters to model parameters. If a certain index of the similarity criterion is 1, this model parameter is fully compliant with the field parameter. Based on the discussion mentioned above, the similarity

coefficients and similarity indices of the physical model are obtained as shown in Tables 5 and 6.

3.2 Model fabrication

Longitudinal sections of the example reservoir were divided into six layers, as shown in Fig. 3. Caves and fractures were carved in each core based on the example reservoir. After that, these six cores were bonded together in a given

Table 6 Similarity indices of the physical model

Similarity condition	Similarity criterion	Physical meaning	Source	Similarity index
Geometric similarity	d/l	The ratio of the cave diameter to the diameter of the area of influence	$\pi 1$	1
Kinematic similarity	$\rho u l/\mu$	Reynolds number (the ratio of inertia force to viscous force)	$1/\pi 8$	1
	$P/(\rho g l)$	The ratio of the injection pressure to gravity	$\pi 9/\pi 10$	1.01–1.04
Dynamic similarity	$Q/(u l^2)$	The ratio of the liquid production rate to the injection rate	$\pi 6$	1.01–1.04
	$t u/l$	The ratio of the production duration to the injection duration	$\pi 7$	1
Characteristic parameter similarity	ξ	Pseudo coordination number	$\pi 11$	1
	η	Filling percentage	$\pi 12$	1

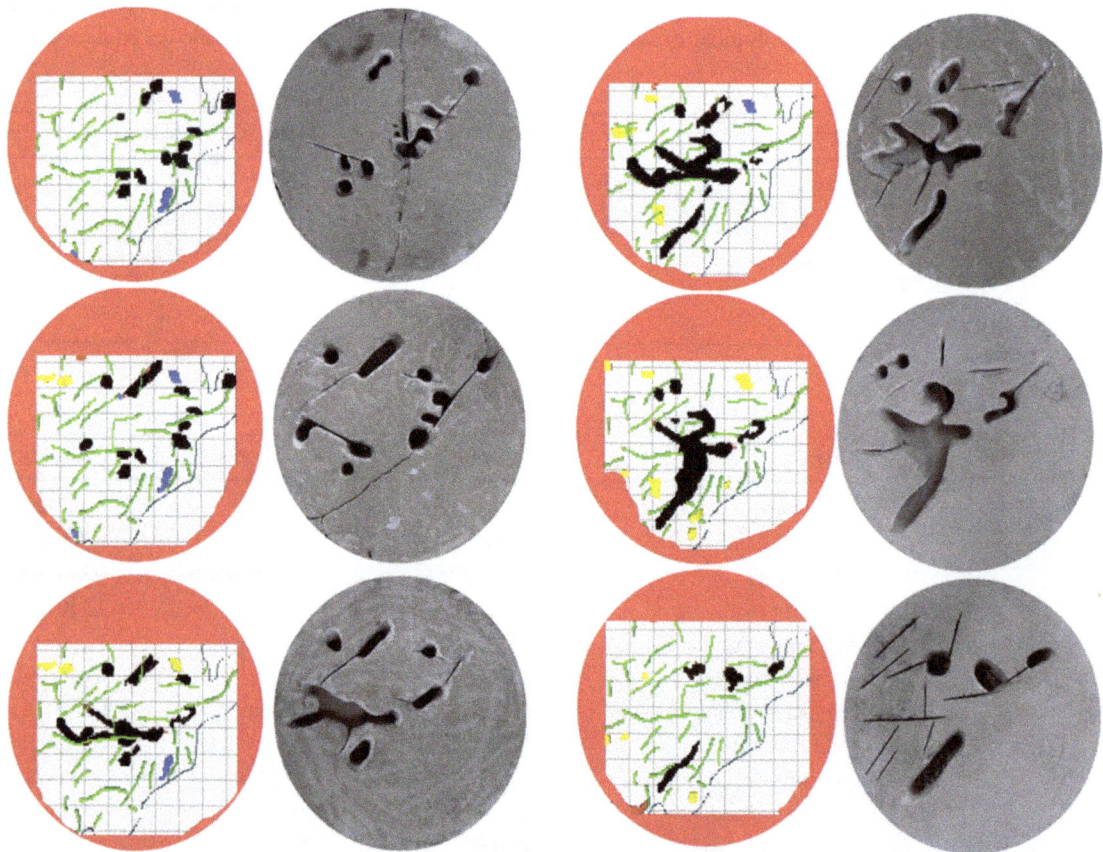

Fig. 3 Comparison of the fractured-vuggy structures between the physical model and the real core for each layer

order. The flow paths of the bottom water and well locations were designed, and a three-dimensional model was fabricated. Finally, the model was put into a cylindrical mold and it was encapsulated and fixated with epoxy resin, as presented in Fig. 4. More details of the physical model were described elsewhere (Hou et al. 2014).

3.3 Fracture and cave structures and well locations

A fractured-vuggy unit with a complex structure was formed in a 3D space, as shown in Fig. 5. In order to make the model similar to the actual formation, caves and fractures were filled with gravel in a certain order of completely filled, half-filled, and unfilled from bottom to top.

Five wells were drilled in the model according to the designed depths and locations. These wells were drilled in different reservoir bodies, i.e., cave and fracture. Wells were divided into two types: one is cave-well which is marked by "D" and the other is fracture-well marked by "L." Figure 6 illustrates the well locations and Table 7 shows the well parameters.

4 Experimental

4.1 Experimental apparatus and materials

The experimental apparatus (Fig. 7) included a 3D physical model, three intermediate containers, two water injection pumps with a pressure range of 0–30 MPa and a flow rate range of 0.01–10.00 mL/min, several pressure transducers, some six-port valves, etc.

The simulated oil used was a mixture of kerosene and dehydrated crude, with a viscosity of 65 mPa s at 25 °C.

The simulated formation water used had a salinity of 2×10^5 mg/L.

4.2 Experimental procedures

The experimental procedures are as follows:

(1) The physical model was evacuated and then saturated with the simulated formation water at 10 mL/min from pipelines 6 in Fig. 7. The total volume of water injected into the model was measured to calculate the pore volume (PV) of fractures and caves in the physical model.

(2) The model was flooded with the simulated oil at 10 mL/min from pipelines 13 from the top of the model and the water was drained from pipelines 6. This was continued until no more water was drained.

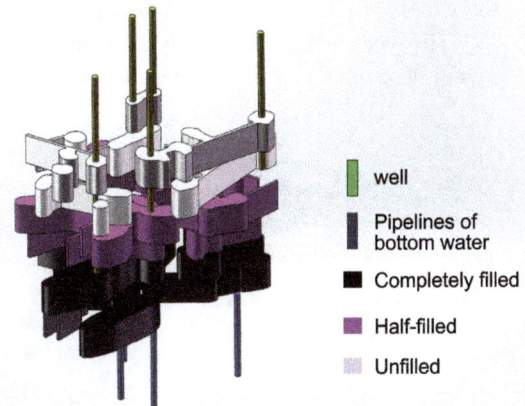

Fig. 5 Fractured-vuggy structure inside the three-dimensional physical model

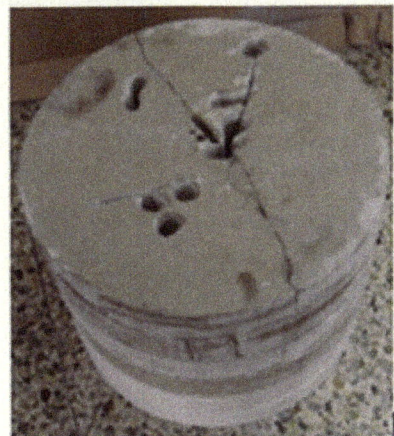

Fig. 4 The real core after combination

The total volume of water displaced from the model was measured, and the initial oil saturation and the water saturation of the model were then calculated.

(3) As shown in Fig. 7a, the simulated formation water was continuously introduced into the bottom of the model; the water injection rate was varied to simulate the stage of depletion of a bottom water drive reservoir. The initial water injection rate was set at 10 mL/min and then decreased to 6 mL/min after 15 min, and to 4 mL/min after another 5 min. The injection rate was kept at 4 mL/min until a 98 % water cut was reached in one of these five production wells. The oil and water production rate was monitored to calculate oil recovery and periodic water cut. Five wells were put into production in the order D1, D2, L2, L3, and L1, as depicted in Table 8.

(4) During the bottom water drive, the water cut of Well L1 reached 98 % first. This well was then switched to be an injection well. After the bottom water drive, the experimental apparatus was set up as shown in Fig. 7b. Water was subsequently introduced into the model through Well L1; water injection rates at

different waterflood projects are listed in Table 9. Each production well was shut in once its water cut reached 98 %. The experiment was stopped when the four production wells were all shut in. The oil and water production rates were monitored to calculate the oil recovery and periodic water cut.

5 Experimental results and discussion

5.1 Performance of the bottom water drive model

The production performance of the bottom water drive model constructed in this study is shown in Fig. 8. The oil recoveries of five production wells increased slowly at this stage, i.e., the oil production rate declined, and the water cut of fluids produced from Well L1 increased as the bottom water energy was depleted.

Water breakthrough occurred immediately in Well L1 and its water cut rose rapidly after this well was put into production. This is because Well L1 was the deepest well (Table 7). Experimental results are consistent with actual production situation. The experiment was stopped when the water cut of fluids produced from Well L1 reached 98 %. The total oil recovery was about 22.1 %, which is in agreement with the actual oil recovery in the S48 unit. Thus, experimental results verified that the 3D physical model had similarity to the example reservoir (S48 unit).

5.2 Optimization of water injection schemes

Due to the viscosity difference between water and oil in the water/oil interface, the flow rate of water is higher than that of oil under the same differential pressure. When the differential pressure is high enough, water can probably flow vertically upward. Therefore, a conical water/oil interface was formed, which is known as bottom water coning (Xiao et al. 2010).

By injecting water into the reservoir through a drowned well to supply energy, bottom water coning could be mitigated and the intrusive water phase could be suppressed toward the opposite direction (Lu et al. 2009). However, different water injection modes exert different effects on

Fig. 6 Well locations in the three-dimensional physical model

Table 7 Well parameters of the models

Number	Well	Well depth, cm	Well type	Filling percentage
D1	S48	14.8	Cave-well	Half-filled
D2	T401	12.2	Cave-well	Half-filled
L1	TK467	18.0	Fracture-well	Completely filled
L2	TK411	17.8	Fracture-well	Completely filled
L3	TK426CH	9.2	Fracture-well	Unfilled

(a)

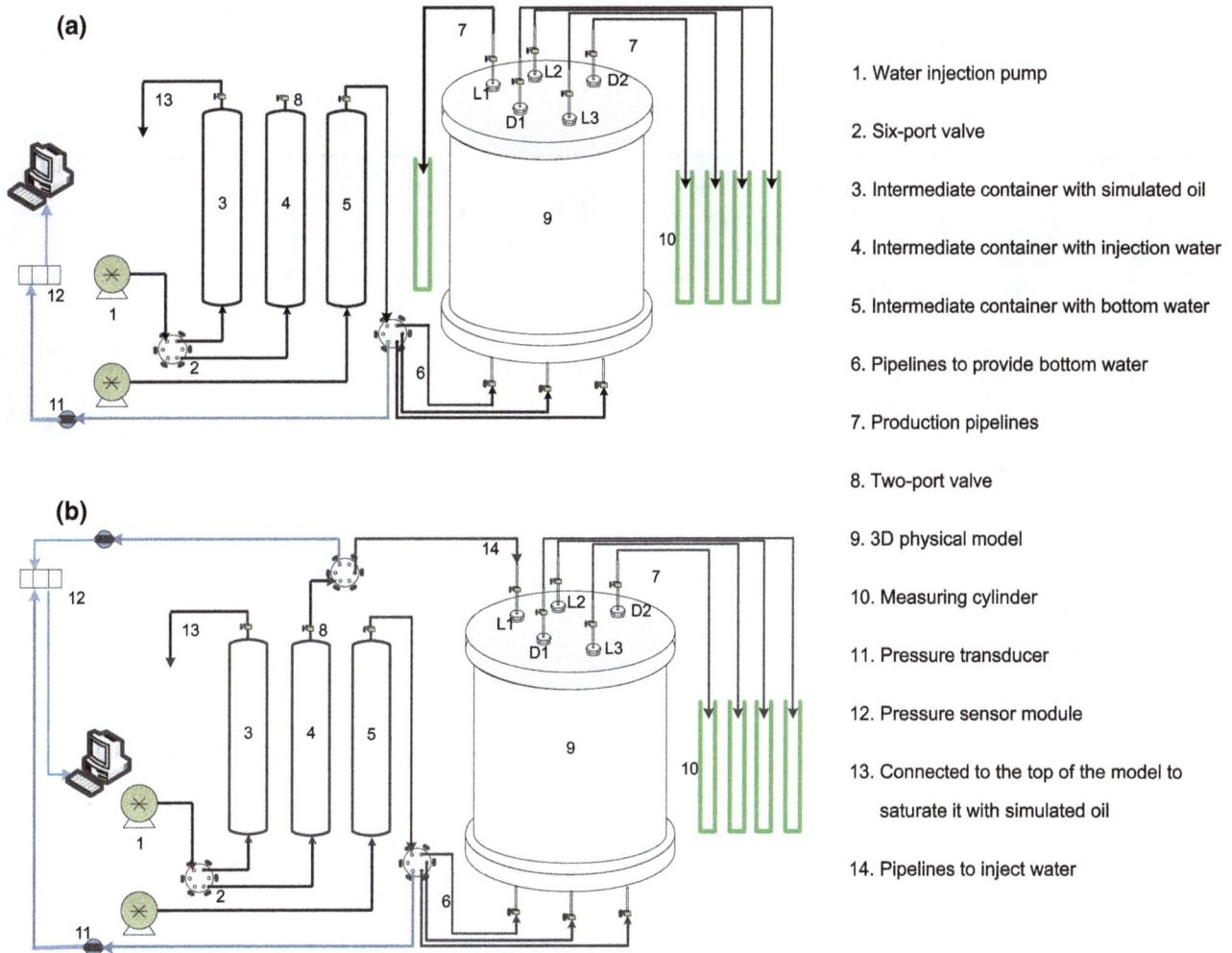

1. Water injection pump

2. Six-port valve

3. Intermediate container with simulated oil

4. Intermediate container with injection water

5. Intermediate container with bottom water

6. Pipelines to provide bottom water

7. Production pipelines

8. Two-port valve

9. 3D physical model

10. Measuring cylinder

11. Pressure transducer

12. Pressure sensor module

13. Connected to the top of the model to saturate it with simulated oil

14. Pipelines to inject water

(b)

Fig. 7 Schematic diagram of experimental setups for the bottom water drive (**a**) and the water injection mode optimization (**b**)

Table 8 Well startup order in the bottom water drive model

Timing, min	Well startup order
0	D1
5	D2
10	L2
15	L3
25	L1

mitigating water coning. Therefore, the objective of the following work is to optimize water injection modes, including continuous injection (constant rate), intermittent injection, and pulsed injection of water.

5.2.1 Overall effect of subsequent water injection

Figures 9 and 10 show the dynamic curves of oil recovery and differential pressure for three water injection schemes, respectively.

No matter which water injection scheme was performed, the oil recovery increased substantially at the beginning and then continued to rise smoothly until the oil recovery became relatively stable as no more oil was being produced. The highest oil recovery of 47.0 % was obtained after pulsed injection of water. Similar oil recoveries, 44.7 % and 44.3 %, respectively, were obtained after continuous and intermittent injection of water.

The differential pressure fluctuated significantly during intermittent and pulsed injection of water (Fig. 10). The amplitude of pressure fluctuations was relatively large when water was intermittently injected into Well L1 every 25 min, while the frequency of pressure fluctuations was relatively high during pulsed injection of water. This indicates that unsteady-state water injection, including intermittent injection and pulsed injection, might promote the instability of fluid flow rate, flow direction, and pressure distribution in the model. In this way, the water displacement efficiency was improved. However, the

Table 9 Experimental schemes for optimizing water injection

Scheme	Injection rate of bottom water, mL/min	Subsequent water injection		Diagrams of injection rate
		Injection scheme	Injection rate	
A	4	Continuous injection	4 mL/min	
B	4	Intermittent injection	8 mL/min for 25 min and 0 mL/min for another 25 min in one cycle	
C	4	Pulsed injection	2 mL/min for 15 min, 4 mL/min for 15 min, 6 mL/min for 15 min, and 4 mL/min for 15 min in one cycle	

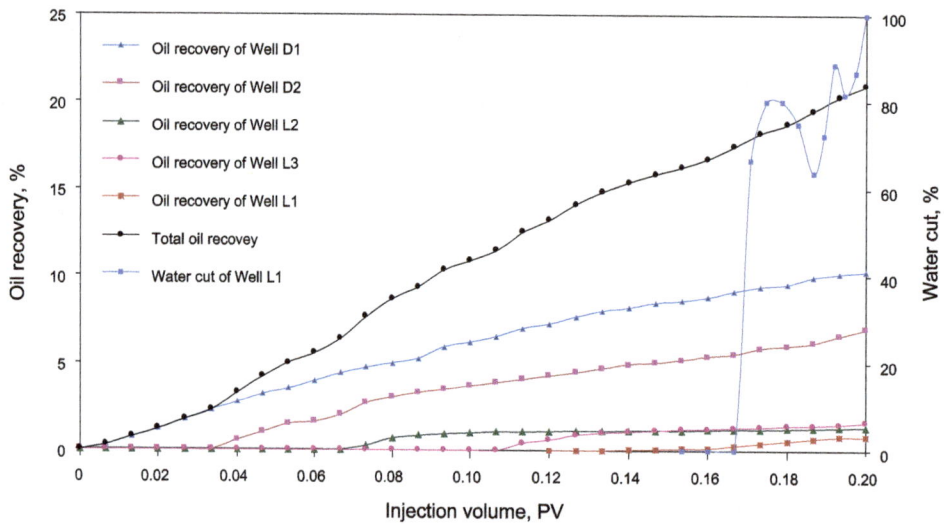

Fig. 8 Production performance of the bottom water drive at the depletion stage

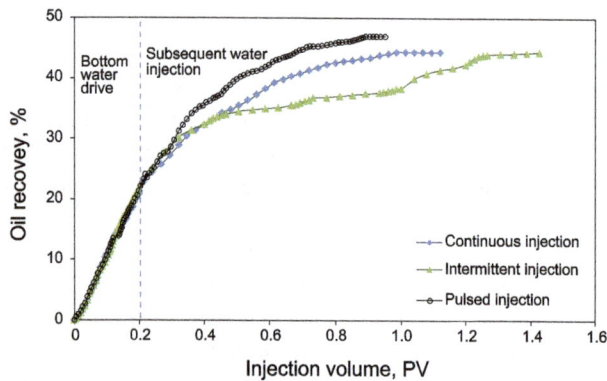

Fig. 9 Dynamic oil recovery curve

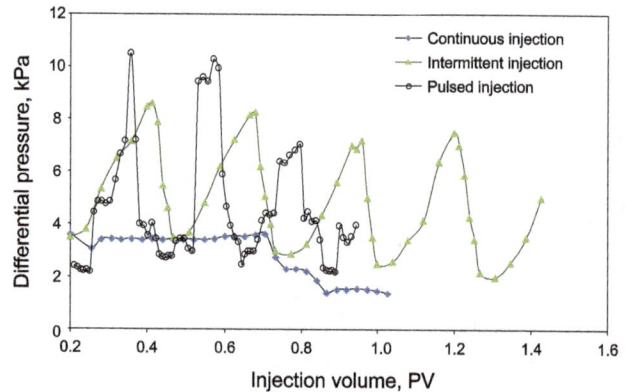

Fig. 10 Injection pressure during water injection

displacement efficiency of intermittent injection was not as good as expected. This is due to the serious heterogeneity and the complex structure of fractures and caves in the fractured-vuggy reservoir model.

5.2.2 Production performance at the subsequent water injection stage

Figures 11, 12, 13, and 14 present the oil recovery and water cut of four production wells at the subsequent water injection stage, respectively.

Well L2 was the nearest to the injection well L1, and its well depth was similar to Well L1. Therefore, water breakthrough occurred immediately in Well L2 when the water was injected into Well L1. Correspondingly, the water cut of Well L2 increased to 100 % rapidly. The incremental oil recovery was only about 0.1 %–0.2 %.

Because Well D1 was drilled deeply into a cave that was near to injection well L1, water breakthrough occurred easily during subsequent water injection. The final oil recovery of Well D1 after intermittent injection was the highest among these three water injection modes. During intermittent injection, the injection pressure of Well D1 was high enough to suppress bottom water coning. Therefore, the oil/water co-production duration of Well D1 was long and its oil recovery was relative high. However, the incremental oil recovery was still at a low level, less

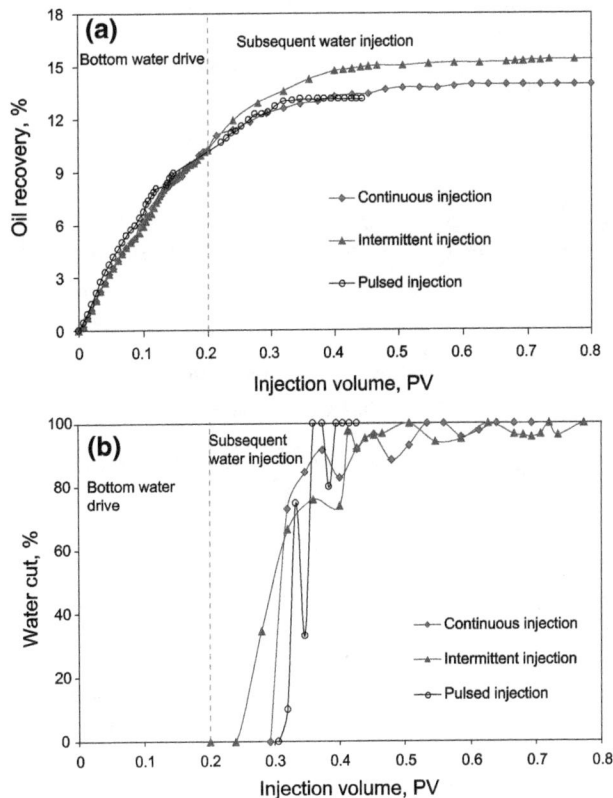

Fig. 12 Dynamic curves of oil recovery and water cut of Well D1

than 5.5 %. All these phenomena can be attributed to the well location, well depth, and the early water breakthrough.

Well D2 was comparatively far from Well L1 and the gravity effect in this well was relatively strong due to a bottom cave. The injection pressure was very low and only supplied bottom water energy during continuous water injection. So, the oil recovery of this well was not high after continuous water injection. On the other hand, the advantage of intermittent and pulsed injection of water was apparent. By changing the injection rate, pressure fluctuations were generated. The formation pressure distribution was changed and the oil–water flow was promoted between fractures and caves. In this way, the sweep efficiency and the displacement efficiency were improved.

Well L3 was also comparatively far away from injection well L1. Unlike Well D2, Well L3 was a fracture-well, and the well depth was the shallowest among those five wells. Therefore, the production performance of Well L3 was completely different from others. During intermittent injection, the injection pressure increased rapidly and water breakthrough occurred easily. However, after the water injection ended, water and oil were hard to redistribute due to the high viscosity resistance and slight gravity effect in fractures. This also led to a relative low oil recovery in Well L3 after immediate injection of water. During pulsed

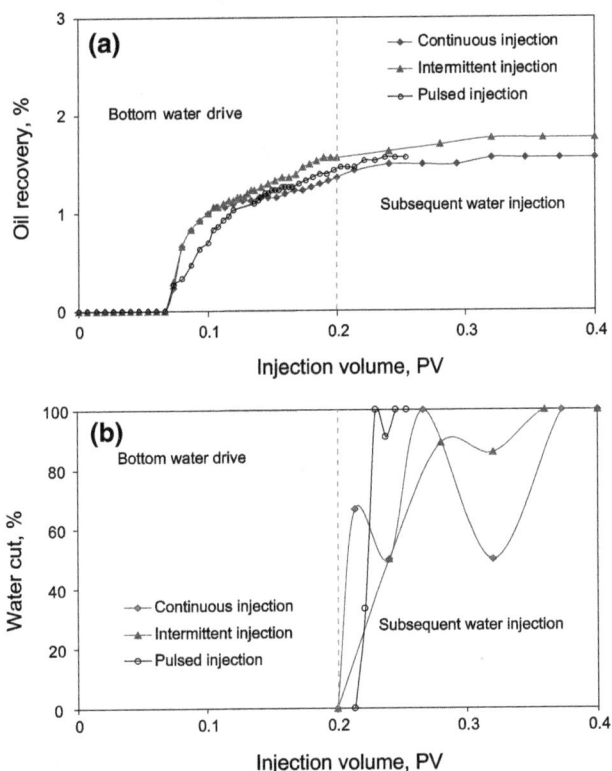

Fig. 11 Dynamic curves of oil recovery and water cut of Well L2

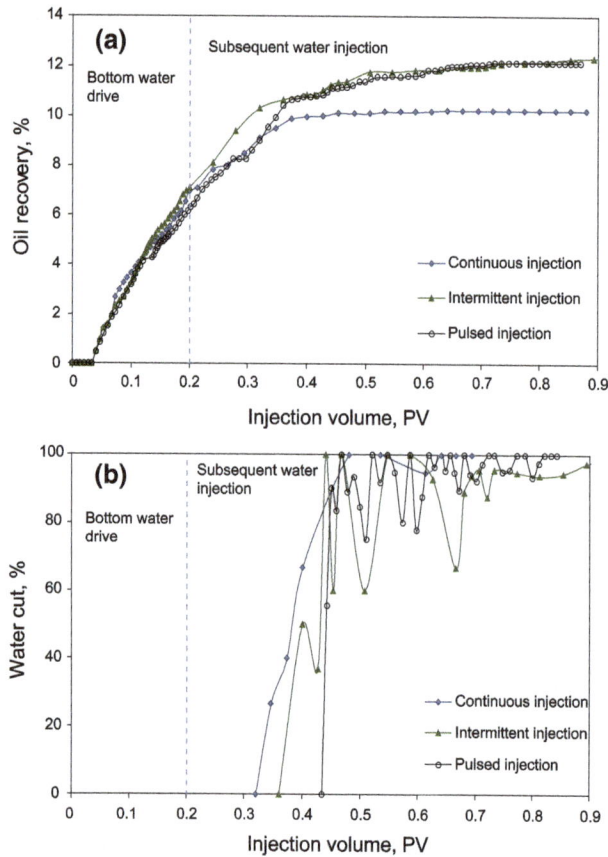

Fig. 13 Dynamic curves of oil recovery and water cut of Well D2

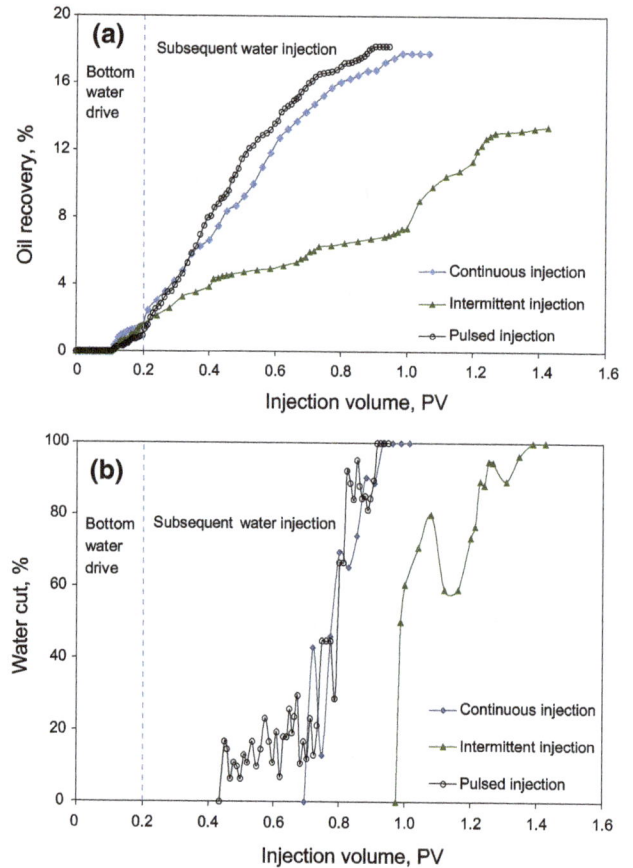

Fig. 14 Dynamic curves of oil recovery and water cut of Well L3

water injection, the high injection rate led to a high pressure increase and caused an early water breakthrough; however, the frequently fluctuated pressure induced a long duration of oil and water co-production, and led to a high final oil recovery.

Due to the complicated fractured-vuggy structure, well location, and well depth, not all the wells exhibited effective response to these three water injection modes. As a whole, the rate of increase of water cut during intermittent and pulsed injection of water was higher than that during continuous water injection. Unsteady-state injection of water indeed enhances water displacement efficiency to some extent compared with steady-state water injection.

6 Conclusions

(1) Based on a multi-well fractured-vuggy geological example unit in the Tahe Oilfield, a large-scale three-dimensional physical model was designed and constructed to simulate the fluid flow according to similarity criteria. Several factors were considered in the design, including storage types, filling types, and filling percentages of storage space.

(2) The depletion production was conducted by changing the bottom water injection rate. Oil recovery of the physical model was similar to that of the field cases, which verified the reliability of the physical model.

(3) During pulsed injection of water, unbalanced formation pressure was caused, and the swept volume was expanded, and consequently the highest oil recovery increment was achieved. Intermittent injection presented the same effect as pulsed injection. However, similar to continuous injection, intermittent injection was influenced by factors including fractured-vuggy connected manner, well depth, and injection–production relationship. Continuous injection and intermittent injection finally get a comparatively low oil recovery.

(4) If a well is deep and close to the injection well, water breakthrough may occur earlier and water cut rises more rapidly. Besides, production performances of the fracture-well and the cave-well are completely different due to the difference in gravity effect and viscosity resistance.

Acknowledgments This work was supported by China National Science and Technology Major Project (2011ZX05009-004, 2011ZX05014-003), National Key Basic Research and Development Program (973 Program), China (2011CB201006), and Science Foundation of China University of Petroleum, Beijing (2462014YJRC053).

References

Bünger F, Herwig H. An extended similarity theory applied to heated flows in complex geometries. Zeitschrift fur angewandte Mathematic und Physik. 2009;60(6):1095–111. doi:10.1007/s00033-009-8076-8.

Cruz HJ, Islas JR, Perez RC, et al. Oil displacement by water in vuggy fractured porous media. In: SPE Latin American and Caribbean petroleum engineering conference, 25–28 March, Buenos Aires, Argentina; 2001. doi:10.2118/69637-MS.

Elkaddifi K, Shirif E, Ayub M, Henni A. Bottomwater reservoirs: simulation approach. J Can Pet Technol. 2008;47(2):38–43. doi:10.2118/08-02-38.

Guo JC, Nie RS, Jia YL. Dual permeability flow behavior for modeling horizontal well production in fractured-vuggy carbonate reservoirs. J Hydrol. 2012;464–5:281–93. doi:10.1016/j.jhydrol.2012.07.021.

Haftkhani AR, Arabi M. Improve regression-based models for prediction of internal-bond strength of particleboard using Buckingham's pi-theorem. J For Res. 2013;24(4):735–40. doi:10.1007/s11676-013-0412-3.

Hou JR, Li HB, Jiang Y, et al. Macroscopic three-dimensional physical simulation of water flooding in multi-well fracture-cavity unit. Pet Explor Dev. 2014;41(6):784–9. doi:10.1016/S1876-3804(14)60093-8.

Hu XY, Li Y, Wang YQ, et al. Application of the probability method in a 3D geological model for petroleum geological reserves in fractured-cavity carbonate reservoirs: a case of Tahe-IV district, Tahe Oilfield. Pet Geol Recovery Effic. 2013;20(4):46–8 (in Chinese).

Jia YL, Fan XY, Nie RS, et al. Flow modeling of well test analysis for porous-vuggy carbonate reservoirs. Transp Porous Media. 2013;97(2):253–79. doi:10.1007/s11242-012-0121-y.

Kang YY, Tang DB. An approach to product solution generation and evaluation based on the similarity theory and ant colony optimisation. Int J Comput Integr Manuf. 2014;27(12):1090–104. doi:10.1080/0951192X.2013.855945.

Li HB, Hou JR, Li W, et al. Laboratory research on nitrogen foam injection in fracture-vuggy reservoirs for enhanced oil recovery. Pet Geol Recovery Effic. 2014a;21(4):93–6 (in Chinese).

Li JY, Chen XH. A rock-physical modeling method for carbonate reservoirs at seismic scale. Appl Geophys. 2013;10(1):1–13. doi:10.1007/s11770-013-0364-6.

Li M, Lou ZH, Zhu R, et al. Distribution and geochemical characteristics of fluids in Ordovician marine carbonate reservoirs of the Tahe Oilfield. J Earth Sci. 2014b;25(3):486–94. doi:10.1007/s12583-014-0453-3.

Li PL. The development of Ordovician fractured-cavity carbonate reservoirs in the Tahe Oilfield. Beijing: Petroleum Industry Press; 2013. p. 1–6 (in Chinese).

Lu ZY, Yang M, Dou ZL, et al. Study of geological model of pressure coning by water injection in well group TK440 of Ordovician

reservoir in Tahe Oilfield, Tarim Basin, China. J Mineral Petrol. 2009;29(4):95–9 (in Chinese).

Lucia FJ, Kerans C, Jennings JW. Carbonate reservoir characterization. Soc Pet Eng. 2003;55(6):70–2. doi:10.2118/82071-MS.

Lv AM, Yao J, Wang W. Characteristics of oil-water relative permeability and influence mechanism in fractured-vuggy medium. Proc Eng. 2011;18:175–83. doi:10.1016/j.proeng.2011.11.028.

Nam JS, Park YJ, Kim JK, et al. Application of similarity theory to load capacity of gearboxes. J Mech Sci Technol. 2014;28(8):3033–40. doi:10.1007/s12206-014-0710-5.

Popov P, Qin G, Bi LF, et al. Multiphysics and multiscale methods for modeling fluid flow through naturally fractured vuggy carbonate reservoirs. SPE Eval Eng. 2009;12(2):218–31. doi:10.2118/105378-PA.

Presho M, Wo S, Ginting V. Calibrated dual porosity, dual permeability modeling of fractured reservoirs. J Petrol Sci Eng. 2011;77:326–37. doi:10.1016/j.petrol.2011.04.007.

Rong YS, Li XH, Liu XL, et al. Discussion about the pattern of water flooding development in multi-well fractured-vuggy units of carbonate fracture-cavity reservoirs in the Tahe Oilfield. Pet Geol Recovery Effic. 2013;20(2):58–61 (in Chinese).

Wang J, Liu HQ, Ning ZF, et al. Experiments on water flooding in fractured-vuggy cells in fractured-vuggy reservoirs. Pet Explor Dev. 2014;41(1):74–81 (in Chinese).

Wu YS, Di Y, Kang ZJ, et al. A multiple-continuum model for simulating single-phase and multiphase flow in naturally fractured vuggy reservoirs. J Pet Sci Eng. 2011;78:13–22. doi:10.1016/j.petrol.2011.05.004.

Xiao MH, Cao Y, Zhang XB, et al. The research on the ancient Karst of Ordovician reservoir in Block 4 of Tahe Oilfield. Pet Geol Eng. 2010;24(3):31–3 (in Chinese).

Xu X, Tian SS, Xu T. The equivalent numerical simulation of fractured-vuggy carbonate reservoir. Mech Aerosp Eng. 2012;110–116:3327–31. doi:10.4028/www.scientific.net/AMM.110-116.3327.

Xu X, Wei GQ, Yang ZM. The productivity calculation method of a carbonate reservoir. Pet Sci Technol. 2013;31(3):301–9. doi:10.1080/10916466.2010.525586.

Yi B, Cui WB, Lu XB, et al. Analysis of dynamic connectivity of a carbonate reservoir with fractures and caves in the Tahe Field, Tarim Basin. Xinjiang Pet Geol. 2011;32(5):469–72 (in Chinese).

Yousef AN, Behzad T, Abolghasem KR, et al. A combined Parzen-wevelet approach for detection of vuggy zones in fractured carbonate reservoirs using petrophysical logs. J Petrol Sci Eng. 2014;119:1–7. doi:10.1016/j.petrol.2014.04.016.

Zhang D, Li AF, Yao J, et al. A single-phase fluid flow pattern in a kind of fractured- vuggy media. Pet Sci Technol. 2011;29(10):1030–40. doi:10.1080/10916466.2011.553657.

Zhang K, Wang DR. Types of karst-fractured and porous reservoirs in China's carbonates and the nature of the Tahe Oilfield in the Tarim Basin. Acta Geol Sinica. 2004;78(3):866–72. doi:10.1111/j.1755-6724.2004.tb00208.x.

Zhang P, Wen XH, Ge LZ, et al. Existence of flow barriers improve horizontal well production in bottom water reservoirs. In: SPE annual technical conference and exhibition, 21–24 September, Denver, Colorado, USA; 2008. doi:10.2118/115348-MS.

Zheng XM, Sun L, Wang L, et al. Physical simulation of water displacing oil mechanism for vuggy fractured carbonate rock reservoirs. J Southwest Pet Univ (Sci Technol Ed). 2010;32(2):89–92 (in Chinese).

Zhu R, Lou ZH, Jin AM, et al. Fluid distribution and dynamic responses to exploitation in fracture-cave unit S48 in Tahe Oilfield. J Zhejiang Univ (Eng Sci). 2009;43(7):1344–8 (in Chinese).

Zuta J, Fjelde I. Mechanistic modeling of CO$_2$-foam processes in fractured chalk rock: effect of foam strength and gravity forces on oil recovery. In: SPE enhanced oil recovery conference, 19–21 July, Kuala Lumpur, Malaysia; 2011. doi:10.2118/144807-MS.

8

Experimental investigation of shale imbibition capacity and the factors influencing loss of hydraulic fracturing fluids

Hong-Kui Ge[1] · Liu Yang[1] · Ying-Hao Shen[1] · Kai Ren[1] · Fan-Bao Meng[1] · Wen-Ming Ji[1] · Shan Wu[1]

Abstract Spontaneous imbibition of water-based fracturing fluids into the shale matrix is considered to be the main mechanism responsible for the high volume of water loss during the flowback period. Understanding the matrix imbibition capacity and rate helps to determine the fracturing fluid volume, optimize the flowback design, and to analyze the influences on the production of shale gas. Imbibition experiments were conducted on shale samples from the Sichuan Basin, and some tight sandstone samples from the Ordos Basin. Tight volcanic samples from the Songliao Basin were also investigated for comparison. The effects of porosity, clay minerals, surfactants, and KCl solutions on the matrix imbibition capacity and rate were systematically investigated. The results show that the imbibition characteristic of tight rocks can be characterized by the imbibition curve shape, the imbibition capacity, the imbibition rate, and the diffusion rate. The driving forces of water imbibition are the capillary pressure and the clay absorption force. For the tight rocks with low clay contents, the imbibition capacity and rate are positively correlated with the porosity. For tight rocks with high clay content, the type and content of clay minerals are the most important factors affecting the imbibition capacity. The imbibed water volume normalized by the porosity increases with an increasing total clay content. Smectite and illite/smectite tend to greatly enhance the water imbibition capacity. Furthermore, clay-rich tight rocks can imbibe a volume of water greater than their measured pore volume. The average ratio of the imbibed water volume to the pore volume is approximately 1.1 in the Niutitang shale, 1.9 in the Lujiaping shale, 2.8 in the Longmaxi shale, and 4.0 in the Yingcheng volcanic rock, and this ratio can be regarded as a parameter that indicates the influence of clay. In addition, surfactants can change the imbibition capacity due to alteration of the capillary pressure and wettability. A 10 wt% KCl solution can inhibit clay absorption to reduce the imbibition capacity.

Keywords Imbibition · Shale · Fracturing fluid · Capillary pressure · Clay

1 Introduction

Multistage hydraulic fracturing is a critical technology for economic production from shale reservoirs. Large amounts of water-based fracturing fluids are pumped into formations, generating extensive fracture networks and stimulating low-permeability formations. Field operations have demonstrated that large volumes of injected fluids are retained in shale formations, with a flowback efficiency of lower than 30 % (Makhanov et al. 2014). In the U.S. Haynesville shale formation, the flowback rate is even lower than 5 % after fracturing operations (Penny et al. 2006). Besides possibly causing a series of environmental problems, the retention of fracturing fluids in shale formations can greatly enhance the water saturation near fracture surfaces and influence two-phase fluid flow, thus further inhibiting the production of shale gas (Sharma and Agrawal 2013). Furthermore, intense interaction between fluid and shale can dramatically change rock properties and impact on the generation of fracture networks during

✉ Liu Yang
  shidayangliu@126.com

[1] State Key Laboratory of Petroleum Resources and Prospecting, China University of Petroleum, Beijing 102249, China

Edited by Yan-Hua Sun

fracturing (Yuan et al. 2014). Therefore, studying the imbibition capacity and its main controlling factors is essential to understanding reservoir performance and optimizing fracturing operations.

It is generally believed that spontaneous imbibition of fracturing fluids into the shale matrix plays an important role in water loss. Many researchers have focused on the mechanism of fracturing fluid imbibition. Makhanov et al. (2012) found that imbibition rates perpendicular and parallel to the bedding plane are different, and the latter is higher. Hu et al. (2012) considered that the Barnett shale has a poor connectivity, which greatly influences the flow and diffusion of fluid. Roychaudhuri et al. (2013) determined that a surfactant can effectively reduce the imbibition rate of fracturing fluids, and the driving force of imbibition is the capillary pressure. Dehghanpour et al. (2013) mentioned that the amount of imbibition in shale is positively related to mineral composition and physical properties. Fakcharoenphol et al. (2014) investigated the effects of salinity on water imbibition and found that the osmotic pressure can act as the driving force for water intake. Currently, it is well known that the imbibition of fracturing fluids is mainly controlled by the capillary pressure, while the effects of clay absorption have not been studied thoroughly. The imbibition capacity, imbibition rate, and other influencing factors in shale reservoirs have not been investigated systematically.

This paper focuses on the imbibition capacity and the influence of the mineral composition and physical properties of tight rocks. Samples include gas shales from the Sichuan Basin, tight sandstones from the Ordos Basin, and tight volcanic rocks from the Songliao Basin. Experiments can be divided into three groups. In group 1, the imbibition capacity and rate of deionized water uptake are investigated systematically. In group 2, each sample is immersed repeatedly in deionized water several times to address the water sensitivity of different rocks. In group 3, comparative experiments are conducted to explore the effects of different fluids on the imbibition capacity.

2 Experimental

2.1 Rock samples and fluids

Sixty-six shale and tight gas rock samples from the Ordos Basin, Songliao Basin, and the Sichuan Basin were used to conduct comparative imbibition experiments, and reservoir rock properties are presented in Table 1. The mineral composition (in wt%) of the shale and tight gas rock samples and the relative abundance of clay minerals are listed in Table 2. The samples were neither cleaned nor exposed to air beforehand. According to the observed brittleness of rocks, the samples were machined into

Table 1 Tight reservoir properties in this study

Label	Formation	Lithology	Depth, m	Source	Geological age
S	Shihezi	Tight sandstone	2120	Erdos Basin	Early Permian
H	Huoshiling	Tight volcanic	2523	Songliao Basin	Lower Jurassic
UY	Upper Yingcheng	Tight volcanic	3524	Songliao Basin	Lower Jurassic
LY	Lower Yingcheng	Tight volcanic	3557	Songliao Basin	Lower Cretaceous
L	Lujiaping	Shale	1235	Sichuan Basin	Lower Cambrian
LM	Longmaxi	Shale	786	Sichuan Basin	Lower Silurian
N	Niutitang	Shale	895	Sichuan Basin	Lower Cambrian

Table 2 Results of XRD mineralogy analysis

Label	Mineral composition, wt%					Relative abundance, %					TOC, wt%
	Quartz	Feldspar	Calcite	Dolomite	Clay	Smectite	Illite	I/S	Chlorite	Kaolinite	
S	32.2	26.4	5.1	25.8	10.3	0	100.0	0	0	0	0
H	1.3	61.5	3	0	34.2	0	10.5	0	89.5	0	0
UY	13.2	11.9	0	0	74.9	0	16.8	74.9	8.3	0	1.1
LY	40.6	11.6	0	0	47.8	0	7.9	78.0	11.1	2.9	1.2
L	29.4	7.2	24.7	14.9	23.7	7.6	23.6	53.2	8.0	7.6	3.1
LM	40.3	8.8	7.5	6.5	36.9	4.3	15.9	62.3	8.7	8.7	3.6
N	31.2	15.8	11.5	18.2	23.3	3.4	5.2	78.9	12.4	0	2.5

I/S is Illite/smectite mixed-layer

cylindrical or prismatic core plugs, as shown in Tables 3, 4, and 5. The effects of the sample size and shape can be normalized by the scaling method.

Deionized water, 10 wt% KCl solution, and an aqueous solution of cationic surfactant were used as the imbibing fluids. Properties of the test fluids are listed in Table 6. The most commonly used boundary conditions for imbibition are one-end-open (OEO), all-faces-open (AFO), and two-ends-open (TEO). Considering the effect of lamination on the imbibition rate, for 1-D imbibition (OEO and TEO), the open face is parallel to the bedding plane to maintain the same experimental conditions.

2.2 Experimental apparatus and procedure

In group 1 experiments, the experimental results are sensitive to test environments and the instrumental error due to the relatively low imbibition rate in shale and tight rock samples. Therefore, a series of measurements are required to improve the measurement accuracy:

(1) All of the samples were weighted by an analytical balance (Mettler XPE205) with an accuracy of 0.00001 g.
(2) Impermeable and nonelastic strings used to suspend core slugs had a diameter of approximately 0.13 mm, which could avoid the error caused by the reduction in the liquid volume.
(3) The experimental device was placed in a chamber with constant temperature and humidity to lower the effect of variable external temperature and humidity. The experimental device is shown in Fig. 1.
(4) The whole apparatus was placed in the basement of a building to minimize the vibration from the ground surface.

The experimental procedure is as follows:

(1) The initial dimensions and mass of the core slug were measured before experiment.
(2) Impermeable epoxy was used to satisfy TEO and OEO, imbibition boundary conditions.
(3) The core slug was dried in an oven at 105 °C until there was no further change in weight.
(4) After the core slug cooled down, the sample was suspended on the analytical balance. The sample was totally submerged into water by adjusting the liquid level.
(5) The variation of the sample mass with time was measured and then recorded on a computer as water was spontaneously imbibed into the sample.
(6) The mass of the imbibed water was calculated by subtracting the initial mass from the mass recorded

after the imbibition experiments. The experimental data were normalized by the scaling method, which is described in the other sections.

The experimental procedure of the group 2 experiments is the same as that of the group 1 experiments. Each sample was immersed in deionized water and dried repeatedly several times. The basic sample data are presented in Table 4.

In group 3, samples of the same formation were acquired from the same core to reduce the influences of heterogeneity. A total of 18 samples were dried for 24 h and submerged into different fluids until there was no further change in weight. This process lasted approximately 7 days. The basic information about the samples is shown in Table 5.

3 Experimental data and analysis

3.1 Scaling method for experimental data normalization

The samples used had different sizes and shapes. Characterization methods need to be developed to normalize the effects of size and shape and represent the imbibition capacity and rate.

(1) The imbibition capacity can be determined based on the curve of water volume gain per pore volume versus time.
(2) Handy (1960) established a famous gas–water imbibition model, which is given by the following equation:

$$V_{imb}/A_c = \sqrt{\frac{2P_c \phi K S_{wf}}{\mu_w}} \sqrt{t}, \qquad (1)$$

where V_{imb} is the volume of imbibed water, cm^3; P_c is the capillary pressure, MPa; ϕ is the porosity, %; K is the permeability, mD; S_{wf} is the water saturation, %; A_c is the imbibition cross-sectional area, cm^2; μ_w is the fluid viscosity, mPa s; and t is the imbibition time, s.

The slope of the volume of imbibed fluid per sectional area versus the square of time, A_i, can be used to represent the imbibition rate, which can be obtained from experimental data (Makhanov et al. 2014).

According to Eq. (1), the imbibition rate can be given as follows:

$$A_i = \sqrt{\frac{2P_c \phi K S_{wf}}{\mu_w}}.$$

Table 3 Basic properties of core plugs used in group 1

No.	Shape	Cross-sectional area A_c, cm^2	Length L, cm	Permeability[a] K, mD	Porosity[b] Φ, %	Boundary condition	Imbibition rate[c] A_i, cm/h$^{0.5}$	Imbibed volume per sample volume[d] C, %
S-1	Cylinder	5.1	5.1	2.1	12.3	OEO	0.1069	8.6
S-2	Cylinder	5.0	5.1	2.1	13.0	TEO	0.1123	8.9
S-3	Cylinder	5.1	5.0	2.2	12.8	OEO	0.1173	9.5
H-1	Cylinder	4.9	1.1	0.0028	12.5	TEO	0.0382	11.4
H-2	Cylinder	5.1	1.9	0.0045	8.4	OEO	0.0467	8.0
H-3	Cylinder	5.0	0.9	0.0031	9.6	TEO	0.0403	9.6
H-4	Cylinder	4.9	0.8	0.0069	9.7	TEO	0.0356	8.6
H-5	Cylinder	5.0	1.6	0.0083	13.6	TEO	0.0368	11.1
H-6	Cylinder	5.0	1.6	0.0034	14.1	TEO	0.0466	12.0
H-7	Cylinder	4.9	1.0	0.0096	10.8	TEO	0.0375	10.3
H-8	Cylinder	5.1	0.9	0.0069	10.1	TEO	0.0381	9.1
UY-1	Cylinder	4.9	0.6	0.0012	0.3	TEO	0.0017	2.7
UY-2	Cylinder	4.9	0.6	0.0013	0.4	TEO	0.0099	2.8
UY-3	Prism	9.0	0.5	0.0023	0.4	TEO	0.0021	3.2
UY-4	Cylinder	4.9	0.6	0.0012	0.5	TEO	0.0060	2.8
LY-1	Cylinder	4.9	0.5	0.0032	3.3	TEO	0.0044	4.7
LY-2	Cylinder	4.9	0.5	0.0025	1.6	TEO	0.0055	4.5
LY-3	Cylinder	4.9	0.8	0.0007	2.2	TEO	0.0053	4.7
LY-4	Cylinder	4.9	0.8	0.0012	2.1	TEO	0.0029	3.9
LY-5	Cylinder	4.9	0.5	0.0007	1.9	TEO	0.0037	3.4
LY-6	Cylinder	4.9	0.5	0.0031	0.9	TEO	0.0016	2.6
LY-7	Prism	10.7	0.5	0.0011	0.6	TEO	0.0008	1.8
L-1	Cylinder	40.4	0.8	0.0021	1.29	TEO	0.0015	2.7
L-2	Cylinder	40.4	1.0	0.0035	1.5	TEO	0.0017	3.5
L-3	Cylinder	40.4	1.1	0.0042	2.0	TEO	0.0020	4.1
L-4	Cylinder	29.2	1.1	0.0014	1.8	TEO	0.0015	2.2
L-5	Cylinder	29.0	0.9	0.0027	0.9	TEO	0.0016	1.9
L-6	Cylinder	29.4	1.0	0.0032	2.4	TEO	0.0018	4.8
L-7	Cylinder	29.0	1.1	0.0018	1.1	TEO	0.0012	2.0
L-8	Cylinder	29.5	0.6	0.0021	2.1	TEO	0.0012	3.8
LM-1	Prism	6.2	0.5	0.0046	2.0	TEO	0.0037	5.7
LM-2	Cylinder	30.1	0.7	0.0053	1.4	TEO	0.0035	1.4
LM-3	Cylinder	31.3	0.9	0.0038	1.5	TEO	0.0025	4.5
LM-4	Cylinder	31.3	0.9	0.0062	2.3	TEO	0.0020	6.5
LM-5	Prism	6.1	0.7	0.0071	2.2	TEO	0.0028	5.0
LM-6	Prism	6.0	0.4	0.0031	1.9	TEO	0.0041	6.1
N-1	Cylinder	29.8	0.7	0.0026	2.1	TEO	0.0008	2.0
N-2	Cylinder	29.9	0.7	0.0038	1.4	TEO	0.0006	1.3
N-3	Cylinder	29.9	0.7	0.0035	1.9	TEO	0.0007	1.7
N-4	Cylinder	29.3	0.6	0.0025	2.4	TEO	0.0008	2.9
N-5	Cylinder	29.0	0.6	0.0019	2.7	TEO	0.0010	3.5
N-6	Cylinder	29.4	0.6	0.0029	3.1	TEO	0.0011	3.2

[a] Permeability was obtained using a nitrogen pressure pulse decay permeability porosimeter

[b] Porosity was measured in a helium porosimeter

[c] Imbibition rate was obtained from the experimental data

[d] Imbibed volume per sample volume was obtained from the experimental data

The log–log relationship can be given by the following equation:

$$\log A_i = \frac{1}{2} \log \frac{2P_c K S_{wf}}{\mu_w} + \frac{1}{2} \log \phi, \tag{2}$$

$$\log A_i = \frac{1}{2} \log \frac{2P_c \phi S_{wf}}{\mu_w} + \frac{1}{2} \log K. \tag{3}$$

The effective imbibition driving force can be given as follows:

$$P_e = \frac{A_i^2 \mu_w}{2K\phi S_{wf}} = \alpha \frac{\mu_w}{2S_{wf}}, \tag{4}$$

where α is the driving force coefficient, 1/s. α reflects the effect of difference between imbibition driving force and friction resistance, which can be obtained by experiments.

Equations (2)–(4) can be used to analyze contributing factors to the imbibition rate, which is discussed in the following sections.

Table 4 Basic properties of core slugs used in group 2

No.	Shape	Dry mass, g	Boundary condition
S-4	Cylinder	69.5	AFO
H-9	Cylinder	13.9	AFO
UY-5	Cylinder	11.5	AFO
L-9	Cylinder	14.0	AFO
LM-7	Prism	15.0	AFO
N-7	Cylinder	57.5	AFO

3.2 Imbibition curve characteristics

The imbibition characteristics of tight rocks can be characterized by the imbibition capacity, the imbibition rate, and the diffusion rate.

Figure 2 shows plots of the volume normalized by the pore volume versus time. The cumulative imbibed volume increases with time. However, the rate of water intake obviously slows down with increasing time, and the rate approximately reaches zero, which represents the equilibrium condition (Sun et al. 2015). However, some curves may have "upward tails," which demonstrate an obvious diffusion effect and may be related to the complex pore structure in these rock samples, as shown in Fig. 2f. Though the size and shape of core samples vary significantly, the water volume normalized by the pore volume can represent the imbibition capacity well. It is worth noting that the highest points in the curves may not always remain at the same value, which may be explained by the heterogeneity of core samples as shown in Fig. 2c, d, e, f. In addition, the volume gain fluctuates in some samples, which may be due to the large amount of water-sensitive I/S mixed-layer in this tight rock, as shown in Fig. 2c, e, f. This effect will be discussed in detail later.

Figure 3 shows plots of the imbibed volume normalized by the sectional area versus the square root of time. The effect of the sectional area is normalized well. The measurements tend to sit close to a smooth curve that represents the imbibition rate as shown in Fig. 3a, b, d, e. In

Table 5 Basic properties of core slugs used in group 3

No.	Shape	Cross-sectional area A_c, cm^2	Length L, cm	Imbibing fluid	Boundary condition
S-5	Cylinder	5.1	5.0	Deionized water	AFO
S-6	Cylinder	5.1	5.1	2.5 wt% surfactant	AFO
S-7	Cylinder	4.9	5.1	10 wt% KCl brine	AFO
H-10	Cylinder	4.9	1.2	Deionized water	AFO
H-11	Cylinder	4.9	1.2	2.5 wt% surfactant	AFO
H-12	Cylinder	4.9	1.0	10 wt% KCl brine	AFO
UY-6	Cylinder	5.1	1.1	Deionized water	AFO
UY-7	Cylinder	5.0	1.0	2.5 wt% surfactant	AFO
UY-8	Cylinder	5.1	1.1	10 wt% KCl brine	AFO
L-10	Cylinder	40.4	1.2	Deionized water	AFO
L-11	Cylinder	40.4	1.1	2.5 wt% surfactant	AFO
L-12	Cylinder	40.3	1.0	10 wt% KCl brine	AFO
LM-8	Prism	6.1	0.9	Deionized water	AFO
LM-9	Prism	6.0	1.0	2.5 wt% surfactant	AFO
LM-10	Prism	6.1	0.9	10 wt% KCl brine	AFO
N-8	Cylinder	29.4	1.1	Deionized water	AFO
N-9	Cylinder	29.4	1.0	2.5 wt% surfactant	AFO
N-10	Cylinder	29.3	1.1	10 wt% KCl brine	AFO

Table 6 Properties of fluids at 25 °C used in group 3

Fluids	Density, g/cm^3	Viscosity, cP	Surface tension, N/m
Deionized water	0.998	1	0.072
10 wt% KCl	1.06	0.88	0.074
2.5 wt% surfactant	0.96	1.1	0.058

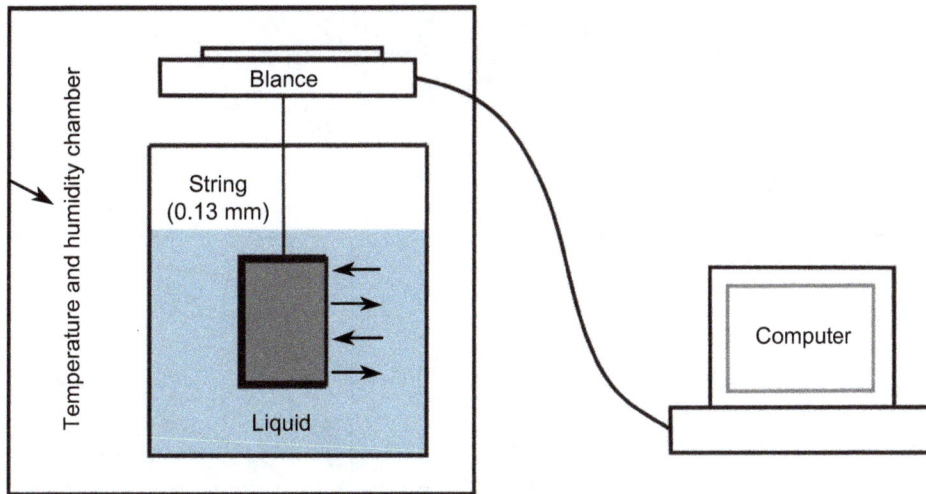

Fig. 1 Schematic for imbibition experiments in the OEO condition

Fig. 3c, e, the departure from a smooth curve can be explained by the strong heterogeneity in these tight rocks.

The imbibition curves generally behave similarly. Each profile is divided into three regions: the initial linear imbibition region (Region 1), the transition region (Region 2), and the diffusion region (Region 3), as shown in Fig. 4 (Lan et al. 2014). However, the curve characteristics of different tight rocks vary significantly, which may be attributed to the pore size distribution and pore connectivity. The slope in Region 1 represents the imbibition rate (A_i), where the capillary pressure may be the primary driving force. As the water saturation increases, the capillary pressure decreases, and the water intake process begins to enter Region 2. In Region 3, diffusion is the primary flow driving force compared with capillary imbibition in Region 1. In tight reservoir rocks, the diffusion effect is more obvious than that in conventional reservoir rocks. It can be represented by the diffusion rate (A_d). However, a study of the diffusion rate is not included in this paper.

3.3 Imbibition capacity and rate

The ratios of the maximum imbibed volume to the pore volume, R exceed 1 in the UY, LY, L, LM, and N formations, as shown in Fig. 5a. In particular, for the UY formation, R is 6–8. In other words, the imbibed volume of the rock in the tight reservoir can exceed the pore volume of the rock by a factor of 6–8. This result is

consistent with the results of Zhou et al. (2014). There is a significant difference between these results and the conventional opinion that the water saturation should be 0–1. Therefore, the capacity for fracturing fluid intake in tight reservoirs may greatly exceed the conventional estimation. The relative relationship of R is UY tight volcanic > LM shale > LY tight volcanic > L shale > N shale > H tight volcanic > S tight sandstone.

Figure 5b presents the distribution of the imbibition rate. The imbibition rates in the S and H formations are approximately 0.05–0.1 cm/h$^{0.5}$, which is obviously higher than the rates for the LY, LM, N, and L formations. The relative relationship of the imbibition rate is S tight sandstone > H tight volcanic > UY tight volcanic > LY tight volcanic > LM shale > L shale > N shale.

It is worth noting that the imbibed volume exceeds the pore volume in rocks, so the water presumably entered into a space where the gas used to measure the porosity could not enter. Another explanation is that absorption of water by the clay can induce cracks and increase the pore volume. The high water imbibition capacity in tight rocks may be related to clay minerals. This is consistent with the well-known phenomenon of wellbore instability in well drilling. The experiments in group 2 were conducted to explore the effects of clay minerals on the imbibition process. A single sample was tested several times repeatedly.

The samples in the S, H, L, and N formations have good reproducibility, which illustrates that the inner pore volume

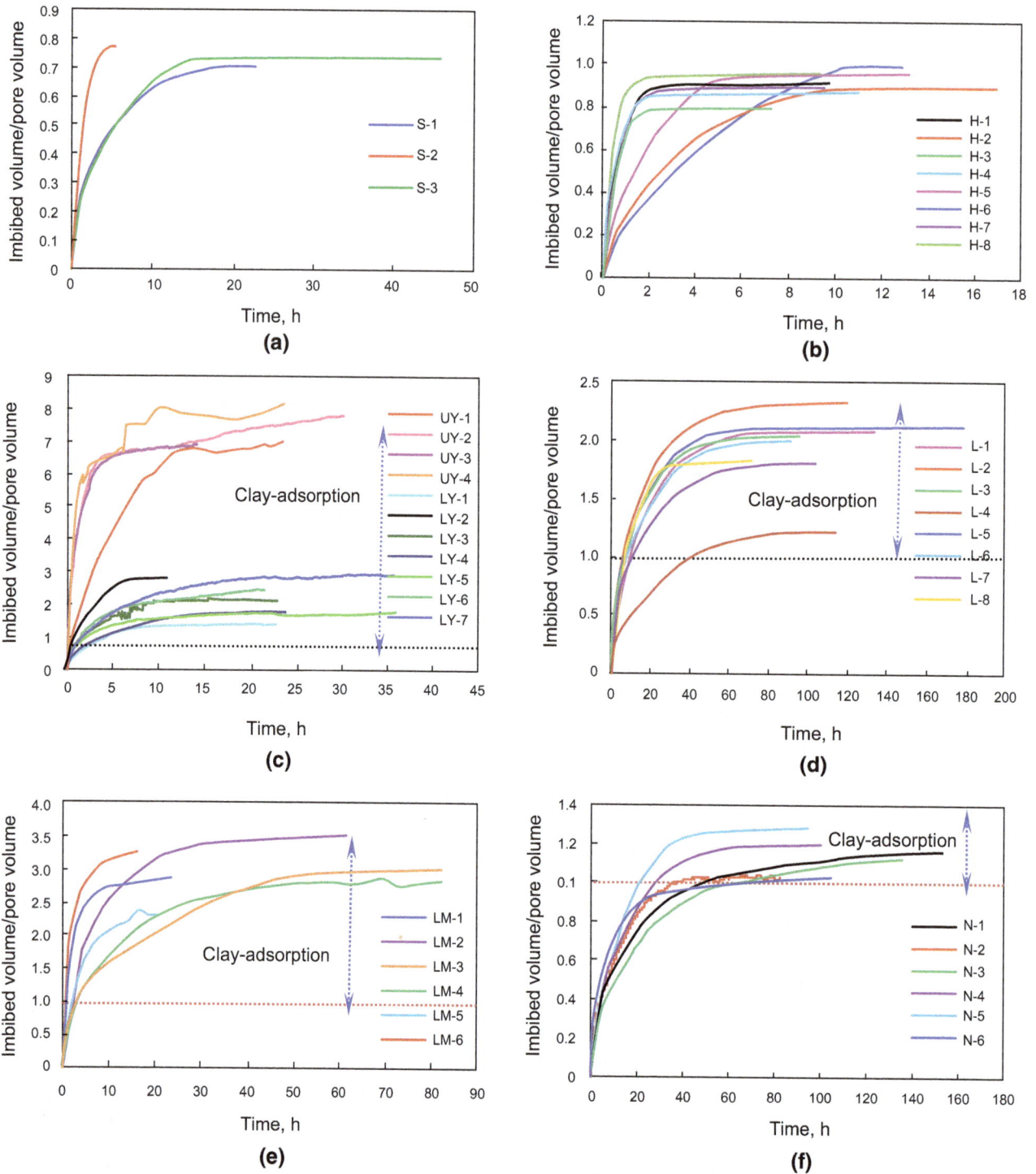

Fig. 2 Normalized cumulative imbibed volume versus time in group 1. **a** S formation; **b** H formation; **c** UY/LY formations; **d** L formation; **e** LM formation; **f** N formation

of the rock does not obviously change after water imbibition, as shown in Fig. 6a, b, d, f. However, the imbibed water volume of the sample from the UY and LM formations obviously increased after the first immersion, as shown in Fig. 6c, e. The rock from the UY formation

disintegrates during the third imbibition, as shown in Fig. 6a. The rock from the LM formation generates micro-fractures, as shown in Fig. 6b. The expansion of clay minerals with water imbibition and the generation of micro-fractures could increase the pore volume, achieving

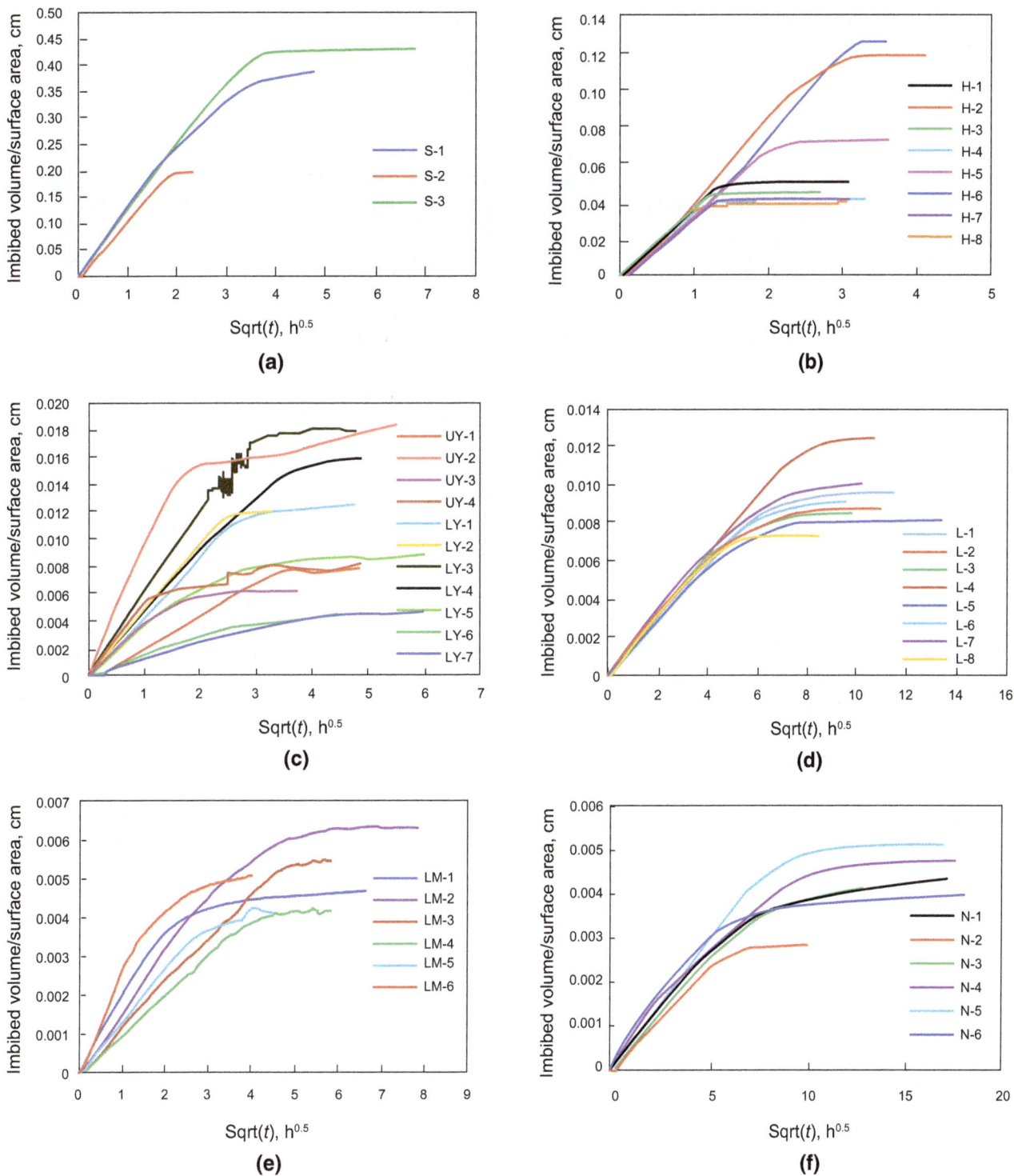

Fig. 3 Imbibed volume per sectional area versus the square of time in group 1. **a** S formation; **b** H formation; **c** UY/LY formations; **d** L formation; **e** LM formation; **f** N formation

an additional imbibition capacity beyond the initial pore volume. In addition, the R values of shales in the L and N formations are higher than 1, though these formations have good imbibition reproducibilities. This suggests that the

water could enter a space that gas could not enter. Therefore, the water imbibed into the tight rocks exits in the space that can be divided into two parts: the pore space and the clay crystal lattice space. In addition, the imbibed water volume per unit pore volume, R can be used as a parameter to indicate the clay content and type.

3.4 Influencing factors

In this section, the authors try to address the factors influencing the imbibition capacity and rate, including the porosity, permeability, clay minerals, and the fluid component.

3.4.1 Porosity

For a ratio of imbibition volume to pore volume of more than 1, the conventional definition of water location is not applicable for tight formations. The imbibition capacity can be defined as the imbibition volume per unit dry sample volume. The parameter is broadly applicable and

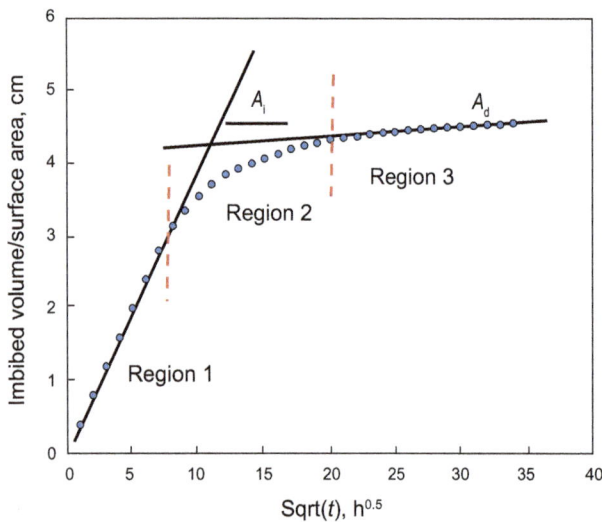

Fig. 4 Schematic of imbibition curve behavior

can be used to determine the fracturing fluid volume intake. A plot of the imbibition capacity versus the porosity is shown in Fig. 7a. The general trend is that the imbibition capacity increases with increasing porosity. The green line in Fig. 7a shows when the imbibition capacity is equal to the porosity. In other words, the imbibed volume is equal to the pore volume. Obviously, when the rocks have high porosities, the points tend to be below the line. One explanation is that the rocks with high porosities have low water saturation. When the rocks have low porosities, the points tend to be above the line. Therefore, the plot can be divided into three regions. In Region I, the low-porosity region, the points are above the line, and R tends to exceed 1. The imbibition capacity is mainly controlled by the clay mineral content. In Region II, the moderate-porosity region, the points are on the line, and R tends to be 1. The imbibition capacity is mainly controlled by the clay mineral content and porosity. In Region III, the high-porosity region, the points are below the line, and R tends to be lower than 1. In this region, the rocks with high porosities tend to have low clay contents. Therefore, the imbibition capacity is mainly controlled by the porosity.

In Fig. 7b, the imbibition rate is positively correlated with the porosity, which is consistent with the prediction of Eq. (2). This result means that porosity is the main controlling factor for the imbibition rate.

3.4.2 Permeability

In Fig. 8, the imbibition capacity and rate are positively correlated with permeability. However, the correlation is not strong. In addition, there is an obvious departure from the prediction of Eq. (3). This result means that permeability is not the main controlling factor for the imbibition rate.

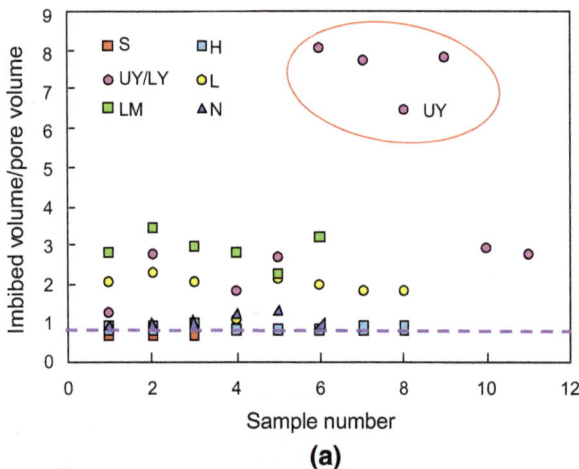

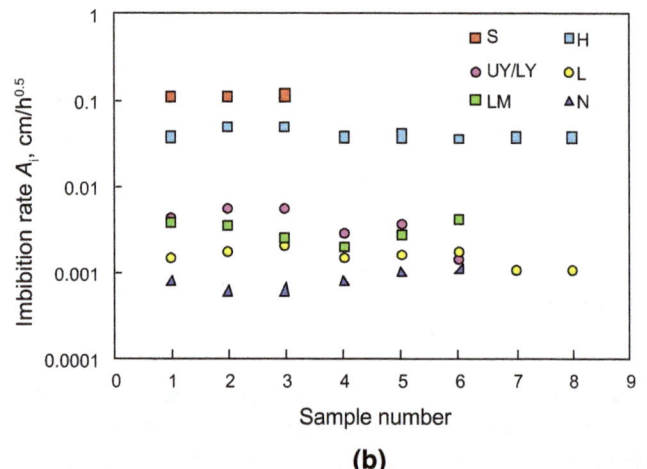

Fig. 5 Distribution of the imbibition capacity normalized by porosity and imbibition rate. **a** Imbibition capacity by porosity; **b** Imbibition rate

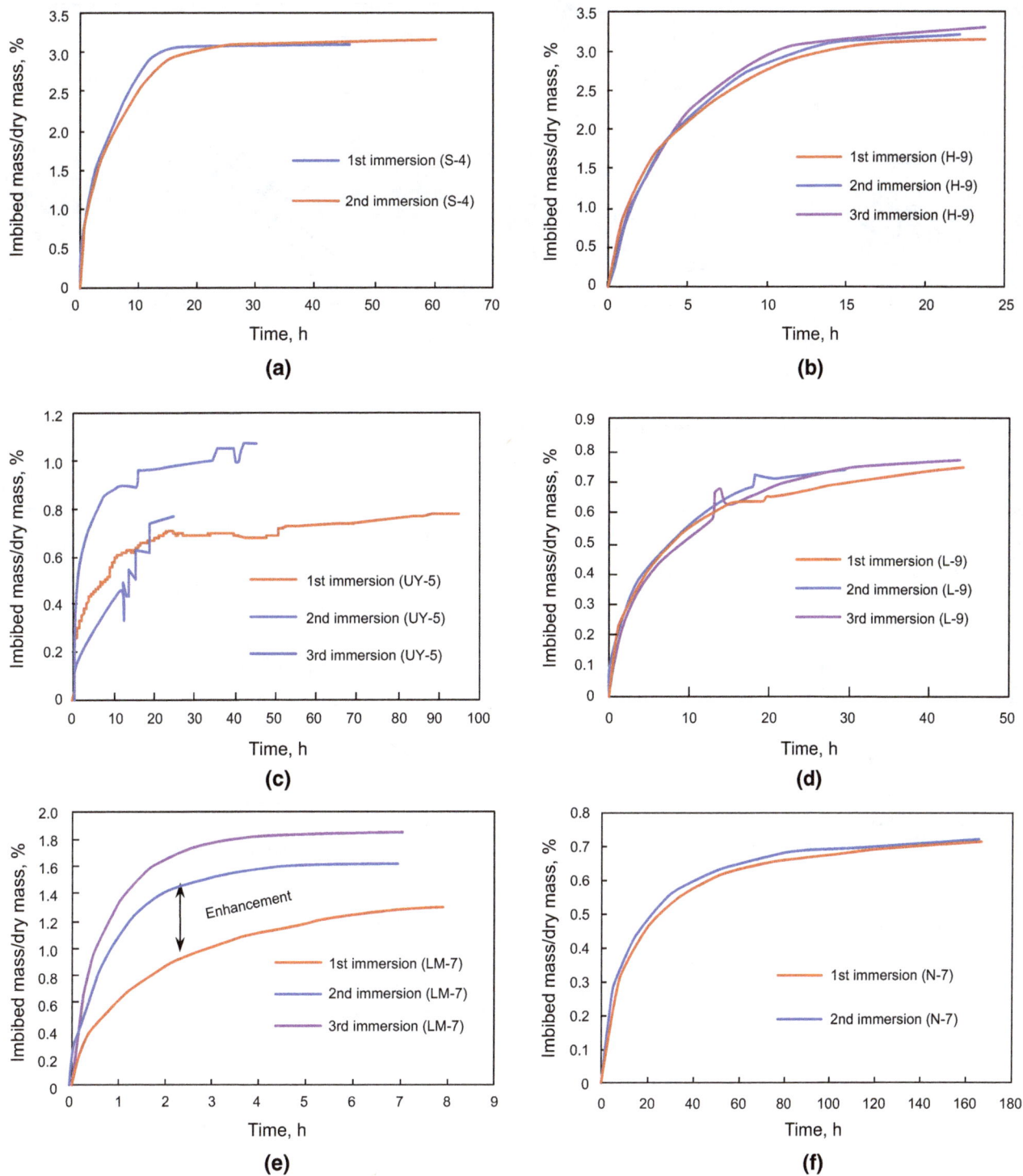

Fig. 6 Repeated immersion tests in group 2. **a** S formation; **b** H formation; **c** UY formation; **d** L formation; **e** LM formation; **f** N formation

3.4.3 Clay minerals

In order to explore the effect of the clay content and type on the imbibition capacity and rate, the effect of the porosity is normalized. Then the water saturation and the driving force coefficient are used to represent the imbibition capacity and rate, respectively.

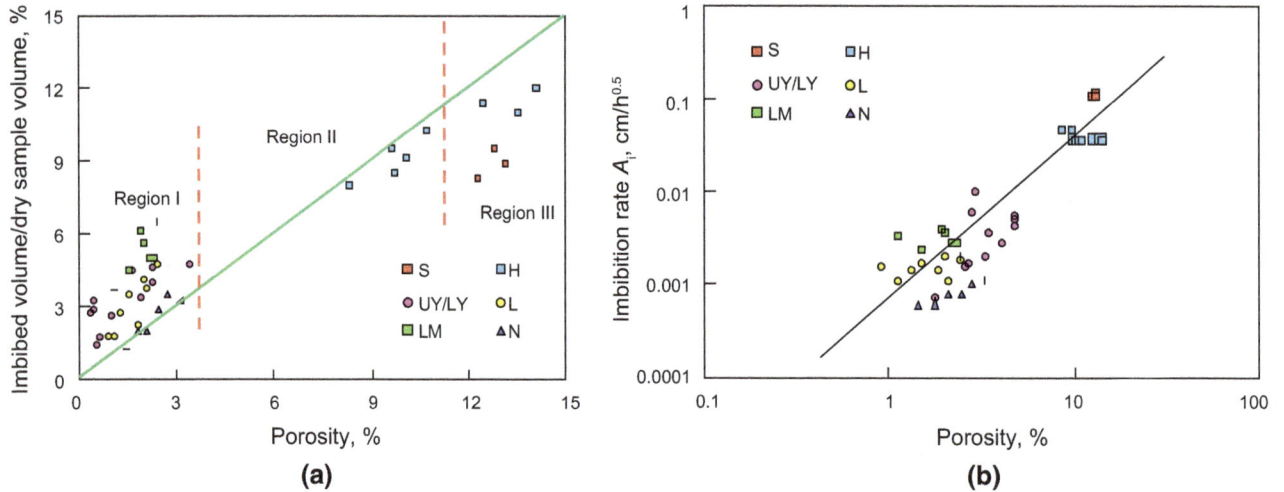

Fig. 7 Imbibition capacity and rate versus porosity. **a** Imbibition capacity; **b** Imbibition rate

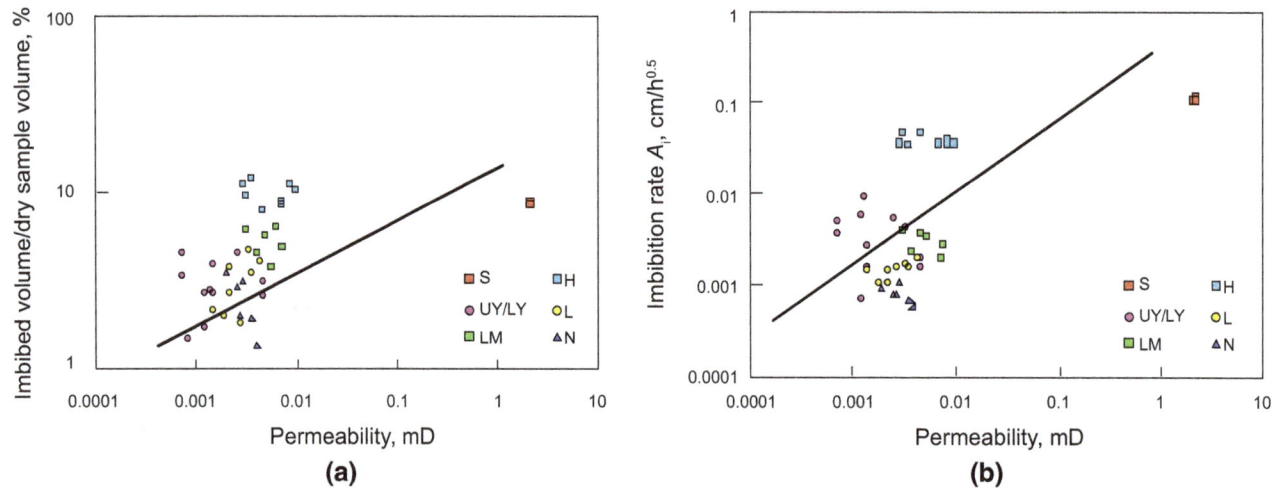

Fig. 8 Imbibition capacity and rate versus permeability. **a** Imbibition capacity; **b** Imbibition rate

The imbibition capacity is positively related to the total clay content, the I/S, and illite concentrations, as shown in Figs. 9a, 10a, and 11a. The imbibition rate is also positively related to the total clay content and the I/S concentration, as shown in Figs. 9b and 10b. However, the correlation between the imbibition rate and the illite concentration is not strong.

In Fig. 9a, the imbibition capacity is closely related to the total clay concentration. However, the H and LY formations deviate from the rule. The H and LY formations have relatively high clay contents, but their imbibition capacities are lower than those of the L and LM formations. The clay type in the H formation is mainly chlorite, while the clay of the L formation is mainly I/S. I/S has a relatively high specific surface area that can absorb larger amounts of water. Unlike the LM formation,

the LY formation does not contain smectite. The specific surface area of smectite is much higher than that of I/S, which is probably the main reason for the deviation from the rule in the LY formation. This proves that the imbibition capacity is related to not only the clay content but also the clay type.

3.4.4 Fluids

In group 3 experiments, a surfactant solution and a KCl solution changed the two driving forces and were used to explore the effect of the fracturing fluid component on the imbibition capacity.

The contact angle on the surface of LM-8 rock was measured with an imaging method, as shown in Fig. 12. The contact angle is 14° before cationic surfactant

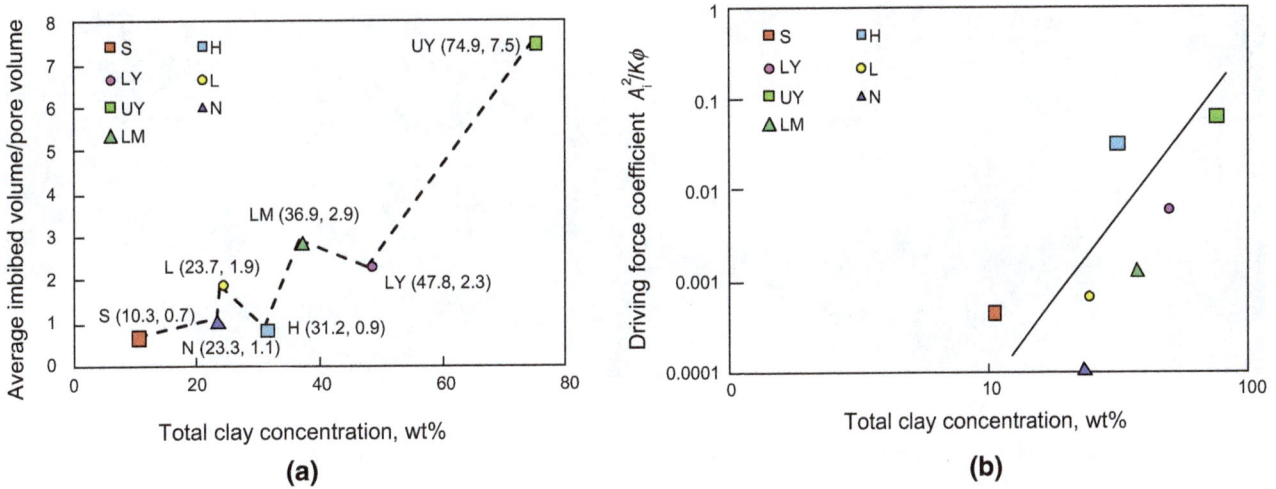

Fig. 9 Imbibition capacity normalized by porosity and driving force coefficient versus total clay concentration. **a** Imbibition capacity normalized by porosity; **b** Driving force coefficient

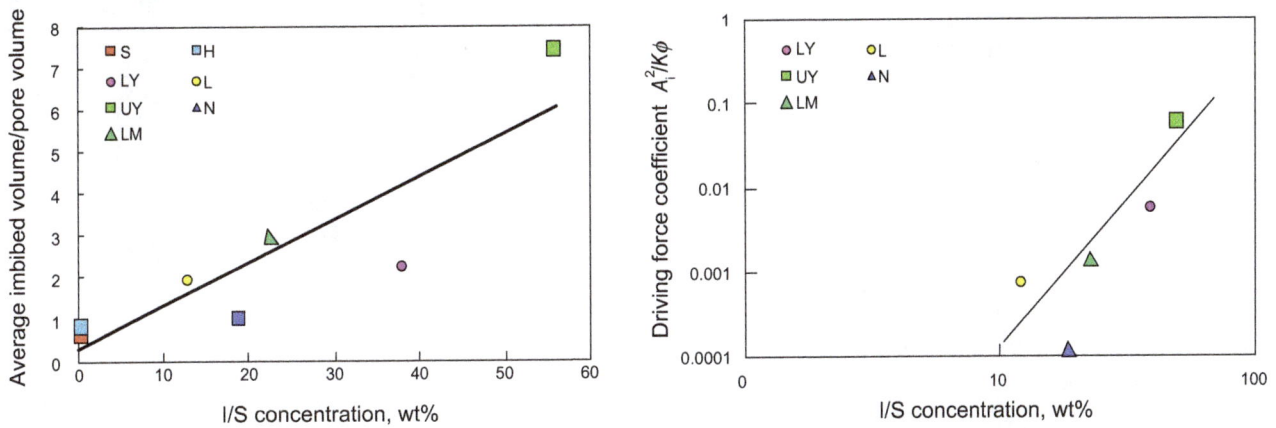

Fig. 10 Imbibition capacity normalized by porosity and driving force coefficient versus I/S concentration. **a** Imbibition capacity normalized by porosity; **b** Driving force coefficient

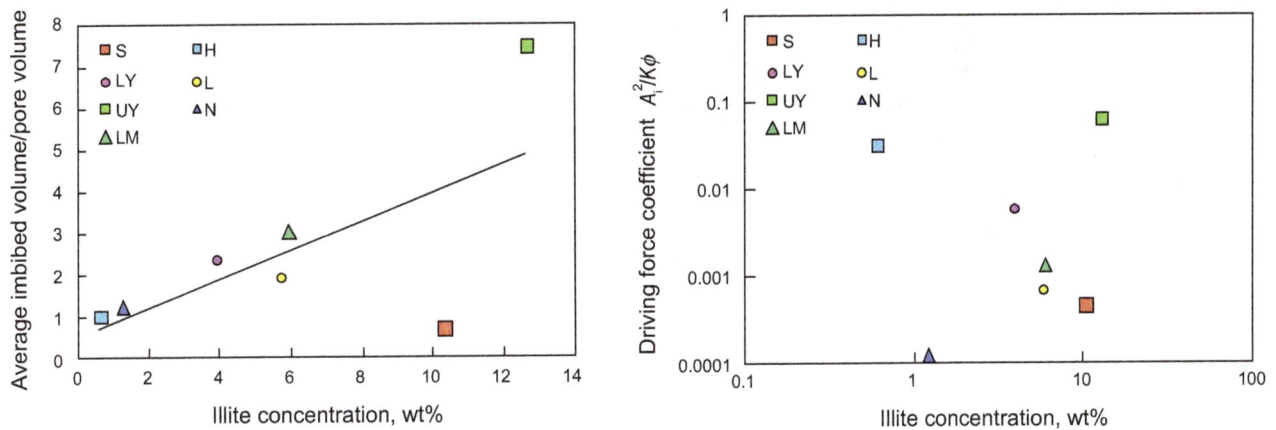

Fig. 11 Imbibition capacity normalized by porosity and rate versus illite concentration. **a** Imbibition capacity normalized by porosity; **b** Driving force coefficient

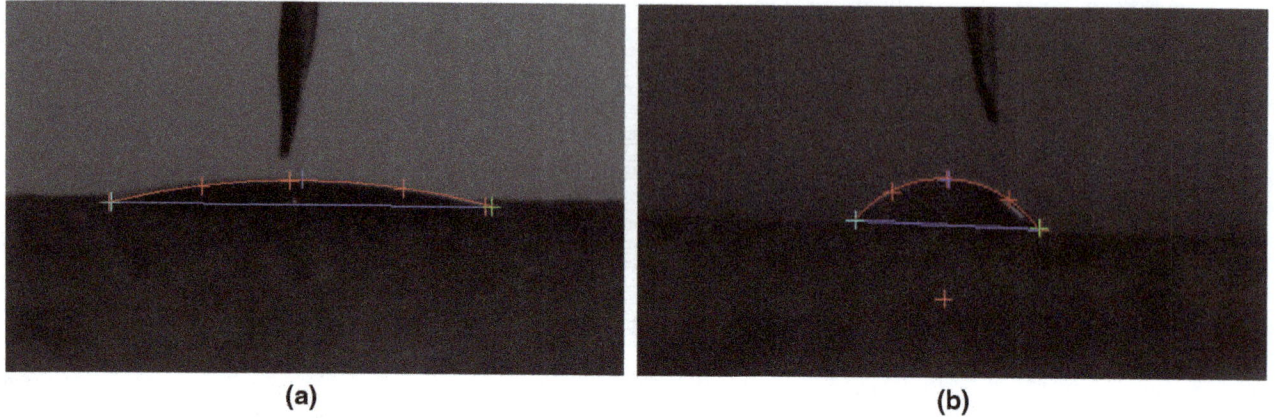

Fig. 12 Contact angle on sample LM-8 with a water drop. **a** Before surfactant treatment (14°); **b** After surfactant treatment (51°)

treatment and 51° after treatment. The surface wettability is altered and tends to be intermediately wet, which may lead to a lower imbibition capacity.

The capillary suction time (CST) tests are quick and easy and can provide information about inhibition characteristics of an additive (Berry et al. 2008). The authors tried to understand the inhibitive ability of a 10 wt% KCl solution on clay expansion and explored its effect on the imbibition capacity. The CST apparatus measures the time required for the fluid to travel a given distance, as shown in Fig. 13. A short CST time interval reflects a poor clay dispersibility. The results of CST tests are shown in Fig. 14. The relative relationship of the CST time in tight formations is UY > LM > N > L > H > S, which is positively related to the I/S concentration. The 10 wt% KCl solution inhibits the clay expansion in tight formations, especially for the UY and LM formations. This effect may lead to a lower imbibition capacity.

The results from group 3 experiments are presented in Fig. 15. The surfactant and 10 wt% KCl solutions reduce

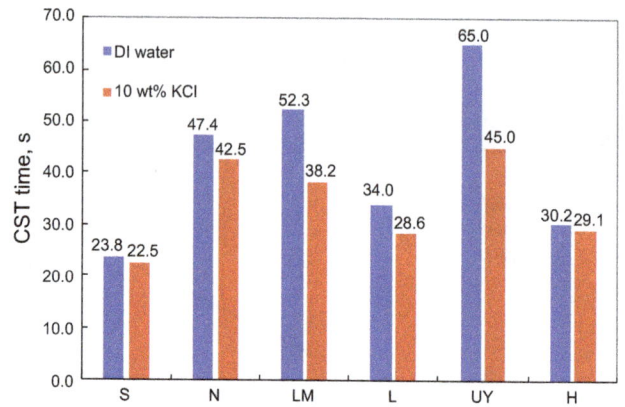

Fig. 14 CST time in different formations with two solutions

the imbibition capacity. For the UY and LM formations, the reduced imbibition capacity is more obvious. Figure 16 shows pictures of the UY and LM formation samples after exposure to the surfactant and 10 wt% KCl solutions. When exposed to deionized water, the sample from the UY

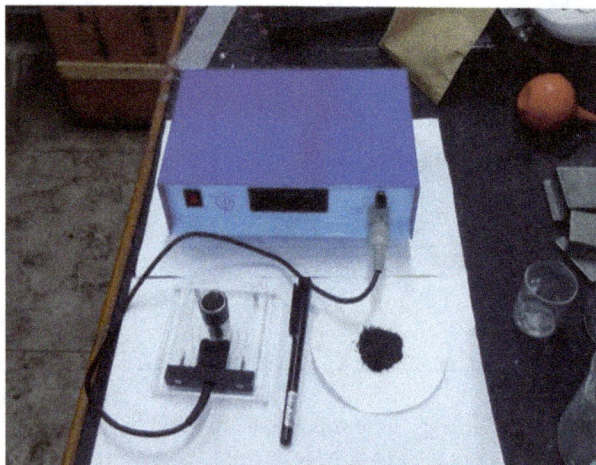

Fig. 13 CST apparatus

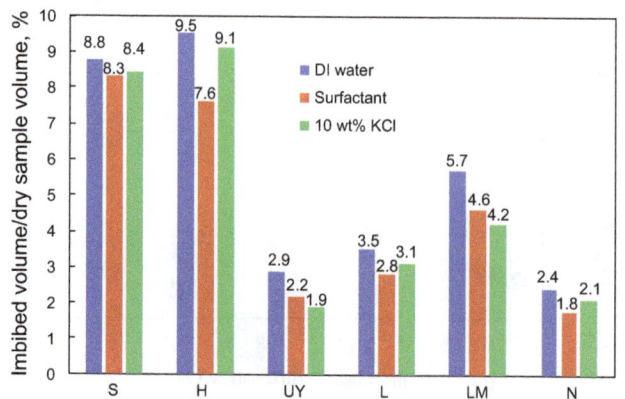

Fig. 15 The effect of imbibed fluids on the imbibition capacity in group 3

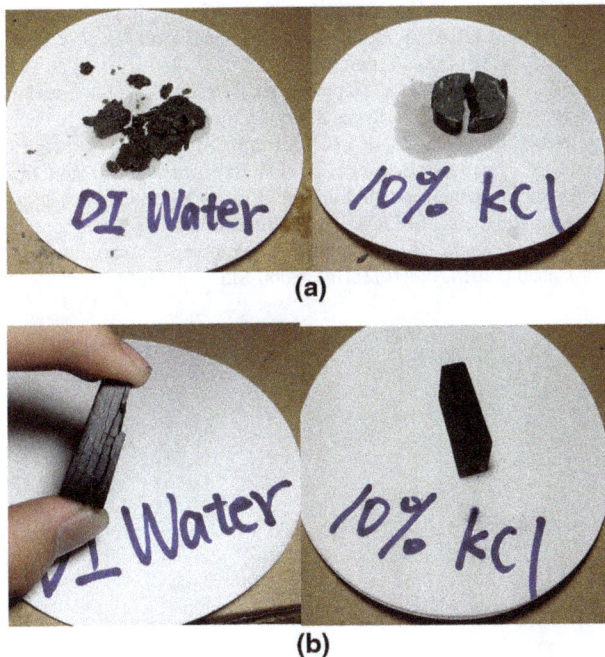

Fig. 16 Pictures of tight rock samples after exposure to different fluids. **a** UY formation; **b** LM formation

formation disintegrates completely, and the sample from the LM formation generates some micro-fractures along the bedding plane. The 10 wt% KCl solution has a good inhibitive effect on disintegration and micro-fractures, which is consistent with the results of CST tests, as shown in Fig. 16.

4 Conclusions

A series of spontaneous imbibition experiments were carried out, and the following conclusions were reached:

(1) Tight rock imbibition can be characterized by the imbibition capacity, the imbibition rate, and the diffusion rate. The imbibition capacity and rate are positively correlated with porosity. The imbibition curves can typically be divided into three regions: the initial linear imbibition region, the transition region, and the diffusion region.

(2) The water-holding space in tight rocks consists of two parts: the pore space and the clay crystal lattice space. The driving forces of spontaneous imbibition in tight rocks are the capillary force and the clay absorption force. A new parameter, the effective driving force of imbibition, is defined to describe the effect of clay absorption.

(3) The clay mineral content significantly influences the imbibition capacity for clay-rich tight rocks. The imbibition capacity normalized by the porosity increases with increasing total clay, I/S, and illite concentrations. Smectite and I/S with a high specific area tend to lead to strong water imbibition.

(4) The water volume imbibed into the clay-rich tight rocks is much greater than the pore volume measured by helium. The imbibed volume per unit pore volume is a parameter that can evaluate the effect of the clay content and type.

(5) Surfactants can change the imbibition capacity of tight rocks by altering the interfacial tension and wettability. A 10 wt% KCl solution can inhibit the clay expansion, which also reduces the imbibition capacity.

Acknowledgments This research program was financially supported by the National Basic Research Program of China (973 Program) Granted No. 2015CB250903 and the National Natural Science Foundation of China Granted No. 51490652. The Chongqing Institute of Geology and Mineral Resources supported this field work.

References

Berry SL, Boles JL, Brannon HD, et al. Performance evaluation of ionic liquids as a clay stabilizer and shale inhibitor. In: SPE international symposium and exhibition on formation damage control, 13–15 February, Lafayette, Louisiana, USA. 2008. doi:10.2118/112540-MS.

Dehghanpour H, Lan Q, Saeed Y, et al. Spontaneous imbibition of brine and oil in gas shales: effect of water adsorption and resulting microfractures. Energy Fuels. 2013;27(6):3039–49. doi:10.1021/ef4002814.

Fakcharoenphol P, Kurtoglu B, Kazemi H, et al. The effect of osmotic pressure on improve oil recovery from fractured shale formations. In: SPE unconventional resources conference, 1–3 April, The Woodlands, Texas, USA. 2014. doi:10.2118/168998-MS.

Handy LL. Determination of effective capillary pressures for porous media from imbibition data. Trans AIME. 1960;219:75–80.

Hu Q, Ewing RP, Dultz S. Low pore connectivity in natural rock. J Contam Hydrol. 2012;133:76–83. doi:10.1016/j.jconhyd.2012.03.006.

Lan Q, Dehghanpour H, Wood J, et al. Wettability of the Montney tight gas formation. In: SPE/CSUR unconventional resources conference—Canada, 30 September–2 October, Calgary, Alberta, Canada. 2014. doi:10.2118/171620-MS.

Makhanov K, Dehghanpour H, Kuru E. An experimental study of spontaneous imbibition in Horn River shales. In: SPE Canadian unconventional resources conference, 30 October–1 November, Calgary, Alberta, Canada. 2012. doi:10.2118/162650-MS.

Makhanov K, Habibi A, Dehghanpour H, et al. Liquid uptake of gas shales: a workflow to estimate water loss during shut-in periods after fracturing operations. J Unconv Oil Gas Resour. 2014;7:22–32. doi:10.1016/j.juogr.2014.04.001.

Penny GS, Dobkins TA, Pursley JT. Field study of completion fluids to enhance gas production in the Barnett Shale. In: SPE gas technology symposium, 15–17 May, Calgary, Alberta, Canada. 2006. doi:10.2118/100434-MS.

Roychaudhuri B, Tsotsis TT, Jessen K. An experimental investigation of spontaneous imbibition in gas shales. J Pet Sci Eng. 2013;111:87–97. doi:10.1016/j.petrol.2013.10.002.

Sharma M, Agrawal S. Impact of liquid loading in hydraulic fractures on well productivity. In: SPE hydraulic fracturing technology conference, 4–6 February, The Woodlands, Texas, USA. 2013. doi:10.2118/163837-MS.

Sun Y, Bai B, Wei M. Microfracture and surfactant impact on linear cocurrent brine imbibition in gas-saturated shale. Energy Fuels. 2015;29(3):1438–46. doi:10.1021/ef5025559.

Yuan W, Li X, Pan Z, et al. Experimental investigation of interactions between water and a lower Silurian Chinese shale. Energy Fuels. 2014;28(8):4925–33. doi:10.1021/ef500915k.

Zhou Z, Hoffman BT, Bearinger D, et al. Experimental and numerical study on spontaneous imbibition of fracturing fluids in shale gas formation. In: SPE/CSUR unconventional resources conference—Canada, 30 September–2 October, Calgary, Alberta, Canada. 2014. doi:10.2118/171600-MS.

9

Investigation of asphaltene deposition under dynamic flow conditions

Farhad Salimi[1] · Javad Salimi[2] · Mozafar Abdollahifar[1]

Abstract Asphaltene deposition is one of the most serious problems, which usually occurs in oil wells, petroleum production, oil processing, and transportation facilities. Deposition of heavy organic components, especially asphaltene, can lead to wellbore blockage and impacts well economics due to reduction in oil production. Therefore, it is necessary to pay more attention to finding some solution to overcome this problem. In this study, a pipe-loop apparatus for investigation of oil stability was employed to measure deposition thickness using a thermal method. The effects of many factors such as oil type, oil temperature, oil velocity, inhibitors, and solvents on asphaltene deposition were investigated. The results showed that the deposition increased with the increasing value of the colloidal instability index. Besides, the deposition thickness increased with the decreasing velocity of oil, but did not change with oil temperature. In addition, n-heptane could result in more deposition; however, toluene had no effect on the deposition. Branched dodecyl benzene sulfonic acid (Branched DBSA) and Linear DBSA as inhibitors decreased the rate of asphaltene deposition.

Keywords Asphaltene · Deposition · Dynamic flow · Pipe-loops · Inhibitors

✉ Farhad Salimi
  f.salimi@iauksh.ac.ir; farhadsalimi348@gmail.com

[1] Department of Chemical Engineering, Faculty of Basic Sciences, Kermanshah Branch, Islamic Azad University, Kermanshah, Iran

[2] Young Research and Elite Club, Kermanshah Branch, Islamic Azad University, Kermanshah, Iran

Edited by Xiu-Qin Zhu

1 Introduction

The formation of asphaltene deposits during petroleum production can cause several operational issues including total or partial blocking of pipelines and wellbores, changes in wettability, and damage to equipment (Islam 1994). Asphaltene deposition inside the oil reservoirs and production facilities is known as the main flow assurance problem in the oil industry.

In general, asphaltenes, which are regarded as part of the crude oil, are insoluble in normal alkanes such as pentane and heptane; however, they are soluble in aromatics like toluene and benzene. Also asphaltenes are known to have a high molecular weight and number of polar oil components (Ali and Al-Ghannam 1981; Hirschberg et al. 1984; Speight 1999).

There are many methods for identifying the stability of asphaltene in crude oil. These methods can approximately determine oils with the potential of asphaltene precipitation. Some of the older, but more popular, methods have been employed for the stability of asphaltenes, including asphaltene–resin ratio, Oliensis spot test, light scattering, measurement of particle size, colloidal instability index (CII), microscopic observations, and refractive index (RI). The latter (RI) was first introduced and used by Buckley (1999 for predicting the asphaltene precipitation onset. The results showed that the RI was independent of asphaltene content and occurred at a characteristic RI of 1.42–1.44. Another method- used to predict asphaltene deposition CII was presented by Yen et al. (2001) as CII = (Saturates + Asphaltene)/(Aromatics + Resin). They found that, if CII is less than 0.7, the oil would be stable, and asphaltene precipitation would not occur. However, if this index is higher than 0.9, the oil would be very unstable, and thus asphaltene precipitation would occur. Finally, if it is between 0.7 and 0.9, the oil would be moderately unstable.

Asphaltene composition, structure, and stability depend on its source, type of solvent used for oil extraction, pressure, and temperature (Soorghali et al. 2014; Groenzin and Mullins 2000; Alboudwarej et al. 2002; Akbarzadeh et al. 2002). Asphaltenes as colloidal with various sizes or as individual molecules can be absorbed on solid surfaces by the virtue of their weak phenolic and carboxylic acid groups (Kokal et al. 1995). These materials can also be strongly deposited on mineral surfaces and reservoir rocks, and hence cause formation damage, and hinder oil recovery from the reservoirs (Dubey and Doe 1993; Kord et al. 2012; Zanganeh et al. 2012). Furthermore, the absorption and deposition of asphaltenes on steel surfaces would restrict oil flow in transportation pipelines (Faus et al. 1984; Mochida et al. 1988). The remediation of asphaltene is very costly, which limits the production design of many asphaltenic crude oil reserves (Hammami and Ratulowski 2007; Akbarzadeh et al. 2012; Buckley 2012).

Asphaltenes have inherent positive or negative charges, depending on the composition of crude oil. Resins have a strong desire to link asphaltenes and attract opposite charges; consequently, they make a protective layer for asphaltenes. When this protective layer is removed, asphaltene precipitation occurs. The force balance between the adsorbed resins and asphaltenes is the most important factor for crude oil instability (Mullins et al. 2007). Resins naturally act as inhibitors, group contribution of which is a factor which connects the polar (asphaltenes) and non-polar (oil bulk) media.

In the stabilization of asphaltene micelles, resins can be helpful; however, when a paraffinic solvent is added to the crude oil, the stability between resins and asphaltenes is demolished, and the amount of asphaltene monomers increases in the bulk phase. In some quantities, asphaltene concentration reaches the concentration of the onset point, and deposition occurs (Al-Sahhaf et al. 2002).

Amphiphilic molecules are among the inhibitors, often known as dispersants which are used to prevent the aggregation of the micelles of asphaltenes. The performance of inhibitors is mainly dependent on their structural and chemical characteristics. Therefore, many inhibitors have been investigated as dispersants. To prevent asphaltene precipitation, amphiphils, which are soluble in oil, can be more efficient than aromatics. There are many works that have investigated the effects of various inhibitors on the stability of asphaltenes (Hirschberg et al. 1984; Al-Sahhaf et al. 2002; González and Middea 1991; Chang and Fogler 1994; Shadman et al. 2014; Rocha Jr et al. 2006; Boukherissa et al. 2009; Ghloum et al. 2010; Ramos 2001).

Most of these methods are done under static conditions, and the onset point of asphaltene precipitation is the main

criterion for inhibitor selection. The result obtained from this section is not sufficient to choosing a suitable inhibitor. Therefore, it is important to examine the inhibitors in real flow conditions. Besides, a literature study reveals that there are only few reports on asphaltene deposition in real pipe conditions (De Boer et al. 1995; Alboudwarej 2003; Wang et al. 2004; Zougari et al. 2006; Jamialahmadi et al. 2009; Vargas et al. 2009).

In this work, we report the effects of oil, oil velocity, temperature, and inhibitors on asphaltene deposition in dynamic conditions by flow loop apparatus using a thermal approach.

1.1 Asphaltene deposition measurement

Direct and indirect methods are used to measure asphaltene deposition in the pipe. In the direct method, depositions are measured directly such as the weight method, with ultrasonic sensors, and caliper. In the indirect method, after the formation of a deposition layer on the internal pipe wall, convection heat transfer occurs at the interface between the flowing fluid and the deposited layer. A thermal resistance term is added to the total resistance due to heat conduction through the layer (Fig. 1). Hence, the deposit's thickness can be determined from the measurements of the relevant thermal parameters by solving the heat transfer equations. Heat transfer from the internal flowing fluid to the outside environment is described by the following equation (Chen et al. 1997):

$$\frac{1}{U_b} = \frac{1}{h_o}\frac{r_o}{r_i - \delta_d} + \frac{r_o}{k_{dep}}\ln\frac{r_i}{r_i - \delta_d} + \frac{r_o}{k_p}\ln\frac{r_o}{r_i} + \frac{1}{h_g} \quad (1)$$

where U_b is the overall heat-transfer coefficient, r_i and r_o are the inside radius and outside radius of pipe, respectively, and k_{dep}, k_p, h_o, and h_g are the deposit's thermal conductivity, pipe's thermal conductivity, convection coefficient of oil, and convection coefficient of the bath, respectively. The total heat-transfer rate in the test section is given by

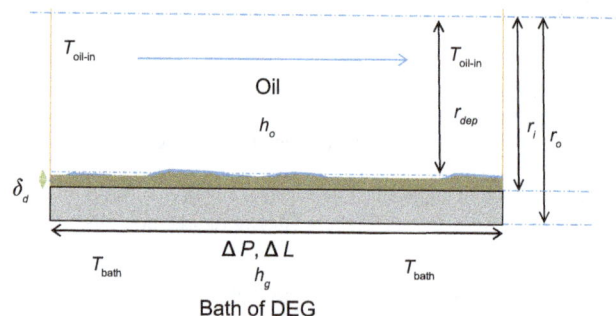

Fig. 1 The pipe cross sections and related parameters

$$Q = U_b A_b \Delta T_{lm} = m_o C_{po}(T_{o,in} - T_{o,out}) \qquad (2)$$

where, $A_b = \pi d_0 L$; and m_o, C_{po}, and ΔT_{lm} are the mass flow rate, specific heat capacity, and logarithmic mean temperature difference in the heat-exchange, respectively. At the beginning of the experiment, asphaltene thickness is zero, and the heat transfer equation is as follows:

$$\frac{1}{U_{bo}} = \frac{1}{h_o}\frac{r_o}{r_i} + \frac{r_o}{k_p}\ln\frac{r_o}{r_i} + \frac{1}{h_g} \qquad (3)$$

If Eq. (3) is subtracted by Eq. (1), the following equation is obtained:

$$Y = \frac{1}{U_b} - \frac{1}{U_{bo}} = \frac{1}{h_o}\frac{r_o \delta_d}{r_i(r_i - \delta_d)} + \frac{r_o}{k_{dep}}\ln\frac{r_i}{r_i - \delta_d} \qquad (4)$$

The overall heat-transfer coefficient, U_b is calculated by Eq. (2), and the deposition thickness is calculated by Eq. (4).

The inside (oil) convective heat-transfer coefficient (h_o) is obtained using the Hausen correlation (Eq. (5)) (Hausen 1943):

$$N_{Nu} = 3.66 + \frac{0.19\left[N_{Pe}\left(\frac{d_w}{L}\right)\right]^{0.8}}{1 + 0.117\left[N_{Pe}\left(\frac{d_w}{L}\right)\right]^{0.467}}, \qquad (5)$$

where $N_{Pe} = N_{Re}N_{Pr}$, $N_{Pr} = \mu C_{po}/k_o$, and d_w and k_o are the diameter of the pipe open to flow and thermal conductivity of oil, respectively. Finally, h_o is obtained by Eq. (6):

$$h_o = \frac{N_{Nu}k_o}{d_w} \qquad (6)$$

1.2 Experimental apparatus

Figure 2 shows a schematic view of the novel designed flow assurance test loop, which was used to measure the thickness of asphaltene deposition as a function of time at different conditions. The apparatus is made of a well-controlled temperature bath containing a long stainless steel tube in coil shape. The temperature of the bath was maintained constant using a heat source, controlling unit and stirrer. The long test tube was equipped with accurate pressure transducers and thermocouples at several intervals, transferring all the information into a data acquisition system. The feedstock was prepared and transferred into the feed storage, and its temperature was maintained at preset temperature prior to flow through the pump into the flowing loop.

In this study, the pipe test section was made of a stainless steel tube (seamless, Fitok Company) of 1 m in length and 3.74 mm in inner diameter, which was coiled and placed inside the bath. The bulk temperatures of the oil were measured with K-type thermocouples, located in the tank and in the mixing chambers, before and after the test section. The temperature of the bath was controlled within ±0.1 °C. The absolute pressure at the outlet of the tube was controlled with a back pressure control regulator (model BP-66). The oil flow rate was controlled by the constant rate pump. A data acquisition system was used to monitor the temperature at various points of both the bath and the tube.

1.3 Error analysis

The experimental error in the measurement of the mass of asphaltene deposition may result from the errors in the measurement of bulk temperature and bath temperature of the test section. The temperatures were measured with PT100 thermocouples located in the bulk of the oil flow before and after the test section. The inaccuracy in temperature measurements due to the calibration errors of the thermocouples may lead to a deviation of approximately ±0.1 °C. The maximum error in the mass of asphaltene deposition measurements was estimated to be less than ±2.5 % by taking into account the temperature and flow rate measurement errors.

1.4 Experimental materials and procedure

Two different samples of Iranian crude oil were used for investigation of asphaltene deposition. The results of SARA test for these oils are shown in Table 1. To establish a relationship between the composition and the stability behavior of the crude oils, their main constituents, namely saturates, aromatics, resins, and asphaltenes, were determined following the procedure that has already been used by Carbognani (Carbognani and Izquierdo 1989). Resins were obtained through chromatographic fractionation of the deasphalted oil. Silica was used as a packing material, and the different fractions were sequentially eluted with hexane (saturates), toluene (aromatics), and a 10 % methanol/toluene mixture (resins). The solvents were removed by distillation, and the resins were further dried under vacuum.

In order to measure the viscosity and density of oil, an Anton Paar SVM 3000 viscometer was used. The solvent used as precipitant in the viscometric method was normal heptane, and toluene, linear dodecyl benzene sulfonic acid (Linear DBSA), and branched dodecyl benzene sulfonic acid (Branched DBSA) were used as the inhibitors. Table 2 presents the chemical structure of inhibitors used. The properties of these inhibitors are presented in Table 3.

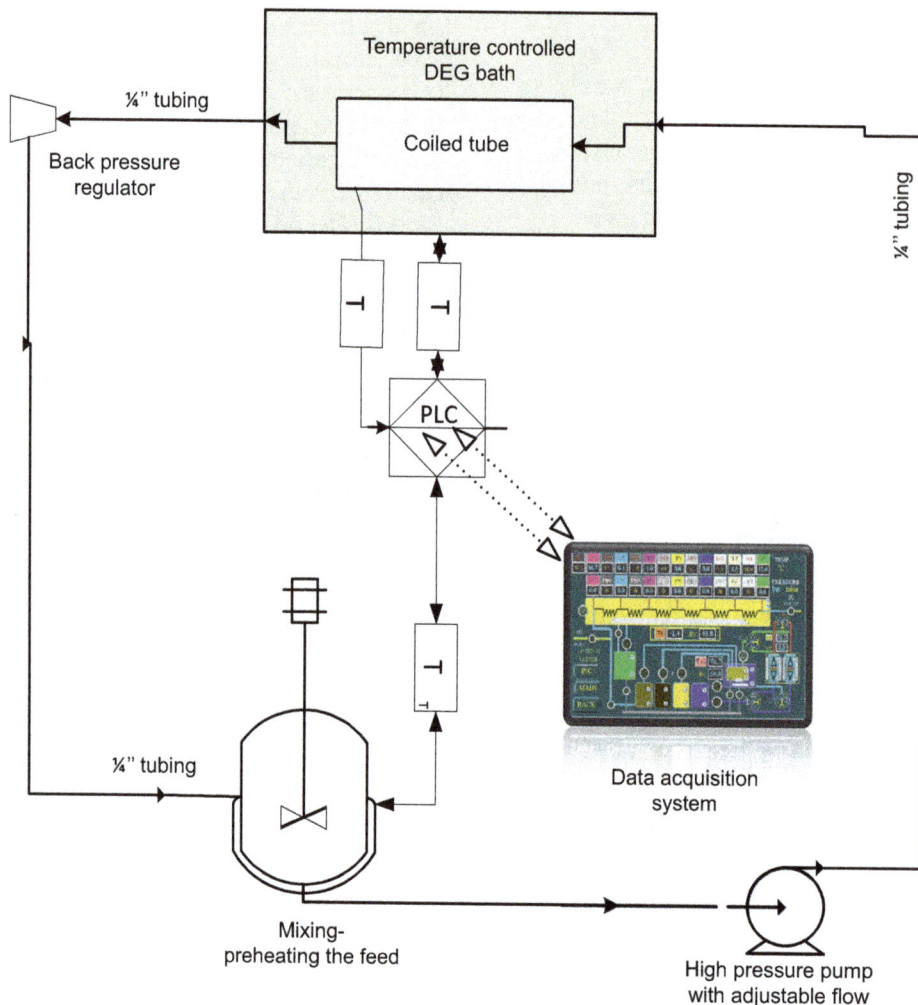

Fig. 2 Schematic of the experimental apparatus

Table 1 Analysis of SARA

	Oil 1	Oil 2
Asphaltenes	13.84	3.75
Resins	13.46	0.49
Saturates	30.31	43.67
Aromatics	42.40	52.09

The oil samples were weighed, mixed with certain amounts of inhibitor, and then put in a closed container (preheating reactor) for 0.5 h.

The experiment was conducted as follows: oil and solvent solutions were poured into the feed tank, then the stirrer was turned on, and the temperature of preheater was set to allow the feed to reach the specified temperature. Simultaneously, the temperature of the bath, which contained the pipe, was set at the test temperature. After reaching the desired temperatures of the bath and preheater, the flow rate of the pump was set and turned on. After the fluid circulated inside the pipe, the pressure of the

Table 2 Chemical structure of the inhibitors

Chemical structure	Inhibitors
	Toluene
	Linear DBSA
	Branched DBSA

experiment was set by a back-pressure control (BPCR) outlet valve. Next, the temperature data were gathered by data-acquisition system. Then, the correlations related to deposition measurement were applied for the collected data.

Table 3 Properties of inhibitors used

Water, %	Molecular weight, g/gmol	Free sulfuric acid, %	Free oil, %	Acidity, mg (KOH)/g	Anionic active, %	Inhibitor
–	92.14	–	–	–	–	Toluene
0.41	322	1.35	1.89	183.9	96.4	Linear DBSA
–	321	1.8	2.2	180	96	Branched DBSA

After the experiments, 2 l of gasoline was added to the feed tank, and the pump was initiated for an hour to wash the pipes. Finally, nitrogen gas was used to dry the pipeline.

2 Results and discussion

The solubility of asphaltene in oil is reduced by adding normal alkanes to the oil. At this point, known as the onset, more solvent will lead the asphaltene to begin to separate from the oil. Further addition of solvent will lead to increased asphaltene precipitation.

The CII index for Oils 1 and 2 is equal to 0.79 and 0.902, respectively, and the results show that the oils are within the moderately unstable range.

According to the CII index calculated for Oil 1 and Oil 2, Oil 2 is in instability zone; however, Oil 1 is in the transition zone. So, the amount of asphaltene deposit formed in Oil 2 is expected to be greater than that in Oil 1. To investigate this issue, two experiments were carried out for both oils at the same flow rate, bulk temperature, and wall temperature. The results (see Fig. 3) showed that more asphaltene deposited from Oil 2 than from Oil 1, indicating Oil 2 was more unstable than Oil 1. Another important point was related to the time of deposition formation, occurring at the later time for Oil 1.

According to the CII definition, n-heptane increases CII, and toluene decreases CII. Comparison between the results of adding n-heptane (n-C7) to Oil 2 indicated that the thickness of asphaltene deposit increased, and the initiation time, when asphaltene deposition occurs, was decreased. Adding n-heptane to the oil led to an increased amount of asphaltene particles in the oil and increased the deposition rate of asphaltene particles. The flow rates (q) of solvent for this test was equal to 220 cc/min, and the temperatures of oil and bath were adjusted to 50 and 70 °C, respectively. Toluene is an aromatic solvent; thus, its addition to the oil led to decreased CII index, and stabilized the oil. To study the effect of toluene, 15 % volume percent of the solvent was added to Oil 2 at the flow rate of 220 cc/min at 50 and 70 °C of oil and bath, respectively, and the obtained results are shown in Fig. 4.

Regarding the toluene solvent, as shown in Fig. 4 addition of toluene had very little effect on the amount of asphaltene deposit. There are two possible reasons for this: Addition of toluene solvent decreased the viscosity of oil and the asphaltene particles in the oil, causing them to easily move and stick to the wall. Besides, reducing the viscosity of oil led to decreased shear stress on the wall, and increased the formation of deposits on it. Although adding toluene reduced the CII index, these two factors could cause toluene to have almost no effect on the amount of asphaltene deposition.

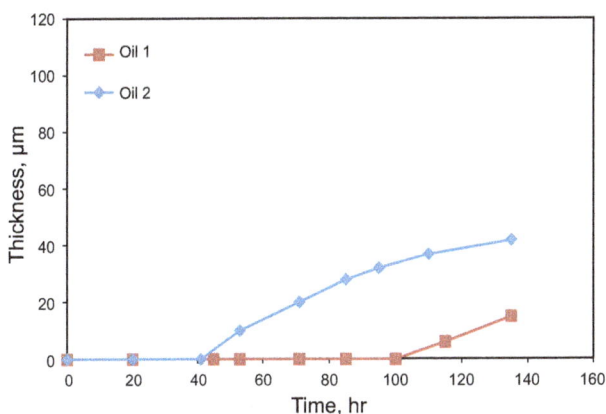

Fig. 3 The effect of oil on the thickness of asphaltene deposit versus time in pipeline for laminar flow. $q = 220$ cc/min, $T_{oil} = 50$ °C, $T_{bath} = 70$ °C

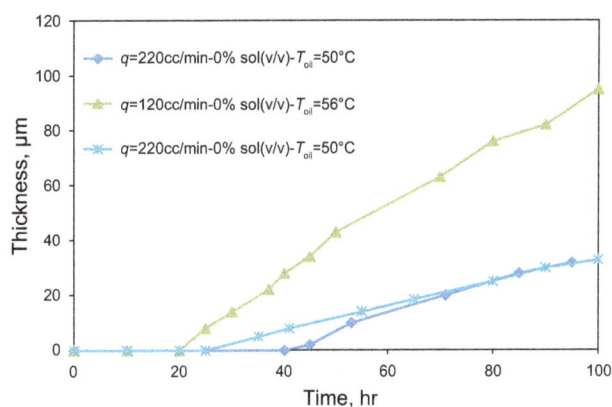

Fig. 4 The effects of flow rate (q) and oil temperature on the thickness of asphaltene deposit versus time for Oil 2, $T_{bath} = 70$ °C

For calculating the laminar flow inside the pipe, the following Hagen Poiseuille equation (Sutera and Skalak 1993) was used:

$$\Delta P = \frac{32L \times \mu \times v}{d^2} \qquad (7)$$

In the above equation, ΔP is the pressure loss, L, d, and v are the length of pipe, diameter, and dynamic viscosity, respectively

If the momentum balance is written for the pipe's cross section, the following equation is derived:

$$\Delta P \times \pi \times r^2 = \tau \times 2\pi \times r \times L \qquad (8)$$

where τ and r are the shear stress and radius, respectively. After inserting Eq. (7) in Eq. (8), the following equation is obtained for the shear stress:

$$\tau = \frac{32\mu \times v}{d} \qquad (9)$$

According to Eq. (9), the increases of velocity and viscosity led to the increase of shear stress on the wall, which could cause a reduction in the amount of asphaltene deposition.

The effect of flow rate (velocity) on asphaltene deposition was another parameter which was studied, and the results are shown in Fig. 5. As indicated, the deposition thickness was strongly dependent on this parameter. When the flow rate reduced to half, the thickness of deposition increases to about 2.5 times after 100 h. Besides, the initiation time for the deposition formation declined from about 40 h to 20 h, because the shear stress on the wall, which was directly proportional to the fluid velocity, might have caused a reduction in the erosion of the deposition layer.

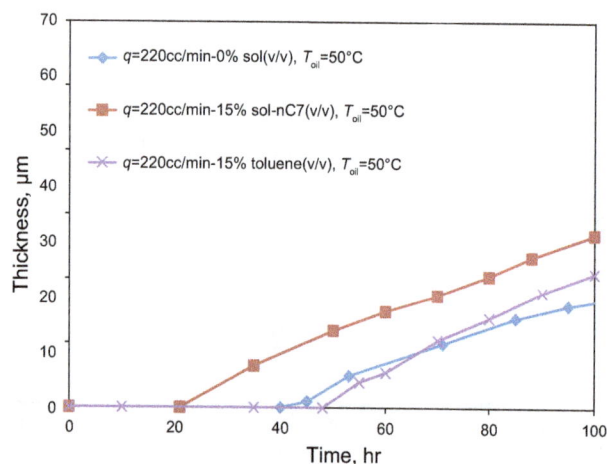

Fig. 5 The effect of solvent on the thickness of deposition versus time for Oil 2, $T_{bath} = 70$ °C

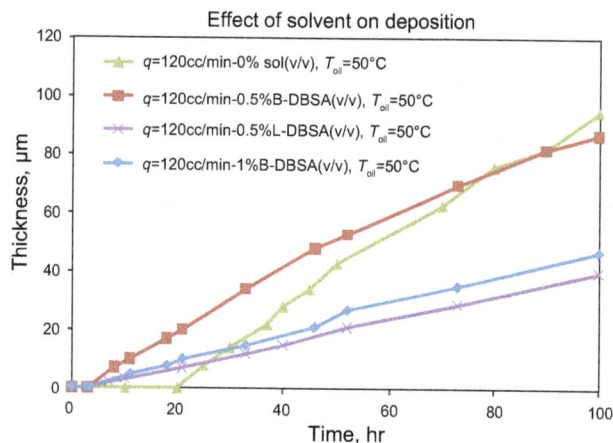

Fig. 6 The effects of inhibitors on the deposit thickness versus time for Oil 2, $T_{bath} = 70$ °C

Effect of oil temperature on asphaltene deposition was studied in this work. The results demonstrated that a rise of oil temperature (up to 56 °C) decreased the initiation time, when the deposition occurs; however, it had no effect on the amount of deposit after 100 h (Fig. 5).

A decreased viscosity of oil was the most important factor for reducing the initiation time of deposition to 5cp, which could decrease the shear rate on the inner pipe surface, and increase the deposit.

Figure 6 shows the effects of type and amount of inhibitors on asphaltene deposition formed on the pipe surface. It can be seen that the presence of inhibitors can decrease the initiation time, when the deposition occurs, and accelerate asphaltene deposition. However, an important point, which was found while adding inhibitors to the crude oil, was the reduction of slope related to the formation of asphaltene deposition or deposition rate on the inner pipe surface. Comparison of different inhibitors indicated that L-DBSA had greater influence on decreasing of deposition rate than B-DBSA. It can be explained that the existence of branches on B-DBSA could lead to the reduced effect of inhibitors in comparison to the inhibitors without branches on its structure. Nevertheless, the deposition rate decreased with increasing concentration of B-DBSA.

3 Conclusions

In this study, asphaltene deposition stability was successfully investigated under dynamic flow conditions. The effects of many factors such as oil, temperature, solvent, inhibitor, and velocity on the rate of deposition inside the pipe were studied. It was found that the presence of oil with more CII led to the increased asphaltene deposition. The deposition thickness increased with the decreasing velocity but did not with oil temperature. More deposition occurred by adding n-heptane; however, toluene had no effect on the deposition. Data revealed that addition of chemical inhibitors had a significant effect on the asphaltene' particle size, prevented further growth of the asphaltene particles, and delayed the onset of asphaltene deposition.

Acknowledgments We would like to thank the supports from Islamic Azad University, Kermanshah Branch.

References

Alboudwarej H, Beck J, Svrcek WY, et al. Sensitivity of asphaltene properties to separation techniques. Energy Fuels. 2002;16:462–9.

Alboudwarej H. Chemical and Petroleum Engineering. Calgary: University of Calgary; 2003.

Ali LH, Al-Ghannam KA. Investigations into asphaltene in heavy crude oils. I. Effect of temperature on precipitation by alkane solvents. Fuel. 1981;6:1043–6.

Al-Sahhaf TA, Fahim MA, Elkilani AS. Retardation of asphaltene precipitation by addition of toluene, resins, deasphalted oil and surfactants. Fluid Phase Equilib. 2002;194:1045–57.

Akbarzadeh K, Ayatollahi S, Nasrifar K, et al. Equations lead to asphaltene deposition prediction. Oil Gas J. 2002;100(44):51.

Akbarzadeh K, Dmitry E, Ratulowski J, et al. Asphaltene deposition measurement and modeling for flow assurance of tubings and flow lines. Energy Fuels. 2012;26:495–510.

Boukherissa M, Mutelet F, Modarressi A, et al. Ionic liquids as dispersants of petroleum asphaltenes. Energy Fuels. 2009;23:2557–64.

Buckley JS. Predicting the onset of asphaltene precipitation from refractive index measurements. Energy Fuels. 1999;13:328–32.

Buckley JS. Asphaltene deposition. Energy Fuels. 2012;26:4086–90.

Carbognani L, Izquierdo A. Preparative and automated compound class separation of Venezuelan vacuum residua by high-performance liquid chromatography. J. Chromatogr. A. 1989;484:399–408.

Chang CL, Fogler HS. Stabilization of asphaltenes in aliphatic solvents using Alkylbenzene-derived amphiphiles. 2. Study of the asphaltene-amphiphile interactions and structures using Fourier transform infrared spectroscopy and small-angle X-ray scattering techniques. Langmuir. 1994;10:1749–57.

Chen XT, Butler T, Volk M, et al. Techniques for measuring wax thickness during single and multiphase flow. In: SPE annual technical conference and exhibition. Society of Petroleum Engineers. Accessed Jan 1997.

De Boer R, Leerlooyer K, Eigner MRP, et al. Screening of crude oils for asphalt precipitation: theory, practice, and the selection of inhibitors. SPE Prod. Facil. 1995;10:55–61.

Dubey ST, Doe PH. Base number and wetting properties of crude oils. SPE Reserv. Eng. 1993;8:195–200.

Faus FM, Grange P, Delmon B. Influence of asphaltene deposition on catalytic activity of cobalt molybdenum on alumina catalysts. Appl. Catal. 1984;11:281–93.

Ghloum EF, Al-Qahtani M, Al-Rashid A. Effect of inhibitors on asphaltene precipitation for Marrat Kuwaiti reservoirs. J. Pet. Sci. Eng. 2010;70:99–106.

González G, Middea A. Peptization of asphaltene by various oil soluble amphiphiles. Colloids Surf. 1991;52:207–17.

Groenzin H, Mullins OC. Molecular size and structure of asphaltenes from various sources. Energy Fuels. 2000;14:677–84.

Hammami A, Ratulowski J. Precipitation and deposition of asphaltenes in production systems: a flow assurance overview. In: Mullins OC, Sheu EY, Hammami A, Marshall AG, editors. Asphaltenes, heavy oils, and petroleomics. Berlin: Springer; 2007. p. 617–60.

Hausen H. Darstellung des Warmeuberganges in Rohren durch verallgemeinerte Potenzbeziehungen. Z. VDI Beih. Verfahrenstech. 1943;4:91–8.

Hirschberg A, DeJong LNJ, Schipper BA, et al. Influence of temperature and pressure on asphaltene flocculation. Soc. Pet. Eng. 1984;24:283–93.

Islam MR. Role of asphaltenes on oil recovery and mathematical modeling of asphaltene properties. Dev. Pet. Sci. 1994;40:249–98.

Jamialahmadi M, Soltani B, Müller-Steinhagen H, et al. Measurement and prediction of the rate of deposition of flocculated asphaltene particles from oil. Int. J. Heat Mass Transf. 2009;52:4624–34.

Rocha LC Jr, Ferreira MS, Da Silva Ramos AC. Inhibition of asphaltene precipitation in Brazilian crude oils using new oil soluble amphiphiles. J. Pet. Sci. Eng. 2006;51:26–36.

Kokal S, Tang T, Schramm L, et al. Electrokinetic and adsorption properties of asphaltenes. Colloids Surf. A. 1995;94:253–65.

Kord S, Miri R, Ayatollahi S, et al. Asphaltene deposition in carbonate rocks: experimental investigation and numerical simulation. Energy Fuels. 2012;26:6186–99.

Mochida I, Zhao XZ, Sakanishi K. Catalyst deactivation during the hydrotreatment of asphaltene in an Australian brown coal liquid. Fuel. 1988;67:1101–5.

Mullins OC, Sheu YE, Hammami A, et al. Asphaltenes, heavy oils, and petroleomics. Berlin: Springer Sci. & Business Media; 2007.

Ramos ACS. Agregação em solventes aromaticos, desenvolvimento de aditivos e estabilização de emulsões. Doctoral dissertation, Universidade Estadual de Campinas (UNICAMP), 2001.

Shadman MM, Dehghanizadeh M, Saeedi Dehaghani AH, et al. An investigation of the effect of aromatic, anionic and nonionic inhibitors on the onset of asphaltene precipitation. J. Oil Gas Petrochem. Technol. 2014;1:17–28.

Soorghali F, Zolghadr A, Ayatollahi S. Effect of resins on asphaltene deposition and the changes of surface properties at different pressures: a microstructure study. Energy Fuels. 2014;28:2415–21.

Speight JG. The chemistry and technology of petroleum. New York: Marcel Dekker; 1999.

Sutera SP, Skalak R. The history of Poiseuille's law. Annu. Rev. Fluid Mech. 1993;25:1–20.

Wang J, Buckley JS, Creek JL. Asphaltene deposition on metallic surfaces. J. Dispers. Sci Technol. 2004;25:287–98.

Vargas FM, Gonzalez DL, Hirasaki GJ, et al. Modeling asphaltene phase behavior in crude oil systems using the perturbed chain form of the statistical associating fluid theory (PC-SAFT) equation of state. Energy Fuels. 2009;23:1140–6.

Yen A, Yin YR, Asomaning S. Evaluating asphaltene inhibitors: Laboratory tests and field studies. In: SPE international symposium on oilfield chemistry. Society of Petroleum Engineers. Accessed Jan 2001.

Zanganeh P, Ayatollahi S, Alamdari A, et al. Asphaltene deposition during CO_2 injection and pressure depletion: a visual study. Energy Fuels. 2012;26:1412–9.

Zougari M, Jacobs S, Hammami A, et al. Novel organic solid deposition and control device for live oils: design and applications. Energy Fuels. 2006;220:1656–63.

10

Synthesis and self-association of dibenzothiophene derivatives for simulation of hydrogen bonding interaction in asphaltenes

Ying-Hui Bian[1] · Shao-Tang Xu[1] · Le-Chun Song[1] · Yu-Lu Zhou[1] ·
Li-Jun Zhu[1] · Yu-Zhi Xiang[1] · Dao-Hong Xia[1]

Abstract The dibenzothiophene derivatives, namely 2-(dibenzothiophene-2-carbonyl)benzoic acid and 2-(dibenzothiophene-2-carbonyl)alkyl benzoate, were synthesized and characterized by nuclear magnetic resonance (^{1}H NMR), matrix-assisted laser desorption/ionization time of flight mass spectrometry, and elemental analysis. The self-association behavior of these dibenzothiophene derivatives in CH_2Cl_2 and CH_3CN was investigated using UV–visible absorption spectroscopy, fourier transform infrared spectroscopy, and atomic force microscopy. It was found that the carboxylic acid exhibited a strong self-association trend in CH_2Cl_2 solution at a concentration of about 5×10^{-7} M. Hydrogen bonding of carboxyl in the dibenzothiophene derivatives was confirmed to be the main driving force for the formation of the carboxylic acid aggregates.

Keywords Dibenzothiophene derivatives · Association · Hydrogen bonding · Simulation · Asphaltenes

1 Introduction

Petroleum is a complex mixture of hydrocarbons and non-hydrocarbons of various molecular weights. Asphaltene, which is the densest and highest molecular weight fraction of petroleum (Rakotondradany et al. 2006; Rogel 2000; Takanohashi et al. 2003), contains polycyclic aromatic hydrocarbons (PAHs) with alkyl groups, heteroatoms (mostly S, N, and O), and metalloporphyrins (predominantly containing Ni and V) (Strausz and Lown 2003; Tan et al. 2009). It has been found that these components can result in strong self-association via hydrogen-bonding, alkyl–alkyl, and π–π stacking interactions between the molecules (Murgich et al. 1996; Murgich 2002; Sheu 2002). Although two types of molecular models, "island" and "archipelago", have been proposed to describe the structures of molecules present in asphaltenes (Groenzin and Mullins 2000; Zhao et al. 2001; Murgich et al. 1999; Sheremata et al. 2004), the chemical identity of asphaltene remains poorly understood.

Another approach to improving understanding of the molecular structure and liquid-phase association behavior relevant to asphaltenes is to synthesize pure compounds with known chemical structures and similar physico-chemical properties to those of asphaltenes, and then to examine their behavior in solution (Groenzin and Mullins 2000; Zhao et al. 2001; Tan et al. 2009; Akbarzadeh et al. 2005). Recently, a series of compounds which contain polyaromatic π-stacking and heterocyclic hydrogen bonding sites linked by aliphatic tethers have been synthesized for study of the self-association behavior of asphaltenes (Tan et al. 2009; Sheremata et al. 2004; Akbarzadeh et al. 2005). Nordgård and coworkers have investigated the surface-pressure area isotherms and gained more knowledge of their arrangement at an aqueous surface on the basis of a perylene bisimide (PBI) core with acidic groups as asphaltene model compounds (Nordgård and Sjöblom 2008; Nordgård et al. 2010). Gray and coworkers have presented a new supramolecular assembly model that combines cooperative binding by acid–base interactions, hydrogen bonding, π–π stacking, metal coordination,

✉ Dao-Hong Xia
xiadh@upc.edu.cn

[1] State Key Laboratory of Heavy Oil Processing, College of Chemical Engineering, China University of Petroleum (East China), Qingdao 266580, China

Edited by Xiu-Qin Zhu

hydrophobic pockets, porous networks, and host–guest complexes in the aggregation of asphaltene (Gray et al. 2011).

Recently, the topics concerning H-bonds in the self-assembly of carboxylic acids have attracted more attention (Grabowski 2008). Compounds containing –COOH groups can provide two hydrogen bonds, which are particularly promising and reliable in creating and maintaining surface order. The self-assembly aggregates formed in benzoic acid solutions depends on the number and relative placement of carboxylic acid groups (Lackinge and Heckl 2009; Clair et al. 2004; Fuhr et al. 2013; Heininger et al. 2009; Ye et al. 2007). The self-assembly of the derivatives of dibenzothiophene with carboxyl groups affects the physicochemical properties of asphaltenes.

In the present work, the design and synthesis of dibenzothiophene derivatives were studied. The aggregation behavior of carboxylic acids and carboxylic esters in different solvents was comparatively investigated. The results revealed that the formation of aggregates in these carboxylic acid solutions was mainly driven by hydrogen bonding. The dibenzothiophene derivatives could be used to simulate the hydrogen bonding interaction in asphaltenes in solutions.

2 Experimental

2.1 Materials and instruments

2.1.1 Materials

All the chemicals and solvents used in this work are of A.R. grade unless otherwise indicated. 1,2-dichloroethane was purified with standard distillation procedures prior to use. All reactions were performed in dry glassware under a nitrogen atmosphere. The progress of the reactions was monitored by thin layer chromatography (SiO_2, TLC). Column chromatography was carried out on silica gel (Haiyang, Kieselgel 60, 200–300 mesh) with the indicated eluents.

2.1.2 Instruments

Melting points were determined on a Beijing Fukai X-5A apparatus (China). The ^1H NMR (600 MHz) spectra were measured in $CDCl_3$ using a Bruker AVANCE III 600 Spectrometer (Germany) with tetramethylsilane (TMS) as an internal standard. MALDI-TOF mass spectra were taken on a Bruker BIFLEX III (Germany). Elemental analysis was performed on an elemental analyzer, Vario EL III (Elementar, Germany). The FT-IR spectra were recorded with a Nicolet 6700 Fourier transform infrared (FT-IR, Thermo Scientific) spectrometer using potassium bromide

pellets. Atomic force microscopy (AFM) images of the aggregates were obtained with a commercial Bruker Nanoscope IVa MultiMode atomic force microscope (Germany). The UV–visible absorption (UV–Vis) spectra were measured on Hitachi U-3900H spectrophotometer (Japan) at room temperature.

2.2 Experimental methods

2.2.1 Synthesis of dibenzothiophene derivatives

The synthesis routes of 2-(dibenzothiophene-2-carbonyl) benzoic acid (compound 2) and 2-(dibenzothiophene-2-carbonyl)octadecyl benzoate (compound 4), 2-(dibenzothiophene-2-carbonyl)hexyl benzoate (compound 5) and 2-(dibenzo[b,d]thiophene-2-carbonyl)ethyl benzoate (compound 6) were designed as shown in Scheme 1 (Katagiri et al. 2012).

2.2.1.1 Preparation of 2-(dibenzothiophene-2-carbonyl) benzoic acid (compound 2)   Aluminum chloride (3.13 g, i.e., 23.5 mmol) was added to a solution of phthalic anhydride (0.87 g, i.e., 5.88 mmol) in 1,2-dichloroethane (20 mL), at −20 °C using a cooling water circulator and stirred for 1 h. Then, a solution of dibenzothiophene (1.08 g, i.e., 5.88 mmol) and 1,2-dichloroethane (20 mL) was added dropwise. The resultant mixture was stirred for 24 h at

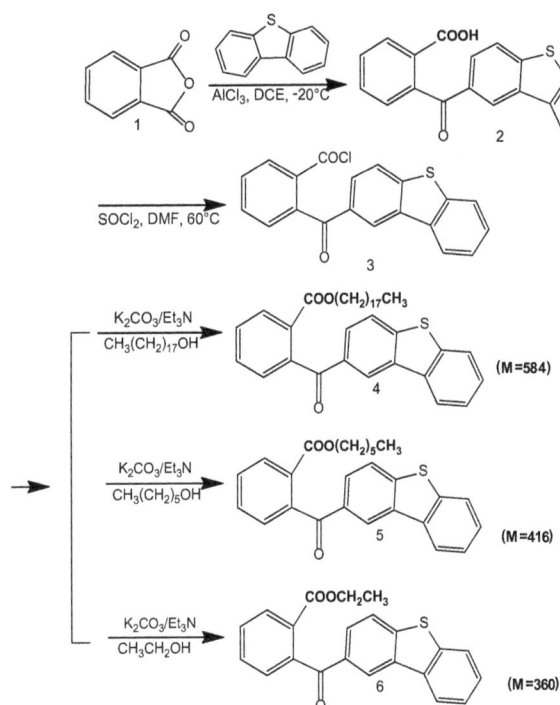

Scheme 1 Synthesis of 2-(dibenzothiophene-2-carbonyl)benzoic acid (compound 2) and octadecyl, hexyl, ethyl 2-(dibenzothiophene-2-carbonyl)benzoate (compounds 4, 5, 6, respectively)

−20 °C under nitrogen and then treated with 30 mL 10 % aqueous hydrochloric acid and stirred for 10 min. Then the mixture was extracted with 200 mL dichloromethane. The organic layer was washed with 200 mL water, and 100 mL saturated salt water, and then dried over anhydrous sodium sulfate. The solvent was removed under reduced pressure, and the residue was purified by column chromatography on silica gel (dichloromethane/methanol = 10/1). Then the product was purified by recrystallization and dried in a vacuum oven to obtain compound **2** as a yellow powder with a yield of 40 % (0.78 g).

2.2.1.2 Preparation of 2-(dibenzothiophene-2-carbonyl) octadecyl benzoate (compound 4)

First, thionyl chloride (5.24 mL) and DMF (1 μL) were added to carboxylic acid (compound **2**) (0.19 g, i.e., 0.57 mmol) at 0 °C under nitrogen and stirred at 60 °C for 3 h, then the excess thionyl chloride was removed under high vacuum to obtain acyl chloride which was used without purification. Then, octadecanol powder (0.15 g, i.e., 0.57 mmol) and ethylenediamine (81 μL) were added to the solution of acyl chloride (0.019 M) in dichloromethane (20 mL), and stirred for 6 h at room temperature. After that, the solvent was removed under reduced pressure, and the residue was purified by column chromatography on silica gel (petroleum ether/acetone = 10/1). The product was purified by recrystallization and then dried in a vacuum oven to give the carboxylic ester (compound **4**) as a pale yellow powder with a yield of 66 % (0.22 g).

2.2.1.3 Preparation of 2-(dibenzothiophene-2-carbonyl)hexyl benzoate (compound 5)

Hexyl alcohol (71.5 μL, i.e., 0.57 mmol) and ethylenediamine (81 μL) were added to a solution of acyl chloride (0.019 mol/L) in dichloromethane (20 mL). The synthesis and post-processing method was similar to that for compound **4**. The carboxylic ester (compound **5**) was a pale yellow powder, and the yield was 72 % (0.17 g).

2.2.1.4 Preparation of 2-[dibenzo(b,d)thiophene-2-carbonyl]ethyl benzoate (compound 6)

Alcohol (33 μL, i.e., 0.57 mmol) and ethylenediamine (81 μL) were added to a solution of acyl chloride (0.019 mol/L) in dichloromethane (20 mL). The synthesis and post-processing method was similar to that for compound **4**. The carboxylic ester (compound **6**) (0.15 g, 73 %) was a pale yellow powder, and the yield was 73 % (0.15 g).

2.2.2 Characterization of dibenzothiophene derivatives

2.2.2.1 2-(Dibenzothiophene-2-carbonyl)benzoic acid (compound 2)

M.p. 202 °C; ^{1}H NMR (600 MHz, CDCl$_3$) δ 8.51 (s, 1H), 8.10 (d, J = 7.3 Hz, 1H), 8.05 (d, J = 7.9 Hz, 1H), 7.86–7.81 (m, 1H), 7.78 (d, J = 8.4 Hz, 1H), 7.71 (dd, J = 8.4, 1.3 Hz, 1H), 7.66 (td, J = 7.6, 0.8 Hz, 1H), 7.55 (td, J = 7.8, 1.0 Hz, 1H), 7.50–7.42 (m, 2H), 7.40 (d, J = 7.4 Hz, 1H), (COOH signal not observed); MALDI-TOF MS calcd for C$_{20}$H$_{12}$O$_3$S m/z = 332.4, found 354.9 [M+Na]$^+$. Anal. calc. for C$_{20}$H$_{12}$O$_3$S: C 72.27, H 3.64, found C 71.96, H 3.63.

2.2.2.2 2-(Dibenzothiophene-2-carbonyl)octadecyl benzoate (compound 4)

M.p. 85.2 °C; ^{1}H NMR (CDCl$_3$) δ 8.35 (s, 1H), 8.22–8.13 (m, 1H), 7.95 (d, J = 7.6 Hz, 1H), 7.86 (t, J = 8.6 Hz, 2H), 7.66 (dd, J = 15.7, 8.1 Hz, 2H), 7.58 (t, J = 7.4 Hz, 1H), 7.53 (d, J = 7.6 Hz, 1H), 7.47 (dd, J = 5.8, 3.2 Hz, 2H), 3.58 (d, J = 8.8 Hz, 1H), 3.39 (d, J = 8.6 Hz, 1H), 1.74–1.60 (m, 2H), 1.37 (s, 2H), 1.24 (s, 28H), 0.88 (t, J = 6.6 Hz, 3H). MALDI-TOF MS calcd for C$_{38}$H$_{48}$O$_3$S m/z = 584.8, found 607.4 [M+Na]$^+$. Anal. calc. for C$_{38}$H$_{48}$O$_3$S: C 78.05, H 8.27, found C 78.11, H 8.25.

2.2.2.3 2-(Dibenzothiophene-2-carbonyl)hexyl benzoate (compound 5)

M.p. 68.2 °C; ^{1}H NMR (CDCl$_3$) δ 8.35 (s, 1H), 8.17 (dd, J = 5.7, 3.3 Hz, 1H), 7.95 (d, J = 7.6 Hz, 1H), 7.85 (dd, J = 6.3, 3.3 Hz, 2H), 7.67 (dd, J = 15.6, 8.0 Hz, 2H), 7.58 (t, J = 7.4 Hz, 1H), 7.53 (d, J = 7.6 Hz, 1H), 7.47 (dt, J = 7.1, 3.6 Hz, 2H), 3.59 (d, J = 8.8 Hz, 1H), 3.40 (d, J = 8.8 Hz, 1H), 1.74–1.60 (m, 2H), 1.44–1.34 (m, 2H), 1.33–1.23 (m, 4H), 0.88 (t, J = 6.8 Hz, 3H). MALDI-TOF MS calcd for C$_{26}$H$_{24}$O$_3$S m/z = 416.5, found 439.1[M+Na]$^+$. Anal. calc. for C$_{26}$H$_{24}$O$_3$S: C 74.97, H 5.81, found C 74.86, H 5.79.

2.2.2.4 2-(Dibenzo[b,d]thiophene-2-carbonyl)ethyl benzoate (compound 6)

M.p. 51.1 °C; ^{1}H NMR (CDCl$_3$) δ 8.37 (s, 1H), 8.19 (dd, J = 6.0, 3.1 Hz, 1H), 7.95 (d, J = 7.6 Hz, 1H), 7.91–7.81 (m, 2H), 7.71–7.63 (m, 2H), 7.62–7.52 (m, 2H), 7.48 (dd, J = 5.9, 3.1 Hz, 2H), 3.81–3.58 (m, 1H), 3.58–3.41 (m, 1H), 1.29 (dd, J = 17.8, 10.8 Hz, 3H); MALDI-TOF MS calcd for C$_{22}$H$_{16}$O$_3$S m/z = 360.4, found 361[M+H]$^+$ and 383[M+Na]$^+$. Anal. calc. for C$_{22}$H$_{16}$O$_3$S: C 73.31, H 4.48, found C 73.01, H 4.47.

3 Results and discussion

3.1 Molecular synthesis and characterization

Satisfactory elemental analysis results were obtained for these newly synthesized compounds after column chromatography purification and recrystallization. The MALDI-TOF mass spectra of the compounds clearly showed an intense signal for [M+Na]$^+$. These dibenzothiophene derivatives were also characterized by spectroscopic methods, including ^{1}H NMR, FT-IR, and UV–Vis spectra. The proposed structures of the compounds were confirmed by these analysis results.

Compound **2** was highly soluble in most organic solvents, including dichloromethane (CH_2Cl_2), acetone (CH_3COCH_3), tetrahydrofuran (THF), N,N-dimethyl formamide (DMF), dimethyl sulfoxide (DMSO), chloroform ($CHCl_3$), and acetonitrile (CH_3CN). However, except for dichloromethane and acetonitrile, the UV–Vis absorption of the organic solvents affected the absorption band of compound **2**. Therefore, dichloromethane and acetonitrile were the best solvents for study of the self-aggregation behavior of compound **2**.

3.2 Aggregation studies

The self-aggregation depends on substituents of the compound, concentration of the solution and nature of the solvent (Bayrak et al. 2011; Öztürk et al. 2012). In this work, the self-aggregation of compound **2** and compounds **4–6** was investigated by UV–Vis spectroscopy in CH_2Cl_2 and CH_3CN, respectively.

As shown in Fig. 1, with the concentration of compound **2** in CH_3CN being increased from 10^{-6} to 5×10^{-5} M, the absorption intensity increased and no new bands formed. The inset of Fig. 1 shows the change of absorption intensity of compound **2** at 241 nm with its concentration in CH_3CN. It could be seen that the Beer–Lambert law was obeyed for compound **2** in the concentration range from 10^{-6} to 5×10^{-5} M. These results indicated that

compound **2** was stable and not aggregated in CH_3CN in the range of $10^{-6} \sim 5 \times 10^{-5}$ M (Bayrak et al. 2011; Öztürk et al. 2012). However, in the concentration range of $10^{-8} \sim 8 \times 10^{-6}$ M, the absorption intensity of compound **2** in CH_2Cl_2 was dependent on concentration, as shown in Fig. 2. The inset of Fig. 2 shows the change of absorption intensity at 241.5 nm with concentration of compound **2**. As shown in the inset, with concentration increasing from 10^{-8} to 2×10^{-7} M, the absorption intensity at 241.5 nm increased, but with further increasing concentration from 2×10^{-7} to 5×10^{-7} M, the absorption intensity decreased significantly, and then the absorption intensity increased again with the concentration increasing from 5×10^{-7} to 8×10^{-6} M. It suggests the formation of aggregates of compound **2** in the CH_2Cl_2 solutions (2×10^{-7} to 5×10^{-7} M). The self-assembly behavior of compound **2** in CH_2Cl_2 resulted in a decrease of the absorption intensity in the range of $2 \times 10^{-7} \sim 5 \times 10^{-7}$ M (Bayrak et al. 2011; Öztürk et al. 2012). When the concentration exceeded 10^{-6} M, it was reasonable to assume that there was equilibrium with higher aggregates. These results demonstrated that the compound **2** could be stabilized by polar solvents and was more stable in CH_3CN than CH_2Cl_2.

To investigate the role that hydrogen bonding and $\pi-\pi$ stacking play in the aggregation process of compound **2**, the carboxylic acid group of compound **2** was protected by octadecyl, hexyl, and ethyl ester to obtain compounds **4**, **5**, **6**, respectively. Then the UV–Vis spectra of compounds **4–6** were obtained and studied (Fig. 3). Taking the UV–Vis spectrum of compound **4** as an example, it can be seen from Fig. 3a and the inset (the change of absorption

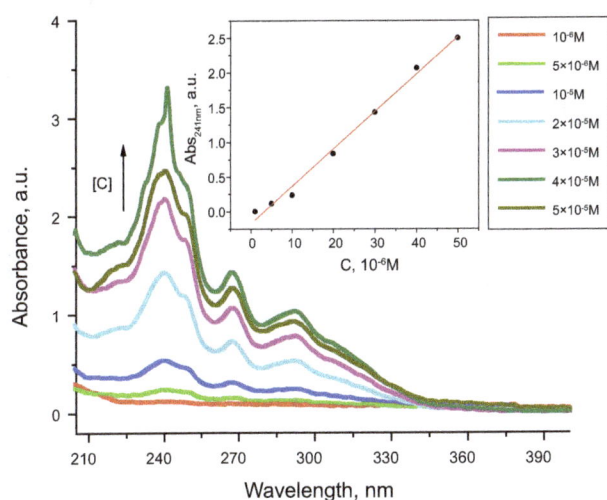

Fig. 1 UV–Vis spectra of compound **2** with different concentrations in CH_3CN. *Inset* shows the absorbance change of compound **2** with its concentration (C) in CH_3CN

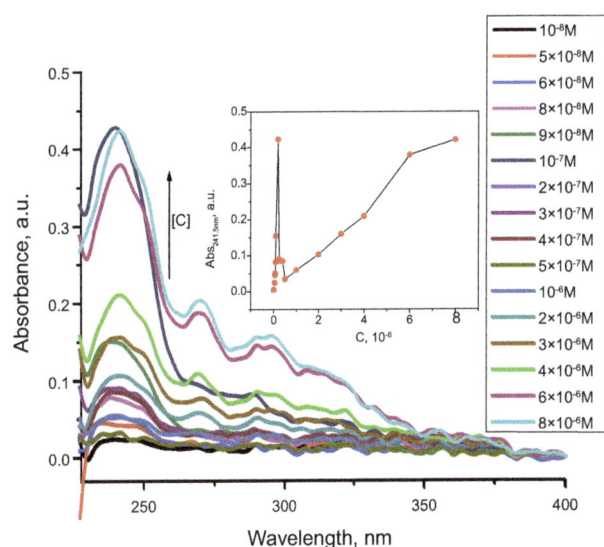

Fig. 2 UV–Vis spectra of compound **2** with different concentrations in CH_2Cl_2. *Inset* shows the absorbance change of compound **2** with its concentration (C) in CH_2Cl_2

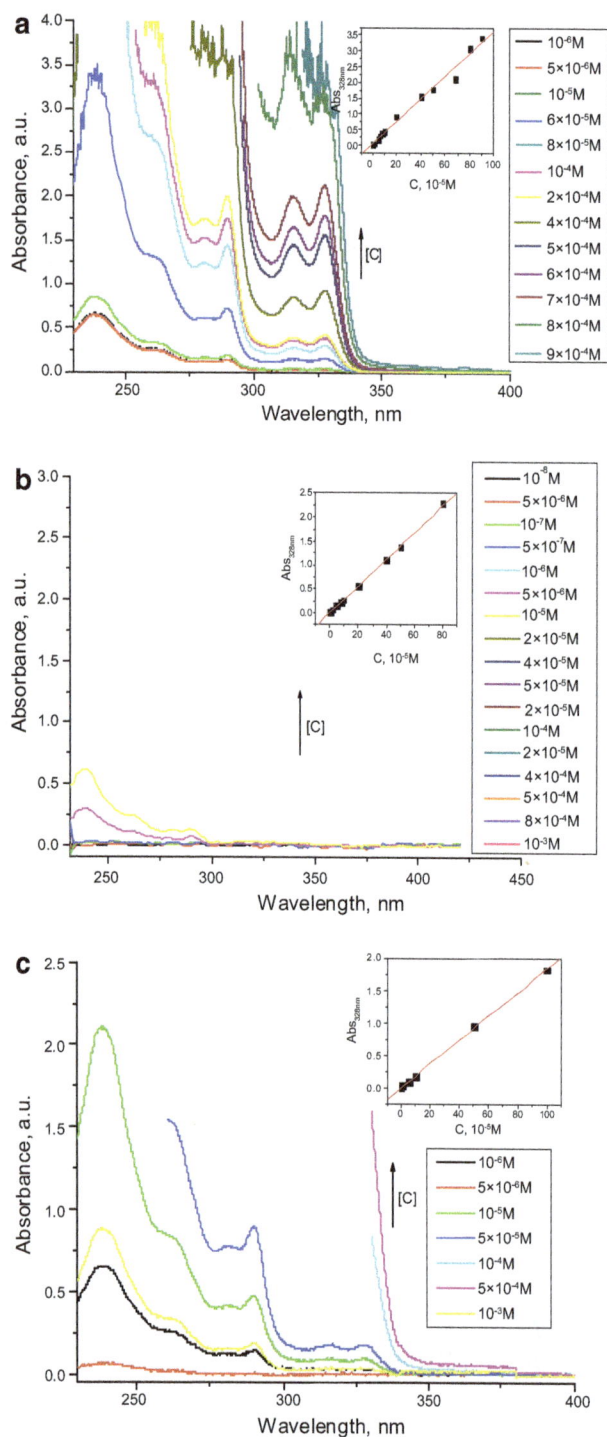

Fig. 3 UV–vis spectra of carboxylic esters with different concentrations in CH$_2$Cl$_2$. **a** compound **4**; **b** compound **5**; **c** compound **6**. *Insets* show the absorbance change of compounds **4**, **5**, **6**, respectively, with their concentration (C) in CH$_2$Cl$_2$

intensity at 328 nm with the concentration of compound **4**), with an increase of concentration, the absorption intensity increased and no new bands are formed. The Beer–Lambert law was obeyed for compound **4** in CH$_2$Cl$_2$ in the concentration range of 10^{-6}–9×10^{-4} M. A similar study was made for compounds **5** and **6** (Fig. 3b, c). The CH$_2$Cl$_2$ solution of compounds **5** and **6** obeyed the Beer–Lambert law in the concentration range of 10^{-8}–10^{-3} M. These results indicated that there were no self-assembled aggregates formed in the solution of these carboxylic esters (compounds **4**, **5**, **6**) in the investigated concentration range of 10^{-8}–10^{-3} M, and π–π stacking did not participate in the formation of aggregates of compound **2**. The results proved the dominant role of hydrogen bonding in the self-aggregation of compound **2**.

Solid state FT-IR can also be useful in showing the ionic nature of hydrogen in the aggregation of compound **2** (Lee and Wang 2010; Das and Baruah 2011). To probe hydrogen bonds among the carboxylic acid groups, the FT-IR spectra of compound **2** and compounds **4–6** were compared. As shown in Fig. 4, the FT-IR spectrum of compound **2** showed a hydroxyl absorption in the region of 3400–3500 cm^{-1}, and the hydroxyl absorption disappeared in the FT-IR spectra of compounds **4–6** after the formation of carboxylic esters. The absorption of hydroxyl of compound **2** was assigned to the absorption region of hydrogen bonding, including the absorption of individual hydroxyl,

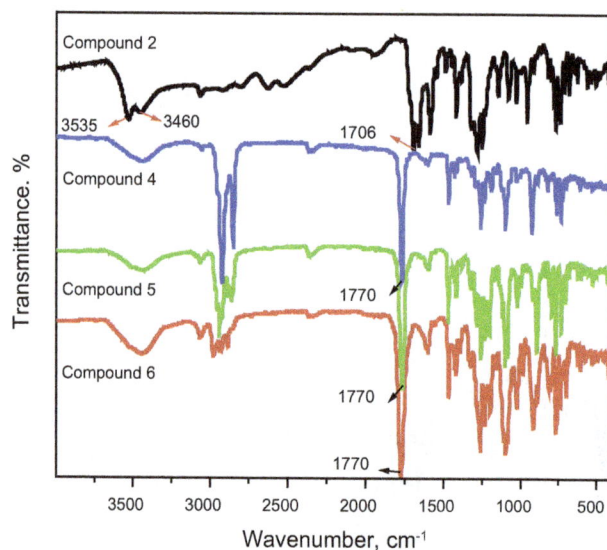

Fig. 4 FT-IR spectra of compounds **2–6**

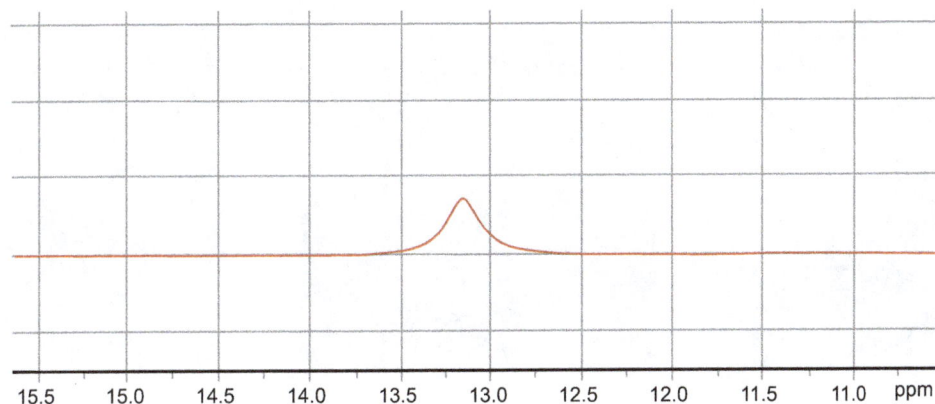

Fig. 5 ^{1}H NMR spectra of a nominally 0.020 M solution of compound **2** in DMSO-d$_6$ at 600 MHz and the horizontal axis is the chemical shift (ppm)

intramolecular hydrogen bonding and hydrogen bonding between molecules (Liu et al. 2010; Fitié et al. 2011). When the carboxylic acid group was protected by octadecyl, hexyl, and ethyl esters, the carbonyl stretching frequency shifted from ≈ 1706 to ≈ 1770 cm^{-1}, which suggested the existence of hydrogen bonds (Molla et al. 2014; Fitié et al. 2011).

The ^{1}H NMR signals at low fields (12–15 ppm) are good reporters of the environment around hydrogen-bonded protons (Suárez et al. 1998; Chin et al. 1997). As shown in Fig. 5, the ^{1}H NMR spectrum of compound **2** in DMSO-d showed one peak in this region (13.3 ppm). The peak is the signal for hydrogen-bonded protons.

Based on UV–Vis, FT-IR spectroscopy and ^{1}H NMR of compound **2** mentioned above, it can be concluded that the H-bonds between different compound **2** molecules could result in the self-assembly of aggregates. A possible growing mode of the compound **2** assembly is shown in Scheme 2. To gain more insight into the self-aggregation of compound **2**, solutions of compound **2** with concentrations of 10^{-8}, 5×10^{-7}, and 10^{-5} M were investigated using AFM. The sample was prepared by dripping a drop of the solutions of compound **2** on a mica sheet, and then the solvent was quickly evaporated under vacuum. The AFM image of Fig. 6a shows no aggregates formed in the 10^{-8} M solution of compound **2**, it was reasonable to assume that the concentration of solution was not high enough to form the aggregates. Nanoparticles of aggregates

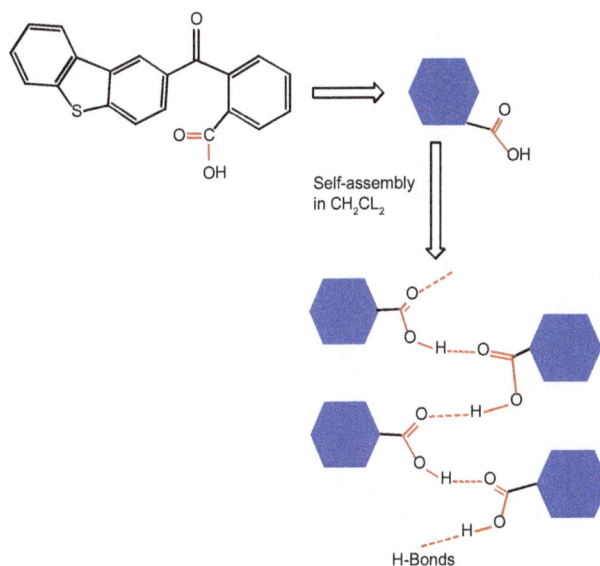

Scheme 2 Schematic presentation for the self-assembly of 2-(dibenzothiophene-2-carbonyl) benzoic acid in CH$_2$Cl$_2$ solution via H-bonds

were observed in the solution of 10^{-5} and 5×10^{-7} M (Fig. 6b, c), and the heights of the particles in the AFM image are about 2 nm (Fig. 6d, e), this phenomenon is in agreement with the result of an equilibrium with higher aggregates when the concentration of compound **2** solution exceeded 10^{-6} M.

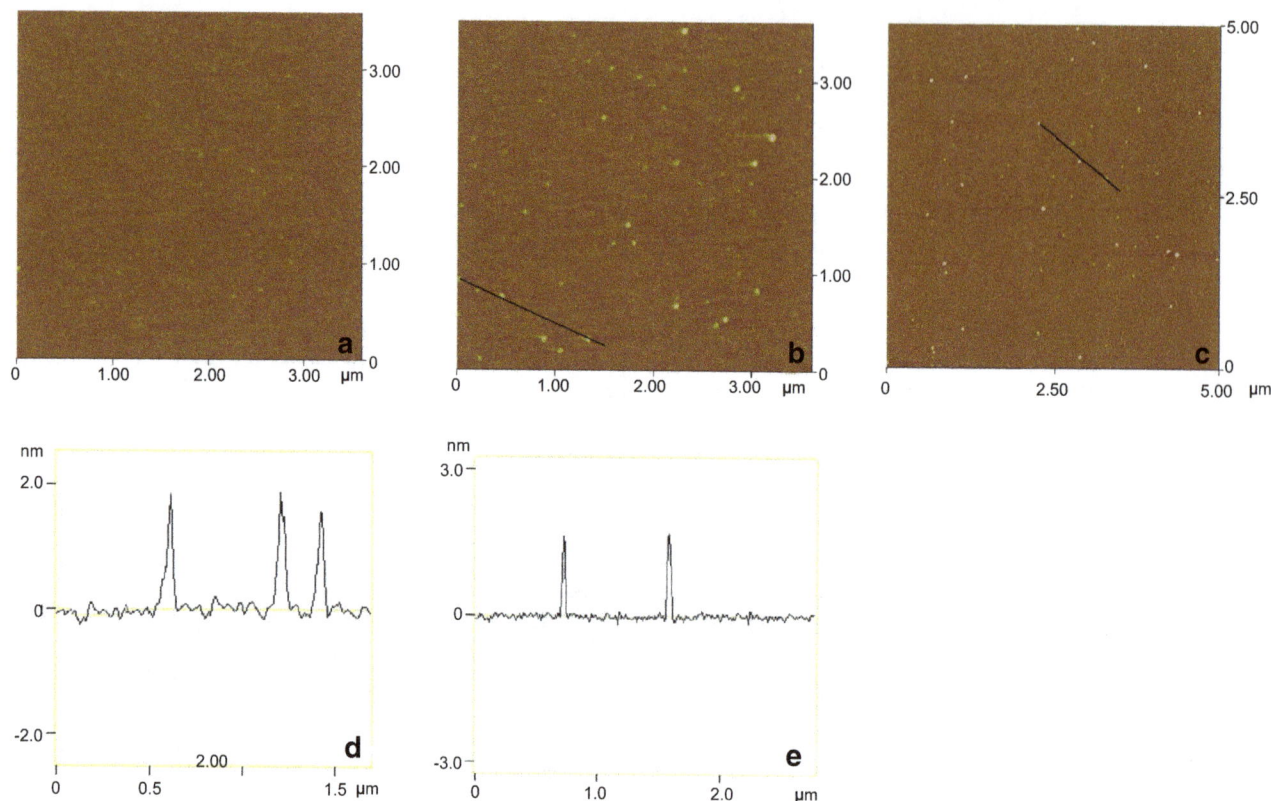

Fig. 6 The AFM representations of compound **2** solution: **a** 10^{-8} M; **b** 5×10^{-7} M; **c** 10^{-5} M; and **d, e** the cross-sectional views of the *black lines* in **b** and **c**, respectively. The *horizontal axis* of images is the scale (μm)

4 Conclusions

The dibenzothiophene derivatives of compound **2** and compounds **4–6** were synthesized successfully and the self-association behavior of these compounds in CH_2Cl_2 and CH_3CN were investigated. Compound **2** showed a strong association tendency in CH_2Cl_2 but no aggregates could be detected in the solution of compounds **4–6**. UV–Vis and FT-IR spectra suggested that hydrogen bonding between the different molecules was the main driving force for the aggregation of compound **2**. In the AFM photos, there were some nano-aggregates formed in the 10^{-5} and 5×10^{-7} M solutions of compound **2**. These results supported the notion that hydrogen bonding played a key role in the aggregation of compound **2**, which could simulate the hydrogen bonding interaction in the aggregation behavior of asphaltenes.

Acknowledgments The authors would like to thank the National Natural Science Foundation of China (No. 21376265) for financial support and the Fundamental Research Funds for the Central Universities (No. 14CX02008A).

References

Akbarzadeh K, Bressler DC, Gray MR. Association behavior of pyrene compounds as models for asphaltenes. Energy Fuels. 2005;19(4):1268–71.

Bayrak R, Akçay HT, Durmuşet M, et al. Synthesis, photophysical and photochemical properties of highly soluble phthalocyanines substituted with four 3,5-dimethylpyrazole-1-methoxy groups. J Organomet Chem. 2011;696(23):3807–15.

Chin DN, Simanek EE, Li X, et al. Computations and ^{1}H nmr spectroscopy of the imide region can distinguish isomers of hydrogen-bonded aggregates. J Org Chem. 1997;62:1891–5.

Clair S, Pons S, Seitsonen A, et al. STM study of terephthalic acid self-assembly on Au(111): hydrogen-bonded sheets on an inhomogeneous substrate. J Phy. Chem B. 2004;108(38): 14585–90.

Das B, Baruah JB. Assemblies of cytosine within H-bonded network of adipic acid and citric acid. J Mol Struct. 2011;1001(1):134–8.

Fitié CFC, Mendes E, Hempenius MA, et al. Self-assembled superlattices of polyamines in a columnar liquid crystal. Macromolecules. 2011;44(4):757–66.

Fuhr JD, Carrera N, Murillo-Quirós N, et al. Interplay between hydrogen bonding and molecule-substrate interactions in the case of terephthalic acid molecules on Cu(001) surfaces. J Phys Chem C. 2013;117(3):1287–96.

Gray MR, Tykwinski R, Tan XL, et al. Supramolecular assembly model for aggregation of petroleum asphaltenes. Energy Fuels. 2011;25(7):3125–34.

Grabowski SJ. Hydrogen bonds assisted by π-electron delocalization–the influence of external intermolecular interactions on dimer of formic acid. J Phys Org Chem. 2008;21(7–8):694–702.

Groenzin H, Mullins OC. Molecular size and structure of asphaltenes from various sources. Energy Fuels. 2000;14(3):677–84.

Heininger C, Kampschulte L, Heckl W, et al. Distinct differences in self-assembly of aromatic linear dicarboxylic acids. Langmuir. 2009;25(2):968–72.

Katagiri H, Yamamoto K, Tairabune H, et al. Synthesis and properties of naphthobisbenzo[b]thiophenes: structural curvature of higher acene frameworks for solubility enhancement and high-order orientation in crystalline states. Tetrahedron Lett. 2012;53(14):1786–9.

Lackinger M, Heckl W. Carboxylic acids: versatile building blocks and mediators for two-dimensional supramolecular self-assembly. Langmuir. 2009;25(19):11307–21.

Lee T, Wang PY. Screening, manufacturing, photoluminescence, and molecular recognition of co-crystals: cytosine with dicarboxylic acids. Cryst Growth Des. 2010;10:1419–34.

Liu D, Kong X, Li MY, et al. Study on the aggregation of residue-derived asphaltene molecules. Energy Fuels. 2010;24(6):3624–7.

Molla MR, Roy L, Kamm V, et al. Self-assembly of carboxylic acid appended naphthalene diimide derivatives with tunable luminescent color and electrical conductivity. Chem Eur J. 2014;20(3):760–71.

Murgich J, Rodríguez MJ, Aray Y. Molecular recognition and molecular mechanics of micelles of some model asphaltenes and resins. Energy Fuels. 1996;10(1):68–76.

Murgich J. Intermolecular forces in aggregates of asphaltenes and resins. Pet Sci Technol. 2002;20(9–10):983–97.

Murgich J, Abanero JA, Strausz OP. Molecular recognition in aggregates formed by asphaltene and resin molecules from the Athabasca oil sand. Energy Fuels. 1999;13(2):278–86.

Nordgård EL, Sjöblom J. Model compounds for asphaltenes and C_{80} isoprenoid tetraacids. Part I: synthesis and interfacial activities. J Dispers Sci Technol. 2008;29(8):1114–22.

Nordgård EL, Sørland G, Sjöblom J. Behavior of asphaltene model compounds at W/O interfaces. Langmuir. 2010;26(4):2352–60.

Öztürk C, Erdoğmuş A, Durmuş M, et al. Highly soluble 3,4-(dimethoxyphenylthio) substituted phthalocyanines: synthesis, photophysical and photochemical studies. Spectrochim. Acta A. 2012;86:423–43.

Rakotondradany F, Fenniri H, Gray MR, et al. Hexabenzocoronene model compounds for asphaltene fractions: synthesis & characterization. Energy Fuels. 2006;20(6):2439–47.

Rogel E. Simulation of interactions in asphaltene aggregates. Energy Fuels. 2000;14(3):566–74.

Sheremata JM, Gray MR, Dettman HD, et al. Quantitative molecular representation and sequential optimization of Athabasca asphaltenes. Energy Fuels. 2004;18(5):1377–84.

Sheu EY. Petroleum asphaltene properties, characterization, and issues. Energy Fuels. 2002;16(1):74–82.

Strausz OP, Lown EM. The chemistry of Alberta oil sands, bitumens and heavy oils. Calgary: Alberta Energy Research Institute (AERI); 2003.

Suárez M, Lehn JM, Zimmerman SC, et al. Supramolecular liquid crystals. self-assembly of a trimeric supramolecular disk and its self-organization into a columnar discotic mesophase. J Am Chem Soc. 1998;120(37):9526–32.

Takanohashi T, Sato S, Tanaka R. Molecular dynamics simulation of structural relaxation of asphaltene aggregates. Pet Sci Technol. 2003;21(3–4):491–505.

Tan XL, Fenniri H, Gray MR. Water enhances the aggregation of model asphaltenes in solution via hydrogen bonding. Energy Fuels. 2009;23(7):3687–93.

Ye Y, Sun W, Wang YF, et al. A unified model: self-assembly of trimesic acid on gold. J Phys Chem C. 2007;111(28):10138–41.

Zhao S, Kotlyar LS, Woods JR. Molecular transformation of Athabasca bitumen end-cuts during coking and hydrocracking. Fuel. 2001;80(8):1155–63.

11

Effects of alternating current interference on corrosion of X60 pipeline steel

Yan-Bao Guo[1] · Cheng Liu[1] · De-Guo Wang[1] · Shu-Hai Liu[1]

Abstract With rapid economic development in China, demand for energy and transportation is growing. Due to the limitations of factors such as terrain and traffic, a large number of buried oil and gas pipelines are parallel to high-voltage transmission lines and electrified railways over long distances. Alternating current (AC) corrosion of pipelines is very serious in such cases. In this work, laboratory experiments were carried out with an electrochemical method in a simulated soil solution at various AC current densities from 0 to 200 A/m² and AC frequencies from 10 to 200 Hz. Experimental results indicated that with an increase in the AC current density, the corrosion potential of an X60 steel electrode shifted negatively, the anodic current density increased significantly, and the corrosion rate increased. Moreover, with an increase in the AC frequency, the corrosion potential of the X60 electrode shifted positively and the anodic current density decreased, which led to a decrease in the corrosion rate. Furthermore, the morphology of X60 electrodes indicated that uniform corrosion occurred at a low AC current density; while corrosion pits were found on the X60 electrode surface at a high AC current density, and deep corrosion pits seriously damaged the pipelines and might lead to leakage.

Keywords Alternating current interference · X60 pipeline steel · Corrosion · AC current density · AC frequency

✉ Yan-Bao Guo
gyb@cup.edu.cn

[1] College of Mechanical and Transportation Engineering, China University of Petroleum, Beijing 102249, China

Edited by Yan-Hua Sun

1 Introduction

It has been acknowledged that the corrosion of many metallic materials and metal constructions is accelerated in the presence of alternating current (AC) interference (Pagano and Lalvani 1994; Wakelin and Sheldon 2004; Goidanich et al. 2005; Rehim et al. 2008; Eliassen and Hesjevik 2000; Xu and Cheng 2013). In the past two decades, with the rapid development of the electric power, petroleum, and transportation industries, accelerated corrosion of buried pipelines from AC stray current interference has received more and more attention (NACE Standard 1996; Gummow 1999; Movley 2005; Muralidharan et al. 2007; Li and Yang 2011). Furthermore, with the rapid development of the Chinese economy, the demand for energy and transportation is also increasing constantly and rapidly. Then, there is the requirement to build more high-voltage AC transmission lines, AC electrified railways, and a large number of long-distance oil pipelines. Due to the limitation of various factors, more and more pipelines are parallel to or cross high-voltage transmission lines or electrified railways in recent years. The AC stray current caused by high-voltage transmission lines or electrified railways can induce serious corrosion in buried pipelines (Kim et al. 2004; Fu and Cheng 2010; Büchler 2012; Xu et al. 2012; Tang et al. 2013), especially when buried pipelines are parallel to high-voltage transmission lines or electrified railways over long distances. Even with the application of cathodic protection (CP), when the pipeline coating contains microscopic defects or surface damage, corrosion of pipelines could also be very serious from the influence of AC stray current (Gummow 1999; Kajiyama and Nakamura 1999; Ibrahim et al. 2007; Büchler and Schöneich 2009; Ormellese et al. 2011; Fu and Cheng 2012).

Laboratory and field tests indicate that AC stray current can accelerate corrosion of pipelines. For example, Lalvani and Zhang (1995) indicated that AC interference would reduce anodic and cathodic polarizations, lower the passivation of metal, and increase the corrosion rate, which is similar to the influence of a depolarization agent. Song et al. (2002) indicated that the AC voltage has no effect on the AC corrosion rate of carbon steel. On the contrary, the AC current density and AC frequency have significant impacts on the corrosion rate. Pagano and Lalvani (1994) investigated corrosion properties of mild steel in seawater. Their results indicated that the corrosion rate decreased with an increase in AC frequency. Weng and Wang (2011) studied the corrosion properties of bare steel without cathodic protection, and the results showed that the relationship between the AC corrosion rate and the AC interference intensity of carbon steel obeyed a power function in the AC current density from 0 to 250 A/m^2. Yang et al. (2013) investigated the effect of the AC current density on corrosion potential, dynamic parameters (e.g., Tafel slope and corrosion current density), and impedance spectroscopy characteristics of the X70 steel using an electrochemical method. Zhu et al. (2014) investigated the stress corrosion cracking (SCC) behavior of X80 pipeline steel under the influence of various AC current densities and found that an AC current density of 30 A/m^2 was a critical value, above which the SCC susceptibility increased. Although carbon steel corrosion caused by AC stray current has been studied by some scholars, the existing research results cannot effectively forecast and control the AC-induced corrosion of carbon steel.

In this study, in order to investigate the effect of AC current density and frequency on X60 pipeline steel, laboratory experiments are carried out using a self-designed device, which simulates AC-induced corrosion, combining a potentiodynamic polarization technique and scanning electron microscopy (SEM) observation. Firstly, the polarization curves under different AC interference are measured using the potentiodynamic scanning method. Then, the electrochemical corrosion parameters including corrosion potential, Tafel slope, and anodic/cathodic current densities are obtained from the polarization curves. Finally, the AC corrosion mechanism is revealed by the electrochemical corrosion parameters and microscopic surface characterization, in order to provide a new strategy to forecast and evaluate the AC-induced corrosion of buried pipelines.

2 Experimental

2.1 Electrode and solution

X60 pipeline steel was used as the working electrode. Its chemical components (wt%) are C 0.1, Si 0.4, Mn 1.3, P 0.02, S 0.025, and Fe balance. The size of carbon steel samples was 8 mm in diameter and thickness. A coated copper wire was welded on the back and the carbon steel was encapsulated in epoxy resin. The specimen preparation was controlled carefully to avoid any grooves or bubbles occurring at the epoxy–steel interface. The area of the bare working electrode was 0.5 cm^2 after encapsulation. An image of a working electrode is shown in Fig. 1. Before electrochemical tests, the working electrode was prepared by the following processes: (1) The sample surface was gradually polished with #800–1500 waterproof abrasive paper, until the sample surface was mirror-like, without any scratches. (2) The sample was cleaned with acetone and deionized water. (3) A vacuum drying oven was used to dry the sample.

The electrolyte solution was a simulated soil solution. It contained 1200 ppm SO_4^{2-} (1.77 g/L Na_2SO_4) and 200 ppm Cl^- (0.31 g/L $CaCl_2$). The solution was made from analytic-grade reagents and deionized water (≥ 18 MΩ cm, Milipore Mili-Q), of pH 7. Deionized water was used for the preparation of all aqueous solutions and for the rinsing process.

2.2 Electrochemical test

The schematic diagram of equipment is shown in Fig. 2. The equipment consists of two basic units: the direct current (DC) test part and the AC interference part. The DC test was performed using a three-electrode electrochemical workstation (Solartron 1280C). The working electrode was X60 pipeline steel, as shown in Fig. 1. The reference electrode was a saturated calomel electrode (SCE), and a platinum electrode was used as the counter electrode. In the AC interference part, AC interference was directly loaded on the working electrode. A DDS function generator (TFG2006) was used as the AC stray current source. It can control the AC interference current density and frequency while combining with an adjustable resistance box

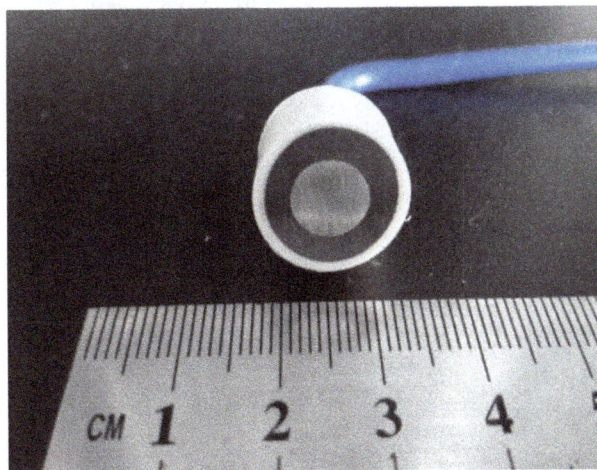

Fig. 1 X60 carbon steel working electrode

Fig. 2 Schematic diagram of the experimental setup for AC corrosion of X60 carbon steel electrode in the simulated soil solution

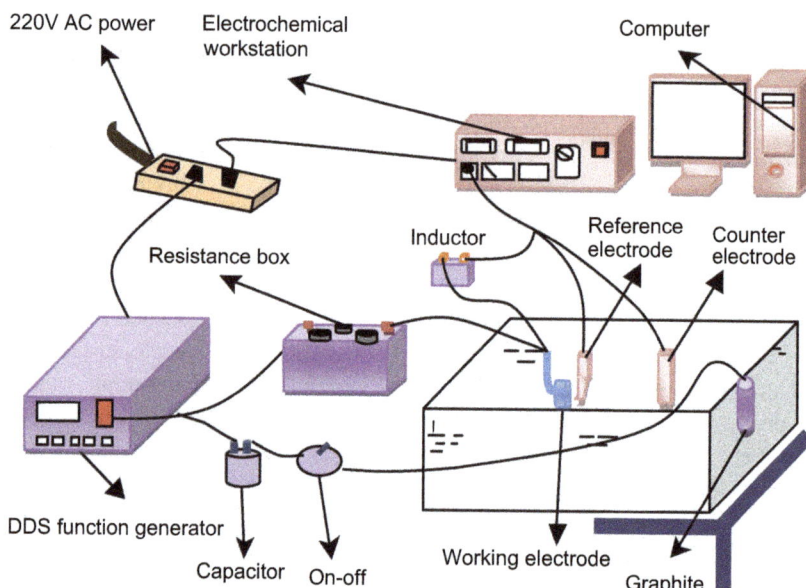

synchronously (0–9999.9 Ω). In our testing, a capacitor (50 V, 500 μF) was used in the AC interference unit to block DC signals. An inductor of 20 H was also used in the DC test unit to eliminate the influence of AC current on the electrochemical workstation. The reference electrode was put in the Luggin capillary, and the tip of the capillary was 2 mm away from the carbon steel electrode surface in order to reduce the ohmic drop influence on the measurement of electric potential.

Before measurement, the carbon steel electrode was immersed in the simulated soil solution for 30 min until the voltage of the open circuit became stable. Then, the potentiodynamic polarization curve was recorded by the electrochemical workstation. The potential scanning rate was 0.5 mV/s. The initial scanning potential was −1300 mV (SCE) and the final was 0 (SCE). The electrochemical parameters were obtained by fitting the polarization curves through software "CView 3.2c" which was supplied by Solartron. All the tests were operated at 22 ± 1 °C.

In order to investigate the corrosion surface, a scanning electron microscope (SEM, FEI Quanta 200F) was used to obtain images of the surface of the carbon steel electrodes before and after corrosion testing.

3 Results and discussion

3.1 Effect of the AC current density on AC corrosion

Figure 3 shows the polarization curves of X60 carbon steel measured at an AC frequency of 50 Hz and various AC

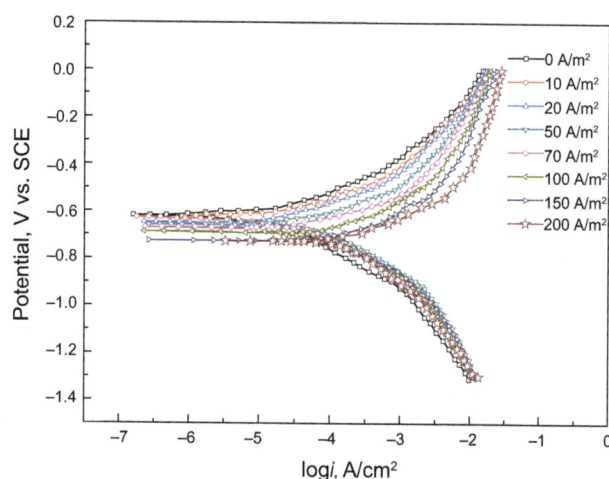

Fig. 3 Polarization curves of X60 carbon steel electrode measured at various AC current densities and a frequency of 50 Hz

current densities. From Fig. 3, it can be seen that the polarization curves were different with an increase in the AC current density from 0 to 200 A/m². It can be also found that the AC current density had a slight effect on the cathodic reaction, but had a significant influence on the anodic reaction. With an increase in the AC interference current density, the anodic current density increased significantly and the cathodic current density also increased, but not as apparently as the anodic one. Furthermore, the increase in the AC current density caused a negative shift of the corrosion potential. The results of fitting the polarization curves from Fig. 3 are summarized in Table 1, which shows the AC current density (i_{AC}), corrosion potential (E_{corr}), corrosion current density (i_{corr}), Tafel slope (β_a, β_c), and the ratio of β_a/β_c.

Table 1 Electrochemical parameters of X60 carbon steel electrode at various AC current densities and a frequency of 50 Hz

i_{AC}, A/m^2	E_{corr}, V versus SCE	i_{corr}, A/cm^2	β_a, mV/decade	β_c, mV/decade	β_a/β_c
0	−0.618	1.426e−5	127.15	168.61	0.75
10	−0.627	1.795e−5	117.10	129.35	0.91
20	−0.647	3.758e−5	141.54	168.82	0.84
50	−0.657	5.902e−5	124.43	157.75	0.79
70	−0.676	6.871e−5	126.85	207.33	0.61
100	−0.690	9.661e−5	145.16	191.21	0.76
150	−0.728	1.390e−4	163.03	167.52	0.97
200	−0.729	1.675e−4	143.27	224.83	0.64

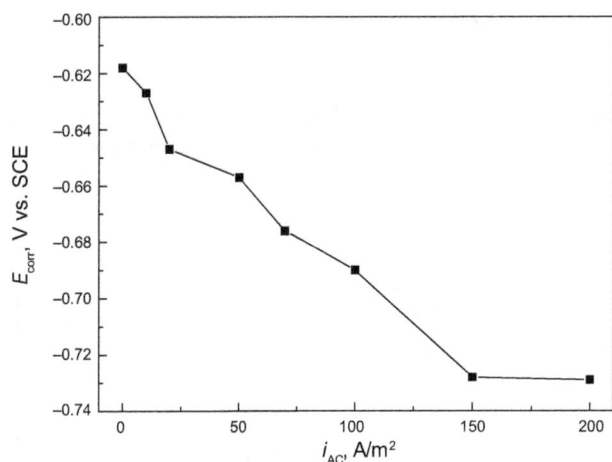

Fig. 4 Changes of corrosion potential of X60 carbon steel electrode at various AC current densities and a frequency of 50 Hz

Figure 4 shows the corrosion potential of X60 carbon steel electrode at various AC current densities and a frequency of 50 Hz. It can be seen that the corrosion potential shifted negatively with an increase in the AC current density. When the AC current acts on the carbon steel electrode, only a very small percentage of current is involved in the electrode reaction (acting as Faradaic current), while the rest of it all takes part in the charging–discharging process (acting as non-Faradaic current) in the electric double layer between the electrode surface and the electrolyte solution. The non-Faradaic current does not participate in electrode reaction, and it almost has no effect on the corrosion of metals. It is known that when the cathodic and anodic reactions of the electrode achieve a balance without AC interference, the electrode potential is its corrosion potential precisely. If AC interference is applied on the carbon steel electrode, the faradic current which is caused by the AC interference can result in polarization of the carbon steel electrode. The polarization of the carbon steel electrode will break the balance of the electrode reaction, meanwhile the corrosion potential of the carbon steel electrode changes. However, the corrosion potential which shifts positively or negatively depends on

the kinetic parameters in the process of electrode reaction (Lalvani and Lin 1994). Lalvani and Lin (1994) provided a mathematical model of the effect of AC interference on the carbon steel corrosion potential:

$$E_{corr,AC} = E_{corr} - \alpha \qquad (1)$$

$$\alpha = \left(\frac{\beta_a}{\frac{\beta_a}{\beta_c} - 1}\right) \ln\left[\frac{\sum_{k=1}^{\infty} \frac{1}{(k!)^2}\left(\frac{E_p}{2\beta_c}\right)^{2k} + 1}{\sum_{k=1}^{\infty} \frac{1}{(k!)^2}\left(\frac{E_p}{2\beta_a}\right)^{2k} + 1}\right], \qquad (2)$$

where $E_{corr,AC}$ is the corrosion potential with AC interference; E_{corr} is the corrosion potential without AC interference; β_a and β_c are the Tafel slopes of the anodic and cathodic reactions, respectively; E_p is the peak voltage of the AC interference; and k is the constant of integration.

From the mathematical model above, it can be seen that the corrosion potential with AC interference ($E_{corr,AC}$) is a function of the Tafel slope ratio ($r = \beta_a/\beta_c$) and peak voltage (E_p). In our experiments, the corrosion surface area of the X60 carbon steel electrode was constant, so the peak voltage E_p increased as the AC interference enhanced. It can be found from Table 1 that the Tafel slope ratios ($r = \beta_a/\beta_c$) are all less than 1 under different AC interference current densities. Then, we know from a combination of Eqs. (1) with (2), when $r = \beta_a/\beta_c < 1$, the $E_{corr,AC}$ is negatively increased with an increase in E_p.

Figure 5 shows the corrosion rate of the X60 carbon steel electrode at various AC current densities and a frequency of 50 Hz. With an increase in the AC current density, the corrosion rate of the X60 carbon steel electrode increased. Corrosion of metals is mainly determined by the anodic dissolution in the electrode reaction, that is to say, it can be characterized by the metal anodic current density. The cathodic and anodic currents were in the same magnitude and opposite direction when the corrosion reaction of the carbon steel electrode reached equilibrium. Furthermore, the electrode corrosion potential was stable. However, the AC interference would affect the electrode reaction. In the positive half-cycle of AC interference, the anodic current density increased with electrode anodic

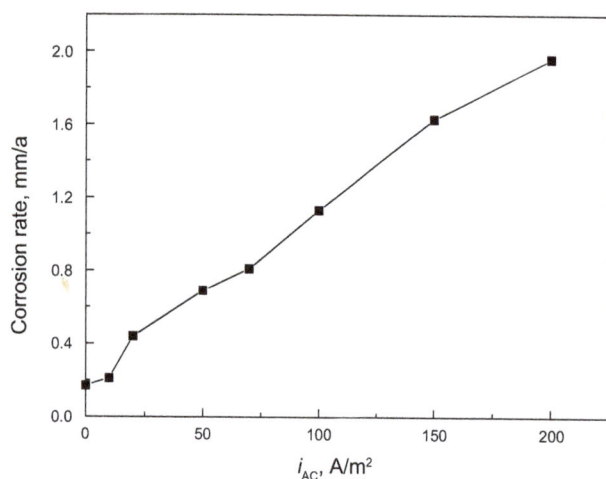

Fig. 5 Changes of corrosion rate of the X60 carbon steel electrode at various AC current densities and a frequency of 50 Hz

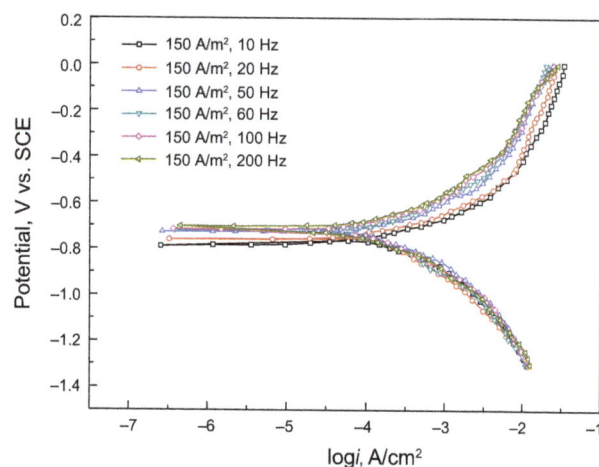

Fig. 6 Polarization curves of the X60 carbon steel electrode measured at various frequencies and an AC current density of 50 A/m^2

polarization. However, in the negative half-cycle of AC interference, the anodic current density decreased with electrode cathodic polarization, but not to the same degree as the increase during the positive half-cycle because of the non-linear polarization curve of the carbon steel anodic reaction. Therefore, the complete AC cycle resulted in a net positive increase in the anodic current density (Goidanich et al. 2010a, b). In fact, from the view of thermodynamics, metal dissolution was the most favored process during the positive half-cycle, while oxygen reduction or hydrogen evolution was favored during the negative half-cycle (Goidanich et al. 2010a, b). It also can be seen from the polarization curves shown in Fig. 3 that the anodic current density increased significantly with an increase in the AC current density. In other words, the enhanced AC current density increased the asymmetry of changes of the anodic current density during the positive half-cycle and negative half-cycle. Thus, the corrosion of carbon steel was accelerated.

3.2 Effect of the AC frequency on AC corrosion

In order to investigate the effect of AC frequency on the X60 carbon steel, polarization curves of the X60 carbon steel electrode were measured at frequencies from 10 to 200 Hz and AC current densities of 50 and 150 A/m^2, respectively.

Figures 6 and 7 show polarization curves of the X60 carbon steel electrode measured at various frequencies and different AC current densities (50 and 150 A/m^2), respectively. The fitting results are presented correspondingly in Tables 2 and 3. It can be found that the anodic reaction was obviously affected by the AC frequency. With an increase

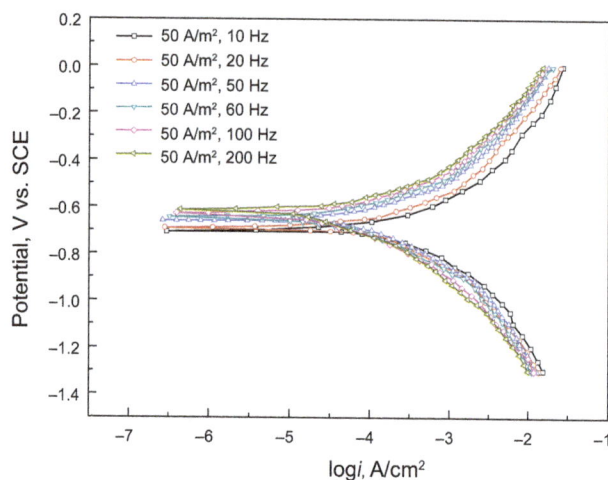

Fig. 7 Polarization curves of the X60 carbon steel electrode measured at various frequencies and an AC current density of 150 A/m^2

in the AC frequency, the anodic current density decreased. Furthermore, the corrosion potential shifted positively with the increase in the AC frequency.

The corrosion process can be expressed as an equivalent circuit which is called a Randle circuit with AC interference. There, the polarization resistance is equivalent to the electrode reaction, the electric double layer is equivalent to an interfacial capacitor, and the polarization resistance is in parallel with the interfacial capacitor. When the equivalent circuit is affected by the same AC current density of different frequencies, Yunovich and Thompson (2004) indicated that the impedance of the electric double-layer capacitor decreased with the increase in the AC frequency. This process may cause the current through the polarization resistance to decrease, which would induce a decrease in the Faradaic current. This also suggested that the current

Table 2 Electrochemical parameters of the X60 carbon steel electrode at various frequencies and an AC current density of 50 A/m^2

f, Hz	E_{corr}, V versus SCE	i_{corr}, A/cm^2	β_a, mV/decade	β_c, mV/decade	β_a/β_c
10	−0.704	1.026e−4	173.91	141.23	1.23
20	−0.687	7.396e−5	129.26	165.43	0.78
50	−0.657	5.902e−5	124.43	157.75	0.79
60	−0.642	5.035e−5	121.76	178.14	0.68
100	−0.627	3.864e−5	129.63	185.79	0.70
200	−0.617	3.548e−5	113.48	186.72	0.61

Table 3 Electrochemical parameters of X60 carbon steel electrode at various current frequencies and an AC current density of 150 A/m^2

f, Hz	E_{corr}, V versus SCE	i_{corr}, A/cm^2	β_a, mV/decade	β_c, mV/decade	β_a/β_c
10	−0.772	1.972e−4	166.72	178.80	0.93
20	−0.757	1.596e−4	157.25	209.12	0.75
50	−0.728	1.390e−4	163.03	167.52	0.97
60	−0.722	1.321e−4	176.81	205.12	0.86
100	−0.718	1.170e−4	164.69	194.18	0.85
200	−0.707	1.081e−4	167.34	208.87	0.80

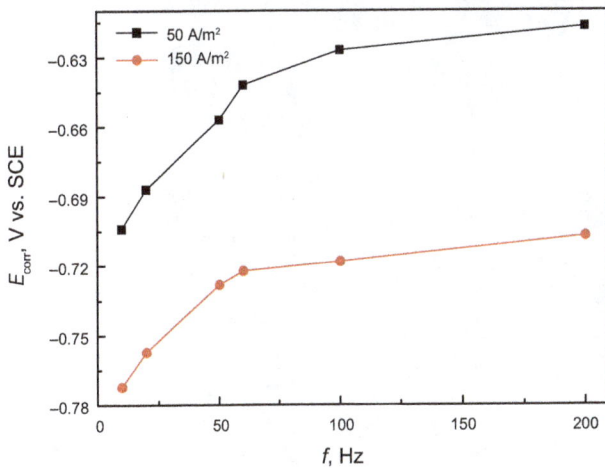

Fig. 8 Changes of the corrosion potential of X60 carbon steel electrode at various AC frequencies

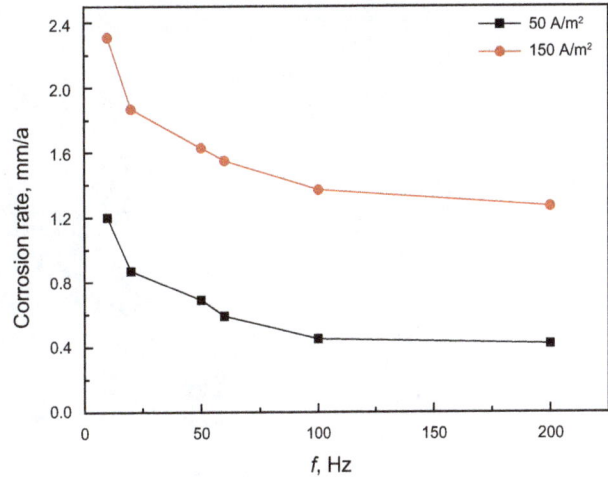

Fig. 9 Changes of the corrosion rate of the X60 carbon steel electrode at various AC frequencies

participating in the electrode reaction decreased. The decrease in the Faradaic current would cause the peak voltage (E_p) to decrease from Eq. (2). From Tables 2 and 3, it can be seen that the Tafel slope ratios ($r = \beta_a/\beta_c$) are lower than 1. Then, a combination of Eqs. (1) with (2) gives that, when $r < 1$, the $E_{corr,AC}$ increased positively with a decrease in E_p, which also can be found in Fig. 8.

With an increase in the AC frequency, the Faradaic current decreases and then the anodic current density decreases. Thus, the asymmetry of changes of the anodic current density decreases during the positive half-cycle and negative half-cycle of AC interference. Based on the above analysis, the corrosion rate of the X60 carbon steel decreased with an increase in the AC frequency, as shown in Fig. 9.

3.3 Morphology of the corrosion surface

In this research, a scanning electron microscope (SEM, FEI Quanta 200F) was used to observe the corrosion surfaces. After 2-h test, the corrosion product formed on the electrode surface was removed carefully by both mechanical and chemical methods. The mechanical methods, including light scraping and scrubbing, were used for removal of tightly adherent corrosion product. In the chemical procedure, a descaling solution containing 100 mL HCl (special gravity 1.19), 0.7 g hexamethylenetetramine, and 100 mL distilled water was used. The SEM images of the X60 electrode surface are shown in Fig. 10. In the absence of AC, the carbon steel surface was relatively smooth except

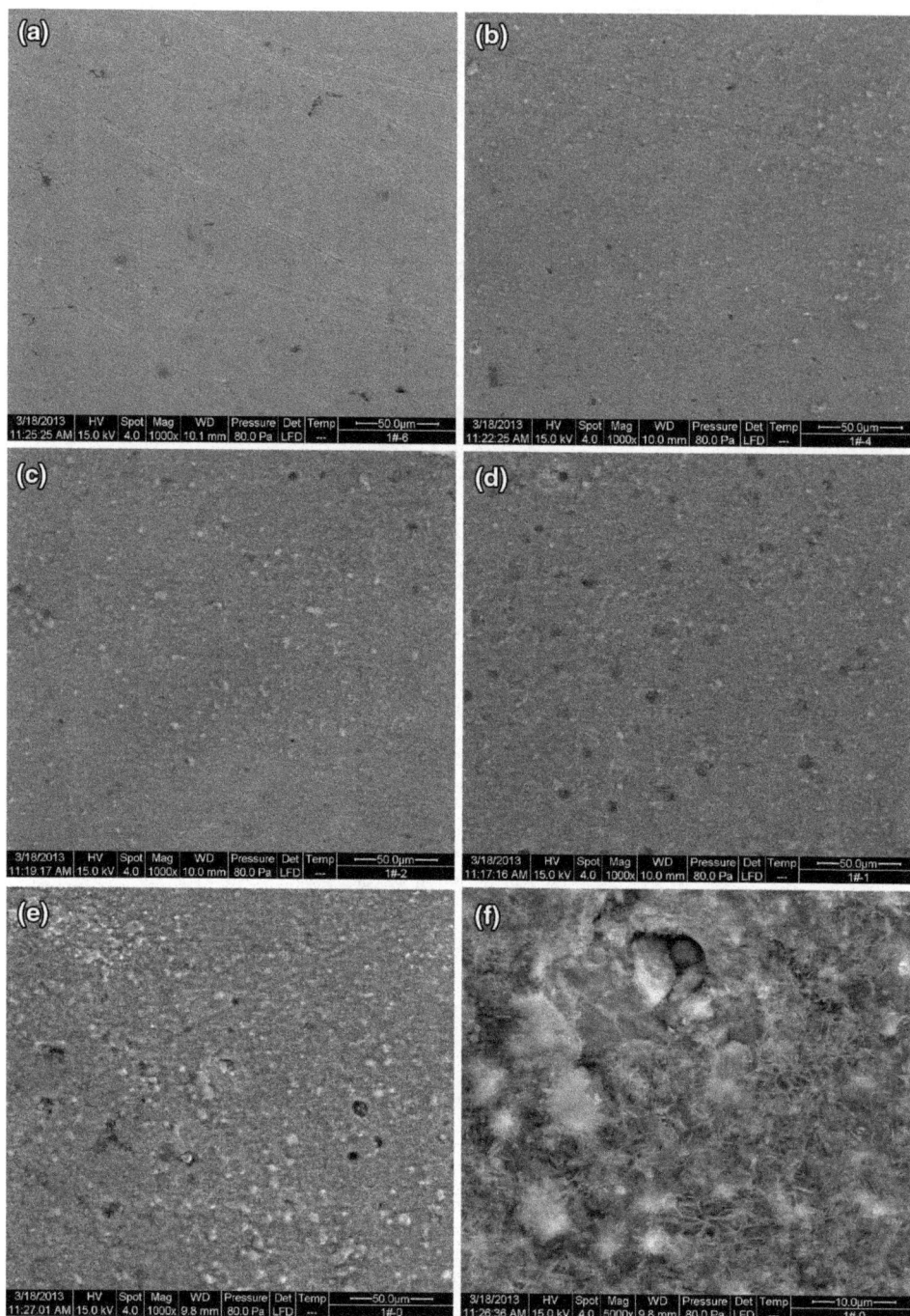

Fig. 10 SEM images of the X60 carbon steel electrode at various AC current densities and a frequency of 50 Hz after 2 h of test. **a** 0 A/m^2; **b** 20 A/m^2; **c** 70 A/m^2; **d** 150 A/m^2; and **e, f** 200 A/m^2

for some slight corrosion marks, as seen in Fig. 10a. When a 20 A/m^2 AC current density was applied, uniform corrosion occurred on the carbon steel surface and the corrosion became serious, as shown in Fig. 10b. With further increase in the AC current density to 70 and 150 A/m^2, the corrosion was promoted and pitting corrosion appeared on the carbon steel surfaces, as seen in Fig. 10c, d. When the AC current density increased to 200 A/m^2, pitting

corrosion became more severe and pitting holes got deeper than before, as shown in Fig. 10e, f. The corrosion form changed from uniform corrosion to pitting corrosion. This suggests that, at a low AC current density, a uniform corrosion occurs, while at a high AC current density, pitting corrosion occurs on the X60 electrode surface. That may be caused by the AC oscillation between the carbon steel surface and electrolyte solution, which leads to the pitting corrosion of carbon steel electrode (Nielsen et al. 2004; Nielsen and Galsgaard 2005).

4 Conclusions

Electrochemical tests on X60 carbon steel in simulated soil solution with AC interference led to the following conclusions:

(1) The presence of AC interference accelerated the corrosion of X60 carbon steel. With an increase in the AC current density, the corrosion potential shifted negatively and the net positive increase in the anodic current density increased during a complete AC cycle, which caused an increase in the corrosion rate of X60 carbon steel.

(2) With an increase in the AC frequency, the impedance of the electric double-layer capacitor decreased, which caused a decrease in the current through the polarization resistance. Thus, the decrease in the Faradaic current caused the anodic current density to decrease, which led to a decrease in the corrosion rate. The increase in the AC frequency also caused a positive shift of the corrosion potential of X60 carbon steel.

(3) Morphology investigations indicated that the corrosion of X60 carbon steel became more and more serious with an increase in the AC interference intensity. At a low AC current density, a uniform corrosion occurred, while at a high AC current density, pitting corrosion occurred on the X60 electrode surface. In general, the AC interference enhanced localized corrosion.

Acknowledgments This research is sponsored by the Projects in the National Science & Technology Pillar Program during the Twelfth Five-year Plan Period (Grant No. 2011BAK06B01).

References

Büchler M, Schöneich HG. Investigation of alternating current corrosion of cathodically protected pipelines: development of a detection method, mitigation measures, and a model for the mechanism. Corrosion. 2009;65(9):578–86.

Büchler M. Alternating current corrosion of cathodically protected pipelines: discussion of the involved processes and their consequences on the critical interference values. Mater Corros. 2012;63(12):1181–7.

Eliassen SL, Hesjevik SM. Corrosion management of buried pipelines under difficult operational and environmental conditions. In: Corrosion 2000, NACE, Paper No. 00724, Orlando. 2000.

Fu AQ, Cheng YF. Effects of alternating current on corrosion of a coated pipeline steel in a chloride–containing carbonate/bicarbonate solution. Corros Sci. 2010;52(2):612–9.

Fu AQ, Cheng YF. Effect of alternating current on corrosion and effectiveness of cathodic protection of pipelines. Can Metall Q. 2012;51(1):81–90.

Goidanich S, Lazzari L, Ormellese M, et al. Influence of AC on corrosion kinetics for carbon steel, zinc and copper. In: Corrosion 2005. NACE, Paper No. 05189, Houston. 2005.

Goidanich S, Lazzari L, Ormellese M. AC corrosion–Part 1: effects on overpotentials of anodic and cathodic processes. Corros Sci. 2010a;52(2):491–7.

Goidanich S, Lazzari L, Ormellese M. AC corrosion. Part 2: parameters influencing corrosion rate. Corros Sci. 2010b; 52(3):916–22.

Gummow RA. Cathodic protection considerations for pipelines with AC mitigation facilities. Pipeline Research Council International. Final Report of Contract PR-262-9809. 1999.

Ibrahim I, Takenouti H, Tribollet B, et al. Harmonic analysis study of the AC corrosion of buried pipelines under cathodic protection. In: Corrosion 2007. NACE, Paper No. 07042, Nashville. 2007.

Kajiyama F, Nakamura Y. Effect of induced alternating current voltage on cathodically protected pipelines paralleling electric power transmission lines. Corrosion. 1999;55(2):200–5.

Kim DK, Ha TH, Ha YC, et al. Alternating current induced corrosion. Corros Eng Sci Technol. 2004;39(2):117–23.

Lalvani SB, Lin XA. A theoretical approach for predicting AC–induced corrosion. Corros Sci. 1994;36(6):1039–46.

Lalvani SB, Zhang G. The corrosion of carbon steel in a chloride environment due to periodic voltage modulation: Part I. Corros Sci. 1995;37(10):1567–82.

Li ZL, Yang Y. Mechanism, influence factors and risk evaluation of metal alternating current corrosion. CIESC J. 2011;62(7):1790–9 (in Chinese).

Movley CM. Pipeline corrosion from induced A.C. In: Corrosion 2005. NACE, Paper No. 05132, Houston. 2005.

Muralidharan S, Kim DK, Ha TH, et al. Influence of alternating, direct and superimposed alternating and direct current on the corrosion of mild steel in marine environments. Desalination. 2007;216(1):103–15.

NACE Standard RP0169–96. Control of external corrosion on underground or submerged metallic piping systems. Houston: NACE; 1996.

Nielsen LV, Nielsen KV, Baumgarten B, et al. AC induced corrosion in pipelines: detection, characterization and mitigation. In: Corrosion 2004. NACE, Paper No. 04211, New Orleans. 2004.

Nielsen LV, Galsgaard F. Sensor technology for on-line monitoring of AC induced corrosion along pipelines. In: Corrosion 2005. NACE, Paper No. 05375, Houston. 2005.

Ormellese M, Goidanich S, Lazzari L. Effect of AC interference on cathodic protection monitoring. Corros Eng, Sci Technol. 2011;46(5):618–23.

Pagano MA, Lalvani SB. Corrosion of mild steel subjected to alternating voltages in seawater. Corros Sci. 1994;36(1):127–40.

Rehim SSA, Hazzazi OA, Amin MA, et al. On the corrosion inhibition of low carbon steel in concentrated sulphuric acid solutions. Part I: chemical and electrochemical (AC and DC) studies. Corros Sci. 2008;50(8):2258–71.

Song HS, Kho YT, Kim YG, et al. Competition of AC and DC current in AC corrosion under cathodic protection. In: Corrosion 2002. NACE, Paper No. 02117, Denver. 2002.

Tang DZ, Du YX, Lu MX, et al. Effect of AC current on corrosion behavior of cathodically protected Q235 steel. Mater Corros. 2013. doi:10.1002/maco.201307234 (published online).

Wakelin RG, Sheldon C. Investigation and mitigation of AC corrosion on a 300 mm diameter natural gas pipeline. In: Corrosion 2004, NACE, Paper No. 04205, New Orleans. 2004.

Weng YJ, Wang N. Carbon steel corrosion induced by alternating current. J Chin Soc Corros Prot. 2011;31(4):270–4 (in Chinese).

Xu LY, Cheng YF. Effect of alternating current on cathodic protection on pipelines. Corros Sci. 2013;66:263–8.

Xu LY, Su X, Yin ZX, et al. Development of a real-time AC/DC data acquisition technique for studies of AC corrosion of pipelines. Corros Sci. 2012;61:215–23.

Yang Y, Li ZL, Wen C. Effects of alternating current on X70 steel morphology and electrochemical behavior. Acta Metall Sin. 2013;49(1):43–50 (in Chinese).

Yunovich M, Thompson NG. AC corrosion: mechanism and proposed model. In: 2004 international pipeline conference. Calgary: ASME; 2004. pp 183–95.

Zhu M, Du C, Li X, et al. Effect of AC current density on stress corrosion cracking behavior of X80 pipeline steel in high pH carbonate/bicarbonate solution. Electrochim Acta. 2014; 117:351–9.

12

Modeling of flow of oil-in-water emulsions through porous media

Ajay Mandal[1] · Achinta Bera[2]

Abstract Formation and flow of emulsions in porous media are common in all enhanced oil recovery techniques. In most cases, oil-in-water (O/W) emulsions are formed in porous media due to oil–water interaction. Even now, detailed flow mechanisms of emulsions through porous media are not well understood. In this study, variation of rate of flow of O/W emulsions with pressure drop was studied experimentally, and rheological parameters were calculated. The pressure drop increases with an increase in oil concentration in the O/W emulsion due to high viscosity. The effective viscosity of the emulsion was calculated from the derived model and expressed as a function of shear rate while flowing through porous media. Flow of O/W emulsions of different concentrations was evaluated in sand packs of different sand sizes. Emulsions were characterized by analyzing their stability, rheological properties, and temperature effects on rheological properties.

Keywords Emulsion · Porous media · Rheology · Modeling · Pressure drop

✉ Ajay Mandal
mandal_ajay@hotmail.com

[1] Department of Petroleum Engineering, Indian School of Mines, Dhanbad 826004, India

[2] Department of Civil & Environmental Engineering, School of Mining and Petroleum Engineering, University of Alberta, Edmonton, AB T6G 2W2, Canada

Edited by Yan-Hua Sun

1 Introduction

Emulsion flow through porous media is an important phenomenon in oil production operations and is also a topic of special interest in many applications of science and engineering, especially petroleum industries (Soo and Radke 1984; Soma and Papadopoulos 1995; Vidrine et al. 2000; Arhuoma et al. 2009; Cobos et al. 2009). It has been suggested that oil migrates through reservoir sands in the form of a fine, dispersed emulsion of oil in water, and that oil accumulations occur where the stream enters finer-grained rock such as silt or shale. Flow behavior of emulsions in both pipelines and reservoirs can be properly described based on emulsion properties and physical laws controlling their flow through porous media. Emulsions are generally liquid–liquid dispersions, in which the dispersed phase interferes with the flow of the continuous phase through the pore space, as they can partially block the already swept, more permeable paths. Thus, the flow of injected water diverts to unswept regions, leading to more efficient reservoir sweep and higher recovery factor (Thomas and Ali 1989; Seright and Liang 1995; Babadagli 2005). These researchers reported that emulsions would have good displacement properties similar to those of a low-viscosity gel. In petroleum production, heavy crude oils are often produced from natural oil reservoirs in the form of water-in-oil (W/O) or oil-in-water (O/W) emulsion with water (Steinhauff 1962). In enhanced oil recovery (EOR) methods a variety of emulsions are observed (Mandal et al. 2010a). During waterfloods for secondary recovery of oil from reservoirs, there is rapid channeling of water from injection to producing wells through the more permeable portions of reservoirs, resulting in low oil recovery. As water flows through tortuous paths in reservoirs in presence of oil, O/W emulsions are often formed. When the O/W emulsion flows through a heterogeneous reservoir, a great amount of emulsion enters the more permeable zones, so water begins to flow into less permeable zones, resulting in higher sweep efficiency. In this study the flow behavior of O/W emulsions through porous media has been discussed.

Different typical configurations of an emulsion may be categorized as O/W, W/O or a more complex dispersion. In complex dispersion phase a dispersed phase is embedded within each droplet already dispersed in a continuous phase, such as the case of oil-in-water-in-oil emulsions. In porous media, it is of great interest for petroleum engineers to know how the emulsions flow in porous media and the flow rate changes with pressure drop during flowing. Emulsions (O/W and W/O emulsions) are formed in situ under adequate conditions in porous media. A third component often present in an emulsion is an emulsifier or surfactant, which can reduce interfacial tension between liquids and can stabilize emulsions. In oil reservoirs, the added chemicals, or crude oil components such as asphaltenes, waxes, resins, and naphthenic acids act as emulsifying agents. These emulsifiers or stabilizers suppress the emulsion breakdown process (Fingas and Fieldhouse 2004). Many researchers carried out experimental investigation on injection of W/O or O/W emulsion as a selective plugging agent to improve oil recovery in water flooding (McAuliffe 1973a, b; Thomas and Ali 1989). During alkali flooding, alkali can penetrate into heavy oil in porous media by the formation of W/O emulsions in situ. In this connection, rheological properties of emulsions play an important role in efficient oil recovery. Due to high viscosity of W/O emulsions, the resistance to water flow in high water saturation zones can be increased significantly to improve sweep efficiency and thus oil recovery. Most of the cases the power-law model is the simplest representation of the viscosity of non-Newtonian fluids.

The objective of this study is to characterize rheological properties of O/W emulsions and their flow behavior through porous media. For the experiments, emulsions were first prepared by so-called agent-in-water technique (Bennett 1967; Becher 2001; Rastogi 2003, Mandal et al. 2010b). After preparation of emulsions, flow tests were performed in sand packs filled with sand of different sizes at different pressures and the corresponding flow rates were calculated and finally experimental results have been analyzed by a proposed model.

2 Mechanism of oil/water emulsion flow in porous media

Oil and water form an emulsion under favorable conditions in the presence of extraneous materials (Gewers 1968; Rupesh et al. 2008). Emulsions are suspensions of droplets (greater than 0.1 μm) of one immiscible fluid dispersed in another fluid. Their kinetic stability is a consequence of small droplet size and the presence of an interfacial film around oil droplets (Murray et al. 2009). Emulsions may be either oil-in-water or water-in-oil and they are transported in reservoir rocks having permeabilities from very low to several Darcys. Many flow models have been reported in the literature to describe flow mechanisms of emulsions through porous media. McAuliffe (1973a, b) studied mechanisms of oil droplets through porous media, and he described the mechanism of flow of a dispersed oil droplet entering a pore throat smaller than the oil droplet. Al-Faris et al. (1994) reported that during flow of the O/W emulsion the permeability of porous media decreases monotonically with time. Alvarado and Marsden (1979) developed a correlation, which uses both capillary and core flow data for describing flow of non-Newtonian macroemulsions through porous media. Jin and Wojtanowicz (2014) described a model considering mass transfer of the oil phase from the produced water to the rock due to capture effects by dispersion, advection, and adsorption inside the rock. Khambharatana et al. (1998) studied physical mechanisms of stable emulsion flow in Berea sandstones and Ottawa sand packs for systems of comparable droplet and pore sizes. They showed that the change in emulsion rheology in a porous medium has an overall trend similar to that in a viscometer for the shear rates of interest.

The better understanding of flow mechanisms of O/W emulsions through porous rock is of utmost importance during secondary and EOR. Although significant works on mechanism of the O/W emulsion flow through porous media have been reported, the detailed mechanisms are still not well understood. In this study, a mathematical model is developed from basic fluid flow equations, and the model is used to describe the flow behavior of the O/W emulsion through sand packs.

3 Experimental

3.1 Materials used

Gear oil (EPX 90) available in the market with sp. gravity of 0.905 and kinematic viscosity of 197 cSt at 40 °C and 17.3 cSt at 100 °C was used for preparation of emulsion in distilled water. Sodium chloride (NaCl) with 98 % purity, procured from Qualigens Fine Chemicals, India, was used for preparation of a brine solution. Sand used in building was sieved into different particle sizes, 12–30, 30–60, and 60–70 mesh and then was used to prepare sand packs.

Table 1 Properties of three types of porous media used for emulsion flow tests

Sand pack	Sand size	Length L, cm	Diameter D, cm	Area A, cm^2	Permeability k, Darcy	Porosity ϕ, %
Sand pack A	IS 12–30	40	3	7.069	0.899	31.82
Sand pack B	IS 30–60	40	3	7.069	1.079	35.36
Sand pack C	IS 60–70	40	3	7.069	1.160	37.14

3.2 Emulsion preparation

Each emulsion was prepared by mixing gear oil with water using a mixer with a standard three-blade propeller. The quality of lubricating oil is often improved by adding detergent, dispersant, etc. (Atkins et al. 1947; Miller 1956), which enables formation of a milky emulsion when properly mixed with water. For O/W emulsions, distilled water was used as the continuous phase and oil was used as the dispersed phase. The different concentrations of emulsions (5 %, 10 %, and 15 %) were prepared by agitating at 3000 rpm for 6 h, which were found to be the optimum conditions for making emulsions from the brine solution and the gear oil. The emulsion was left to stand in a separation flask for 6 h and the bottom part was separated out as emulsion for the present experiments. The separation flask can easily lead to the separation of two different density fluids based on their density.

3.3 Sand pack preparation and property determination

Sand packs were prepared at the ambient temperature of 27 °C, and fresh sand was used for each test to ensure similar wettability. Three major steps involved in preparing each porous medium included seizing the sand, packing the core holder, and determining properties of the porous medium. First, the sand particles were classified into three categories, namely, A, B and C, each having different mesh: IS 12–30, IS 30–60 and IS 60–70, respectively. The sieving process was undertaken using a sieve shaker with different sieves varying from IS 12 to 70 mesh. Second, the core holder was placed in the bench vice in the vertical position and filled with 1.0 wt% NaCl brine, and then the desired sand was gradually added into the core holder. For sand packs of IS 60–70 mesh sand (sand pack C), 90 mL of brine was used for saturation; for sand packs of IS 30–60 mesh sand (sand pack B), 100 mL brine used; and for sand packs of IS 12–30 mesh sand (sand pack A), 105 mL of brine used for complete saturation. In general a perfect packing was needed for preparation of porous media for experimental purposes. Finally, the porosity was measured by two different methods, namely, the weight and volumetric methods, and the permeability measurement

was conducted using Darcy's law. Table 1 summarizes properties of the three types of porous media used for emulsion flow tests.

3.4 Determination of flow rate and emulsion viscosity

The experimental setup (Fig. 1) for flooding tests in sand packs consisted of four components: a sand pack holder, a displacement pump, cylinders for holding brine and emulsion, and fraction collectors. The detailed procedures for determining the flow rate of emulsions are described as follows:

An emulsion flow test was initiated after the properties of the porous medium and emulsions were characterized. First of all the syringe pump is filled with paraffin to maintain the pressure. Once the sand pack was fully saturated with brine, it was placed horizontally. Then, the brine was injected into the sand pack at a pressure of 10 psi. At this point, the flow rate was calculated by measuring the time required to collect 10 mL of brine. Subsequently, the injection pressure of brine was increased to 20 psi and then to 60 psi and the corresponding flow rates were measured. Submitting values of flow rates at different differential pressure (ΔP) into the Darcy Equation, the permeability of sand packs was calculated and verified that the permeability is constant at different ΔP. Accordingly, the same procedure was repeated for emulsions (O/W) of different concentrations (5 %, 10 %, and 15 %) and corresponding flow rates were measured at different ΔP.

Rheological properties of emulsions were calculated theoretically from the core flooding data of emulsion flow tests.

The flow of emulsions through porous media has been studied in laboratory by injecting emulsions through sand packs. Let us consider the steady laminar flow of emulsions through porous sand packs with an uniform size of sand and the cross sectional area of the core holder is A. The sand particles are assumed to be largely uniform and spherical. Suppose the average projected area of sand particles is a and the porosity is ϕ. Therefore, the number of sand particles present in the total projected area of flow is given by

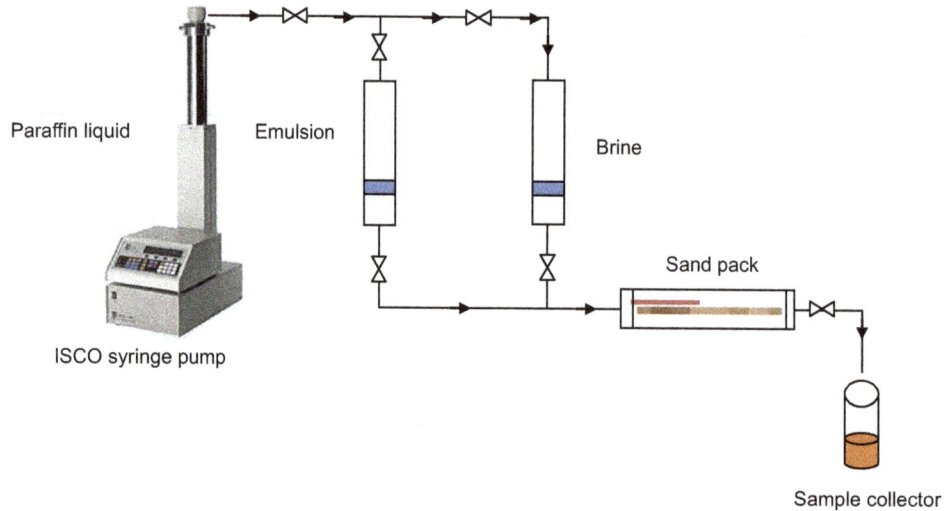

Fig. 1 Schematic diagram of emulsion flow through a sand pack

$$n_p = \frac{A(1-\phi)}{\pi r_a^2}. \tag{1}$$

Therefore, the wetted perimeter is equal to the number of sand particles, i.e.,

$$2\pi r_a = \frac{2A(1-\phi)}{r_a}, \tag{2}$$

where r_a is the average radius of sand particles, cm. Then the hydraulic radius, r_H can be expressed as follows:

$$r_H = \frac{\phi r_a}{2(1-\phi)}, \tag{3}$$

where r_H is the hydraulic radius of sand particles, cm.

Oil/water emulsion is a time-independent non-Newtonian pseudo-plastic fluid and its rheology is described by the Ostwald–Dewaele model or the power-law model as follows:

$$\tau = K\left(\frac{du}{dy}\right)^n, \tag{4}$$

where τ is the shear stress, dynes/cm^2; K and n are constants for a particular fluid and the value of n is less than one. The constant K is known as the consistency index of the fluid, Pa s^n; the higher the value of K, the more viscous the fluid. The constant n is called the flow index and gives a measure of the degree of departure from Newtonian behavior. The detailed rheological properties have been evaluated for three different emulsions with a standard rheometer and it has been found that the viscosity of emulsions decreases with an increase in the shear rate, showing non-Newtonian behavior. In this study, rheology is described by the Ostwald–Dewaele model or the power-law model as reported in a previous study (Al-Fariss et al. 1994).

In order to develop an expression for the effective viscosity, μ_{eff} for pseudo-plastic flow through a porous media let us consider the general equation (Mandal 2010),

$$-\frac{dv}{dr} = f(\tau), \tag{5}$$

where v is the velocity of fluid. For fluid flowing through a porous media, a force balance on a cylindrical fluid element of radius r (imaginary) gives,

$$\tau = \frac{r\Delta P}{2L}, \tag{6}$$

and that at the wall of the core holder and also at the surface of the particle where velocity is zero, τ_w can be written as

$$\tau_w = r_H \frac{\Delta P}{2L}, \tag{7}$$

where L is the length of the core holder, cm; τ_w is the shear stress at the wall, dynes/cm^2.
Therefore,

$$\tau = \tau_w r / r_H. \tag{8}$$

The volumetric flow rate Q through the porous media is given by

$$Q = \int_0^{r_H} 2\pi r v dr, \tag{9}$$

where v is the average velocity, cm/s; Q is the volumetric flow rate of the emulsion through porous media, cm^3/s.

On integration by parts, and applying the condition, at $r = r_H$, $v = 0$, and substitution of Eq. (9) leads to,

$$\frac{Q}{\pi r_{\mathrm{H}}^3} = \frac{1}{\tau_{\mathrm{w}}^3} \int_0^{\tau_{\mathrm{w}}} \tau^2 f(\tau) \mathrm{d}\tau. \tag{10}$$

Since for pseudo-plastic fluids,

$$f(\tau) = (\tau/K)^{\frac{1}{n}}, \tag{11}$$

therefore,

$$\frac{Q}{\pi r_{\mathrm{H}}^3} = \frac{v}{r_{\mathrm{H}}} = \frac{1}{\tau_{\mathrm{w}}^3} \int_0^{\tau_{\mathrm{w}}} \tau^2 \left(\frac{\tau}{K}\right)^{\frac{1}{n}} \mathrm{d}\tau, \tag{12}$$

which on integration gives,

$$\frac{v}{r_{\mathrm{H}}} = \frac{n}{(3n+1)} \left(\frac{\tau_{\mathrm{w}}}{K}\right)^{\frac{1}{n}}. \tag{13}$$

Rearranging one gets,

$$\tau_{\mathrm{w}} = K(2v/D_{\mathrm{H}})^n \left(\frac{3n+1}{n}\right)^n, \tag{14}$$

or,

$$\tau_{\mathrm{w}} = K(8v/D_{\mathrm{H}})^n \left(\frac{3n+1}{4n}\right)^n, \tag{15}$$

where D_{H} is the hydraulic diameter of the sand particles, cm.

From the definition of μ_{eff},

$$\mu_{\mathrm{eff}} = \frac{\tau_{\mathrm{w}}}{8V/D} = K(8v/D_{\mathrm{H}})^{n-1} \left(\frac{3n+1}{4n}\right)^n, \tag{16}$$

$$\mu_{\mathrm{eff}} = 8^{n-1} v^{n-1} D_{\mathrm{H}}^{1-n} K', \tag{17}$$

with

$$K' = K\left(\frac{3n+1}{4n}\right)^n.$$

Therefore, the value of μ_{eff} can be evaluated if K and n are known. It is clear from Eq. (15) that if a logarithmic plot is made between τ_{w} and $8v/D_{\mathrm{H}}$, a linear relationship will obtained and the slope of the line should give the value of n and the intercept $K\left(\frac{3n+1}{4n}\right)^n$ i.e., K'.

4 Results and discussion

4.1 Characterization of emulsions

Emulsions are characterized by analyzing their stability, rheological properties, temperature effects on rheological properties, etc. Figure 2 shows the variation of viscosity with shear rate at different temperatures. Viscosity of emulsions decreases with an increase in the shear rate at lower shear rates, showing non-Newtonian behavior of emulsions. On the other hand, at higher shear rates emulsions show Newtonian behavior. It is also observed from figure that with an increase in temperature the viscosity of the O/W emulsion decreases. This is due to the fact that with an increase in temperature the average speed of the molecules in a liquid increases and the amount of time they spend "in contact" with their nearest neighbors decreases. Thus, as the temperature increases, the average intermolecular forces decrease and hence the viscosity decreases. The variation of viscosity of emulsions of different concentrations is presented in Fig. 3. It may be found that at low shear rates the viscosity of emulsions with higher oil concentration is higher though the variation is only marginal.

Microscopic images of O/W emulsions were taken with a polarizing microscope. A typical microscopic image of

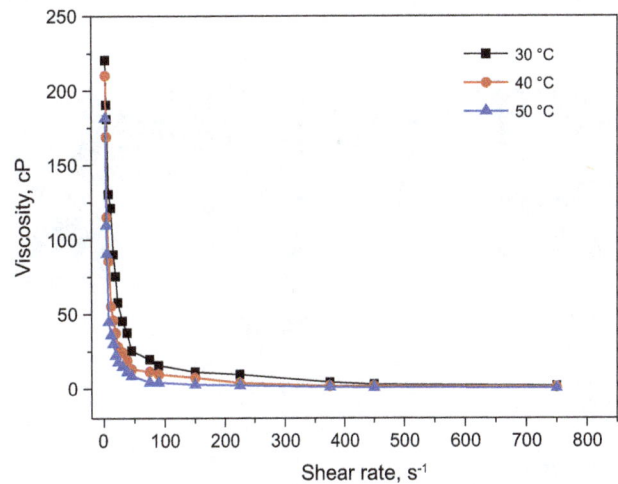

Fig. 2 Viscosity (cP) versus shear rate (s) for 15 % O/W emulsion at different temperatures

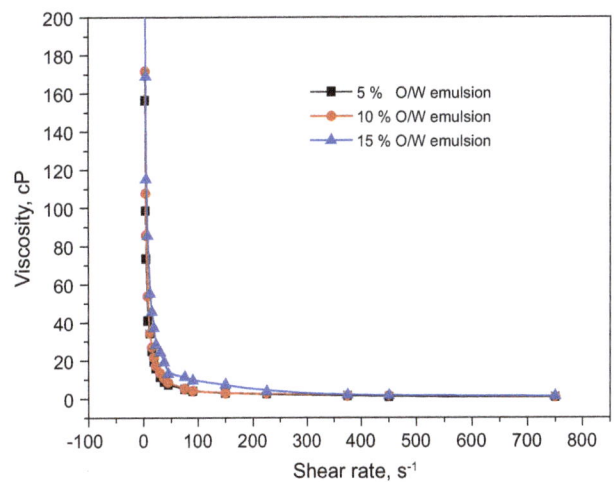

Fig. 3 Viscosity versus shear rate at 30 °C for different O/W emulsion concentrations

10 % O/W emulsion is shown in Fig. 4. The particle size distribution (PSD) of oil in the emulsion was studied using a particle size analyzer. The PSD curve of 10 % O/W emulsion is shown in Fig. 5 and it shows maximum intensity at 466.1 nm.

4.2 Pressure drop of emulsion through a porous media

The emulsion concentration significantly affects the flow rate of emulsions through porous media. In Figs. 6, 7, 8, pressure drop versus flow rate curves are depicted for sand packs A, B and C at different emulsions of varying concentrations. In addition, the rate of brine flow was also measured. In all cases, the flow rate decreases with increasing emulsion concentration. This is due to the differences in viscosity between the different concentrations

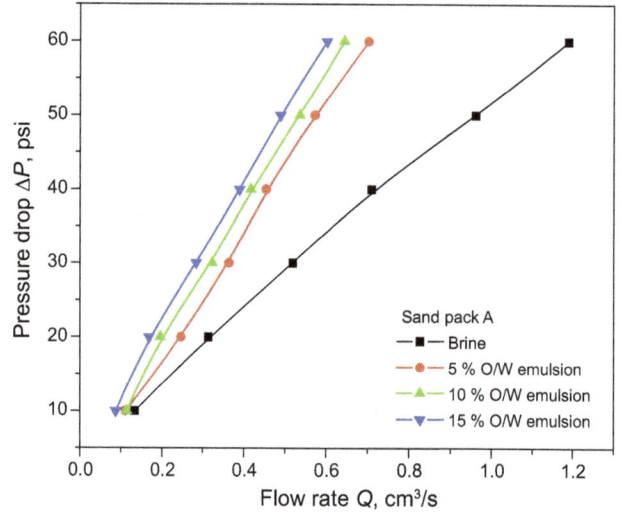

Fig. 6 Plot of pressure drop versus flow rate for O/W emulsions and brine for sand pack A

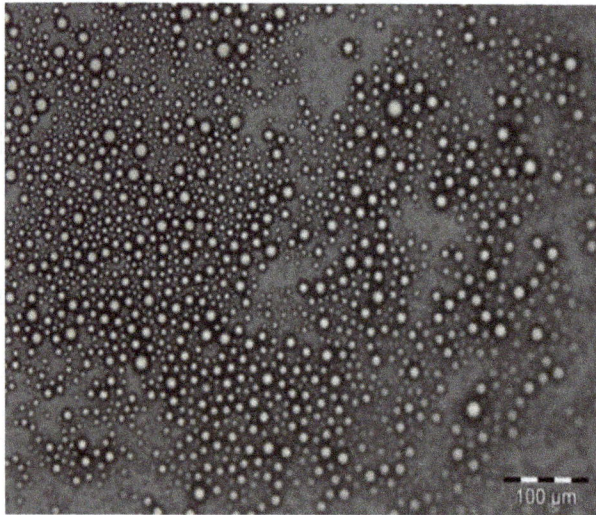

Fig. 4 Microscopic image of the 10 % O/W emulsion

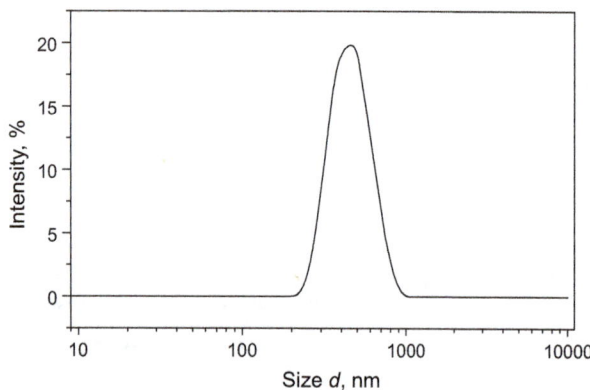

Fig. 5 Particle size distribution of 10 % O/W emulsion

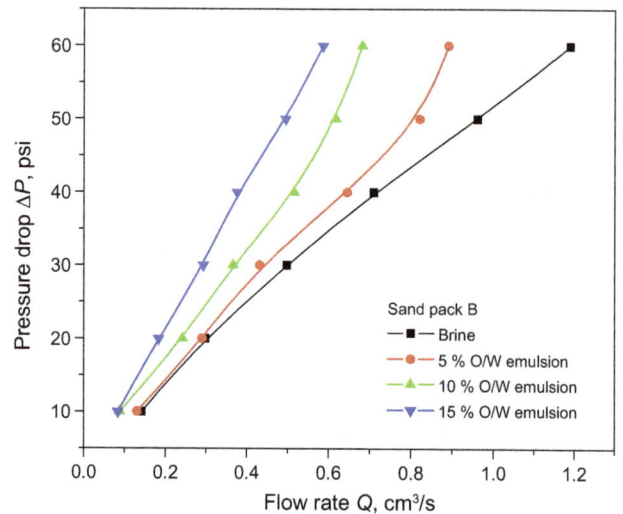

Fig. 7 Plot of pressure drop versus flow rate for O/W emulsions and brine for sand pack B

of injected emulsions. Brine has a higher flow rate compared to emulsions. For a particular pressure gradient, the flow rate decreases with an increase in emulsion concentration. In all cases the relationship between pressure drop and flow rate is linear.

The effect of sand particle size on the flow rate versus pressure drop traces for 10 % O/W emulsion is shown Fig. 9. With an increase in the sand particle size the effective flow area decreases and hence for the same flow rate the higher pressure drop is observed with an increase in the particle size.

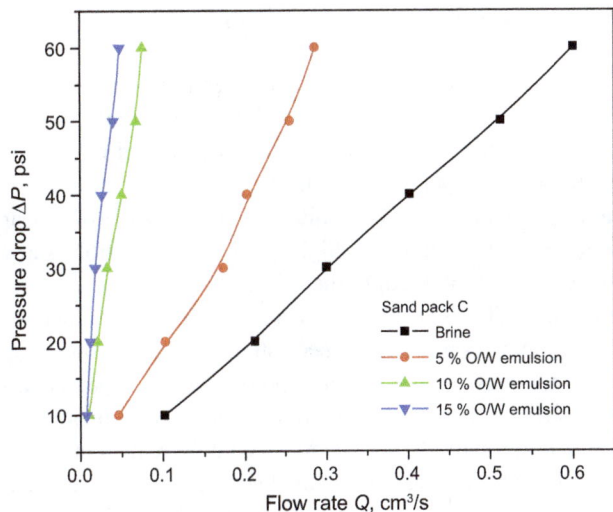

Fig. 8 Plot of pressure drop versus flow rate for O/W emulsions and brine for sand pack C

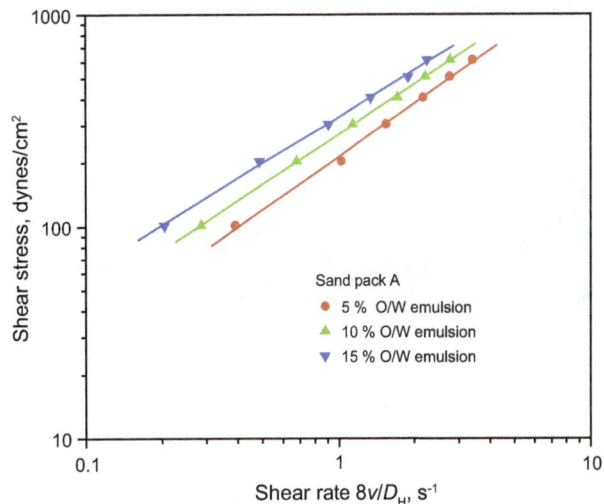

Fig. 10 Rheological behavior of emulsions in sand pack A

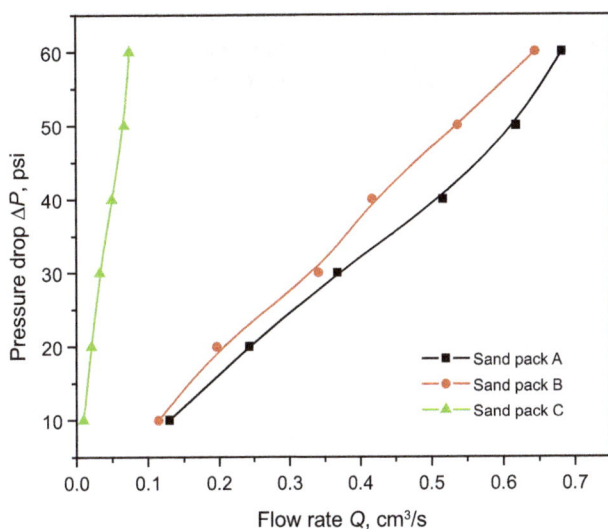

Fig. 9 Plot of pressure versus flow rate for 10 % O/W emulsions with different mesh sizes

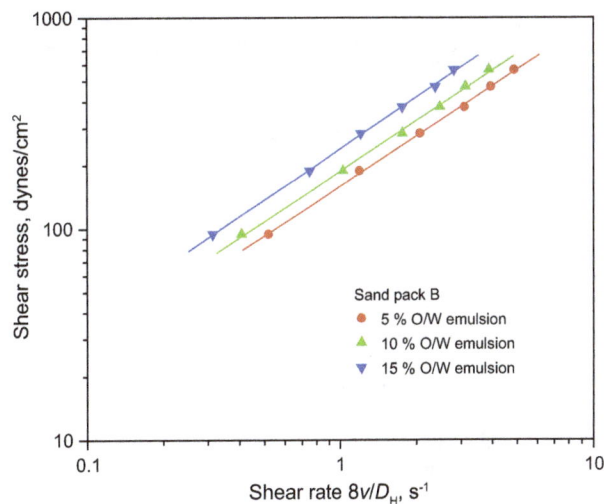

Fig. 11 Rheological behavior of emulsions in sand pack B

4.3 Rheological behavior of emulsions through porous media

The rheological properties of emulsions are explained on the basis of shear stress versus shear rate relationship. The curves of shear stress versus shear rate of 5 %, 10 %, and 15 % emulsions in different sand packs are shown in Figs. 10, 11, 12, respectively. The flow behavior index (n) and consistency constant (K) have been calculated from log–log plots of shear stress versus shear rate. The relationship between "n" and "K" with different emulsion concentrations is shown in Fig. 13. From Fig. 13, it is clear

that with an increase in the emulsion concentration "n" decreases and in all cases the value is less than unity. This indicates that the emulsion is a non-Newtonian fluid in behavior (Uzoigwe and Marsden 1970; Masalova et al. 2011; Saiki et al. 2007). It was also found that "K" increases with an increase in the emulsion concentration. Therefore, the fluid with higher "K" value has a higher viscosity. So 15 % O/W emulsion has the highest viscosity at different shear rates. The effective viscosity has been calculated from the derived model and shown in Fig. 14 as function of shear rate. It has been found that the effective viscosity of emulsions increases with an increase in oil concentration in the O/W emulsion when flowing through porous media at a specified shear rate.

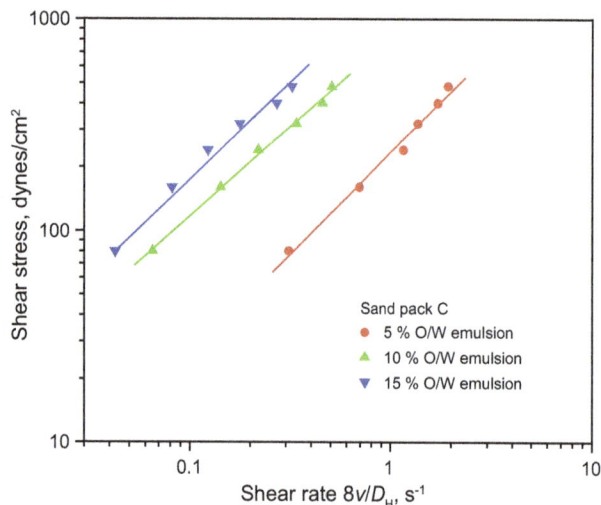

Fig. 12 Rheological behavior of emulsions in sand pack C

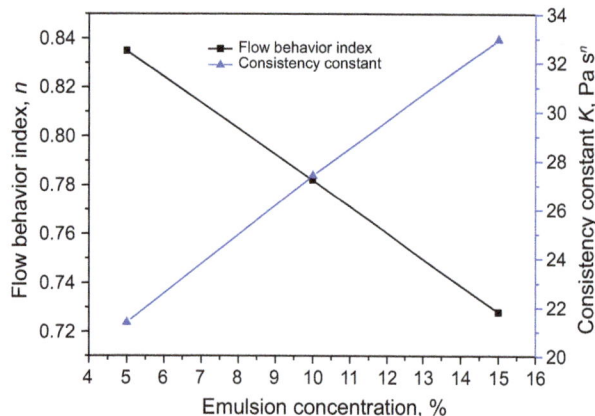

Fig. 13 Rheological parameters (flow index, n and consistency constant, K) as a function of emulsion concentration

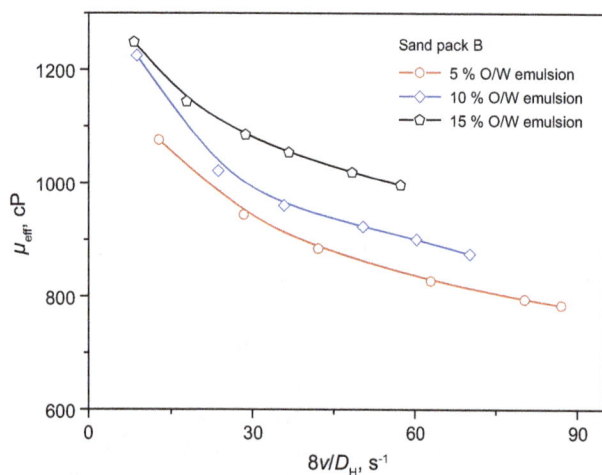

Fig. 14 Effective viscosity of emulsions calculated from the derived model at different shear rates (sand pack B)

5 Conclusions

Oil-in-water emulsions are characterized in terms of their rheological behavior and sizes of the dispersed oil globules in water. It has been found that the viscosity of emulsions decreases with an increase in shear rate at lower shear rates, showing non-Newtonian behavior. At low shear rates, the emulsions of higher oil concentration show higher viscosity though the viscosity variation is only marginal. The flow characteristics of different emulsions through porous media have been studied. The pressure drop for emulsions flowing through porous media depends significantly on the oil concentration in the O/W emulsion and the average particle size of the medium. A mathematical model has been developed to analyze the experimental data of flow of emulsions through porous media. The shear stress–shear rate behavior of different emulsions through porous media shows non-Newtonian behavior and follows the power-law model. The variation of rheological parameters (flow index and consistency constant) as a function of emulsion concentration has also been presented. The effective viscosity of the emulsion calculated from the derived model equation has been presented as a function of shear rate for different sand packs.

References

Al-Fariss TF, Fakeeha AH, Al-Odan MA. Flow of oil emulsion through porous media. J King Saud Univ Eng Sci. 1994; 6(1):1–16.

Alvarado DA, Marsden SS. Flow of oil-in-water emulsions through tubes and porous media. SPE J. 1979;19(6):369–77 (SPE 5859).

Arhuoma M, Dong M, Yang D, et al. Determination of water-in-oil emulsion viscosity in porous media. Ind Eng Chem Res. 2009;48(15):7092–102.

Atkins DC, Baker HR, Murphy CM, et al. Development of additives and lubricating oil compositions. Ind Eng Chem. 1947;39(4): 491–7.

Babadagli T. Mature field development-a review. In: SPE Europec/ EAGE annual conference. Madrid, Spain. 2005 (SPE 93884).

Becher P. Emulsions: theory and practice. Washington, DC: American Chemical Society; 2001.

Bennett H. Practical emulsions. Brooklyn: Chemical Publishing Company; 1967.

Cobos S, Carvalho MS, Alvarado V. Flow of oil-water emulsions through constricted capillary. Int J Multiph Flow. 2009;35(6): 507–15.

Fingas M, Fieldhouse B. Formation of water-in-oil emulsions and application to oil spill modeling. J Hazard Mater. 2004; 107(1–2):37–50.

Gewers CWW. Colloid and surface chemical problems in non-conventional heavy oil recovery. Can J Pet Technol. 1968;7:2–85.

Jin L, Wojtanowicz AK. Progression of injectivity damage with oily waste water in linear flow. Pet Sci. 2014;11(4):550–62.

Khambharatana F, Thomas S, Farouq Ali SM. Macro-emulsion rheology and drop capture mechanism during flow in porous media. In: SPE international oil and gas conference and exhibition. Beijing, China. 1998 (SPE 48910).

Mandal A. Gas-liquid flow in an ejector induced downflow bubble column. KG: LAP Lambert Academic Publishing GmbH & Co; 2010.

Mandal A, Samanta A, Bera A, et al. Characteristics of oil-water emulsion and its use in enhanced oil recovery. Ind Eng Chem Res. 2010a;49(24):12756–61.

Mandal A, Kumar P, Ojha K, et al. Characterization and separation of oil-in-water emulsion. Adv Sustain Pet. Eng Sci. 2010b;1(4): 379–88.

Masalova I, Foudazi R, Malkin AY. The rheology of highly concentrated emulsions stabilized with different surfactants. Colloids Surf A. 2011;375(1–3):76–86.

McAuliffe CD. Crude-oil-water emulsions to improve fluid flow in an oil reservoir. J Pet Technol. 1973a;25(6):721–6 (SPE 4370).

McAuliffe CD. Oil-in-water emulsions and their flow properties in porous media. J Pet Technol. 1973b;25(6):727–33.

Miller A. The chemistry of lubricating oil additives. J Chem Educ. 1956;33(7):308–12.

Murray BS, Dickinson E, Wang Y. Bubble stability in the presence of oil-in-water emulsion droplets: influence of surface shear versus dilatational rheology. Food Hydrocoll. 2009;23(4):1198–208.

Rastogi MC. Surface and interfacial science: applications to engineering and technology. Oxford: Alpha Science; 2003.

Rupesh MB, Prasad B, Mishra IM, et al. Oil field effluent water treatment for safe disposal by electroflotation. Chem Eng J. 2008;137(3):503–9.

Saiki Y, Prestidge CA, Horn RG. Effects of droplet deformability on emulsion rheology. Colloids Surf A. 2007;299(1–3):65–72.

Seright RS, Liang J. A comparison of different types of blocking agents. In: Proceeding of the European formation damage conference. The Hague, the Netherlands. 1995 (SPE 30120).

Soma J, Papadopoulos KD. Flow of dilute, sub-micron emulsions in granular porous media: effect of pH and ionic strength. Colloids Surf A. 1995;101(1):51–61.

Soo H, Radke CJ. The flow mechanism of dilute, stable emulsions in porous media. Ind Eng Chem Fundam. 1984;23(3):342–7.

Steinhauff F. Modern oil field demulsification. Part I Pet. 1962;25:294–6.

Thomas S, Ali SMF. Flow of emulsions in porous media, and potential for enhanced oil recovery. J Pet Sci Eng. 1989;3(1–2): 121–36.

Uzoigwe AC, Marsden Jr SS. Emulsion rheology and flow through unconsolidated synthetic porous media. In: SPE AIME 45th annual fall meeting. Houston, Texas. 1970 (SPE 3004).

Vidrine WK, Willson CS, Valsaraj KT. Emulsions in porous media. I. Transport and stability of polyaphrons in sand packs. Colloids Surf A. 2000;175(3):277–89.

13

Porosity prediction from seismic inversion of a similarity attribute based on a pseudo-forward equation (PFE): a case study from the North Sea Basin, Netherlands

Saeed Mojeddifar[1] · Gholamreza Kamali[1] · Hojjatolah Ranjbar[1]

Abstract The objective of this work is to implement a pseudo-forward equation which is called PFE to transform data (similarity attribute) to model parameters (porosity) in a gas reservoir in the F3 block of North Sea. This equation which is an experimental model has unknown constants in its structure; hence, a least square solution is applied to find the best constants. The results derived from solved equations show that the errors on measured data are mapped into the errors of estimated constants; hence, Tikhonov regularization is used to improve the estimated parameters. The results are compared with a conventional method such as cross plotting between acoustic impedance and porosity values to validate the PFE model. When the testing dataset in sand units was used, the correlation coefficient between two variables (actual and predicted values) was obtained as 0.720 and 0.476 for PFE model and cross-plotting analysis, respectively. Therefore, the testing dataset validates relatively well the PFE optimized by Tikhonov regularization in sand units of a gas reservoir. The obtained results indicate that PFE could provide initial information about sandstone reservoirs. It could estimate reservoir porosity distribution approximately and it highlights bright spots and fault structures such as gas chimneys and salt edges.

Keywords Porosity · Seismic inversion · Tikhonov regularization · Similarity

✉ Saeed Mojeddifar
saeid_miner64@yahoo.com

[1] Department of Mining Engineering, Shahid Bahonar University of Kerman, Kerman, Iran

Edited by Jie Hao

1 Introduction

Porosity is a ignificant criterion in characterizing a reservoir and in determining flow patterns in order to optimize the production of a hydrocarbon field. Also, reliable estimation of porosity is critical for evaluating hydrocarbon accumulations in a basin and to map potential pressure seals in order to reduce drilling risk in the wildcats. Porosity is mostly measured in the laboratory on the cored rocks recovered from the reservoir or could be determined by well-test data. As the well testing and coring methods are expensive and time consuming, all wells in a typical oil or gas field are logged using various tools to measure petrophysical parameters such as porosity and density. However, the spatial distribution of porosity between wells is a very important concern in oil industry (Bhatt and Helle 2002; Tiab and Donaldson 2004). Seismic measurements are often used to delineate the structure of reservoir bodies, but are not often used to estimate the spatial distribution of reservoir and rock properties. In other words, it is very difficult to estimate the porosity directly from seismic data. Inversion was used to improve the prediction of reservoir properties from the 3D seismic. These predictions should become more accurate as wells are added. The past studies showed that inversion of seismic data into acoustic impedance (AI) is widely used in hydrocarbon exploration to estimate petrophysical properties. The acoustic impedance is commonly used for porosity estimation, mostly based on an empirical relationship between acoustic impedance and porosity. However, the relationship differs from area to area because the compaction model varies both laterally and vertically. Thus, in many cases, in a large area, porosity cannot be estimated directly from the acoustic impedance using a single transform function (Anderson 1996). For this reason, Schultz et al. (1994) proposed the idea of using multiple seismic attributes to estimate log properties away from well control. After that, various data integration techniques such as neural networks were used to derive petrophysical properties directly from seismic attributes. The use of artificial neural networks (ANN) in geophysical inverse problems is a relatively recent development and offers many advantages when dealing with the nonlinearity inherent in such applications (Baddari et al. 2009). ANN has been used to predict core properties from well logs (Lim 2005), well log to well log transformations, and seismic properties have been used to predict lithology (Singh et al. 2007; Walls et al. 2000; Calderon and Castagna 2007; Joel et al. 2002), sonic logs and shale content (Liu and Liu 1998), shale stringers in a heavy oil reservoir (Tonn 2002), spontaneous potential (Banchs and Michelena 2002), permeability (Lim 2005; Helle et al. 2001), and porosity (Leite and Vidal 2011; Artun and Mohaghegh 2011; Singh et al. 2007; Calderon and Castagna 2007; Joel et al. 2002; Pramanik et al. 2004; Daniel et al. 2001; Kevin and Curtis 2004; Leiphart and Hart 2001; Russell et al. 1997). Multivariate linear regression (MLR), another technique, is a simple extension of the well-known univariate case. In these circumstances, log properties are estimated from a linearly weighted sum of a number of seismic attributes. This was first demonstrated to yield accurate results by Russell et al. (1997). Although all of these works may show significant advantages compared to impedance-based methods, they have not presented a specific mathematical equation to describe the relationship between attributes and petrophysical properties. To solve this issue, this research attempts to propose a nonlinear mathematical equation to describe the relationship between a seismic attribute (similarity) and the porosity value in a sandstone reservoir. There are several advantages of this mathematical model over the conventional inversion methods: it predicts porosity log rather than acoustic impedance; it uses seismic attribute (similarity) rather than the conventional post-stack volume. It relies on a simple forward model and knowledge of the seismic wavelet is not required that may enhance resolution. In fact, this model which transforms the similarity attribute of a sandstone reservoir to a porosity value is called the pseudo-forward equation (PFE) in this paper. The structure of PFE is implemented based on the dataset of the gas reservoir of the F3 block in the North Sea. This reservoir consists of sand and shale layers, in which shale units are sandwiched between the sand layers. Therefore, the role of PFE in both rock types will be investigated. The initial parameters of PFE are unknown and should be derived from data. This study will use the algebra technique to solve the nonlinear model and finally the quality of the implemented model will be studied. A typical feature of inverse problems is that they are ill-posed and a unique solution may not exist

and small errors in the data may cause prohibitively large variations in the estimations of the quantity sought. To overcome these difficulties one has to regularize the original problem, that is, the original problem has to be replaced by a nearby well-posed problem in order to obtain a stable solution. One of the best known and most used regularization methods is Tikhonov regularization. This work will illustrate how Tikhonov regularization could optimize the PFE acceptably in the North Sea reservoir and ultimately the optimized PFE will be employed in the F3 reservoir to estimate the porosity distribution of the various seismic sections and finally the quality of the implemented model will be compared with results of a conventional method. As mentioned above, most previous studies have used AI to predict porosity. Therefore, cross-plot analysis between AI and porosity derived from density logs is performed to find a regression fit between two datasets. Given a linear relationship provided by regression fit, spatial distribution of porosity is estimated in the F3 block. Thus, the comparison between developed models provides a simple means of testing whether the model is implemented correctly.

2 Geological setting

This research is facilitated by having F3 block data from dGB Earth Sciences. The F3 block is located in the northeastern part of the Dutch sector of the North Sea. During the Cenozoic era, much of this region was a thermally subsiding epicontinental basin, most of which was confined by landmasses (Sørensen et al. 1997). During the Neogene, sedimentation rates exceeded the subsidence rate, and consequently shallowing of the basin occurred. A large fluvio-deltaic system dominated the basin, draining the Fennoscandian High and the Baltic Shield. The Cenozoic succession could be subdivided into two main packages, separated by the Mid-Miocene Unconformity (Fig. 1).

The lower package consists mainly of relatively fine-grained gradational Paleogene sediments (Steeghs et al. 2000), whereas the package above consists of coarser grained Neogene sediments with much more complex geometries. Most of the above package is a progradational deltaic sequence that could be subdivided into three units, corresponding to three phases of delta evolution (Fig. 1). The dominant direction of progradation is toward the west-southwest and is expressed as sigmoid lineaments (clinoforms) in the dip section (Tigrek 1988). Unit 2, containing a conspicuous clinoform package, was chosen as the target zone for gas accumulation, and forms the delta fore set with a coarsening upward sequence. Its age is estimated as Early Pliocene. The coarse sediments are attributed to a

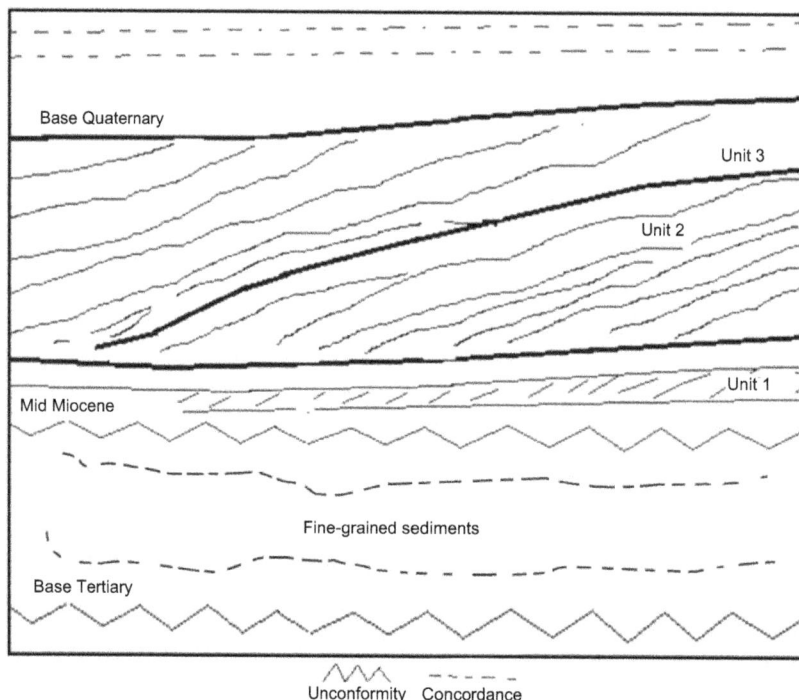

Fig. 1 Sketch of the Neogene fluvio-deltaic system in the south of the North Sea (modified after Steeghs et al. 2000)

regression caused by the Neogene uplift of Scandinavia in the Pliocene (Gregersen 1997).

3 Dataset

A 3D seismic survey in F3 block covering an area of approximately 16×23 km^2 has become publicly available and is provided by a monograph of Aminzadeh and Groot (2006). The data volume consists of 646 inlines and 947 cross-lines. The line spacing is 25 m for both inlines and cross-lines, and the sample rate is 1 ms. A standard seismic data processing sequence was applied to the data. Data from four wells in the area are available, in particular well logs in true vertical depths, including sonic and gamma ray logs. Density logs were reconstructed from the sonic logs using neural network techniques by dGB Earth Sciences. The density logs were also used to calculate porosity logs for all wells. Figure 2 is a seismic cross section of the study area that shows existing well locations (F06-1, F02-1, F03-2, F03-4).

In Fig. 2 gamma ray logs are displayed in every well and one could separate the various shale and sand layers in F3 reservoir. The study area in this paper is the upper package where coarser grained Neogene sediments with much more complex geometries are located (450–1200 ms). In this zone, the presence of a laminated shale and sand sequence is proved especially in well No. F03-4. On the opposite side, the sand layers are the main lithology present in well F03-2.

These sand and shale sequences constitute commercial gas-bearing reservoirs and exhibit an approximate time range of 700–1000 ms. A basic rule for gamma ray log interpretation is that lower values correlate with sandy layers and higher values correlate with the shale-rich layers (Luthi 2001). According to Fig. 2, there are two types of sediments that could be clearly distinguished from the plots: shale-rich sediments with generally higher gamma ray values (>70 API) that mostly belong to the upper and lower target zone (units 1 and 3) and sand-rich sediments with generally low gamma ray values (<70 API) that mostly belong to the middle part of target zone (unit 2). To get better results in the target zone, this research divides the dataset of F3 reservoir into two parts: shale-rich sediments data with gamma ray values more than 70 API and sand-rich sediments data with gamma ray values less than 70 API.

4 Pre-processing

In this work, our objective is to find an operator, possibly nonlinear, which could predict porosity from seismic data. In fact, this paper chooses to apply not the seismic data itself, but attributes of the seismic data. The reason for this choice is that many of attributes are nonlinear; thus, the predictive power of the method is increased. The next reason is that there is often advantage in breaking down the input data into component parts. It means that when the

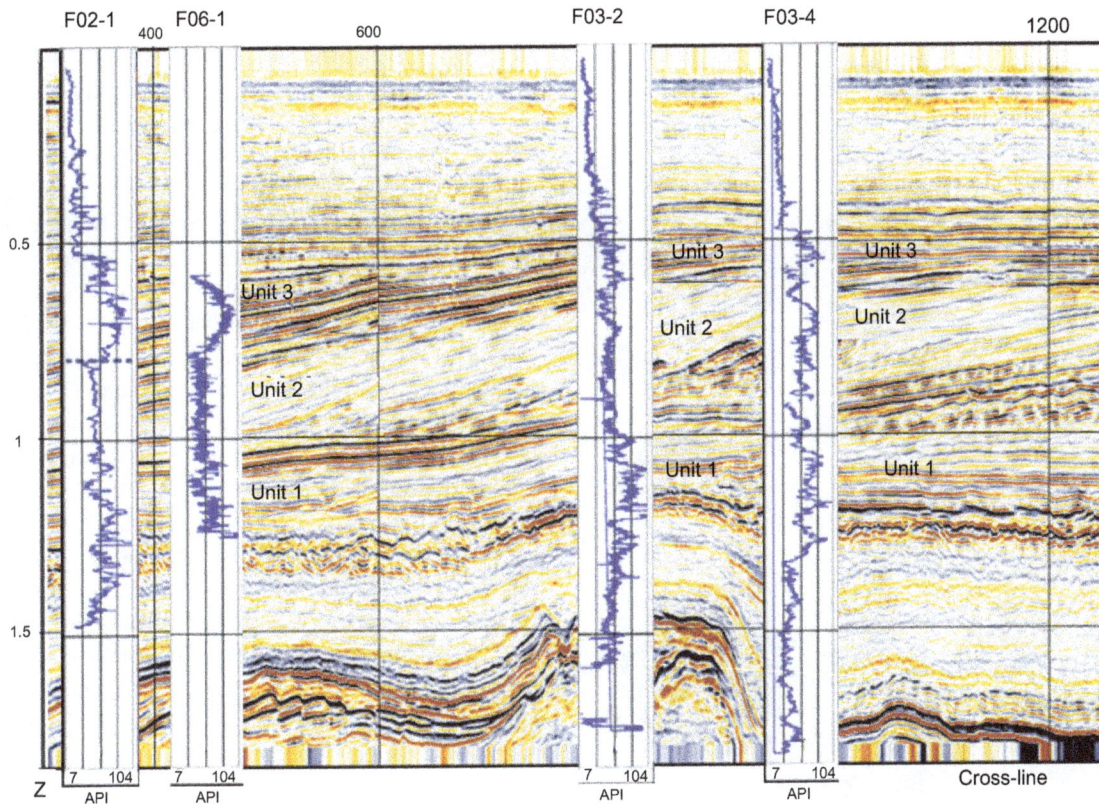

Fig. 2 The seismic section driven from original seismic data (inline 425) and it shows the location of wells and presents the gamma ray logs in every well

Table 1 The list of used attributes and their correlation coefficients with porosity values

Studied attributes	Number of points	Minimum	Maximum	Average	Standard deviation	Correlation coefficient	Sig.[a]
Energy	927	425,250	16,411,000	4,163,775	3,456,700.9	−0.23	0.000
Envelope	927	139.5344	8602.5	2411.765	1748.5479	−0.26	0.000
Spectral decomposition	927	304.8516	13,142	4835.486	2861.6049	−0.307	0.000
Similarity	927	0.6861	0.9488	0.871362	0.0608897	−0.45	0.000

[a] Sig statistical significance of the result. The result is significant if it is smaller than 5 %

raw seismic trace is divided into several mathematical functions (attributes), one could study the behavior of every attribute in contrast to the petrophysical properties and if there is a well-determined linear correlation between seismic attributes and reservoir properties, they will be considered in analysis to predict the unknown properties. Hundreds of seismic attributes have been developed but only some of them are well enough understood to be quantitative, and many are redundant. Also, a seismic trace is the result of complicated interrelationships between bed thickness, porosity, fluid saturation, lithological boundaries, and other rock properties (Kevin and Curtis 2004; Satinder and Kurt 2008). In the present research, the authors have considered statistically the behavior of more than 15 attributes in four wells and they have found that similarity is the fundamental attribute which shows more

correlation than other attributes. A detailed list of the attributes used and their statistical parameters besides the correlation coefficients of the extracted attributes are given in Table 1.

According to Table 1, the similarity attribute is considered to be the optimal one to predict porosity as the output in linearity and nonlinearity mode. In practice, it is not too frequent to have greater correlation than 50 %–60 % between seismic attributes and well log data, so this work is satisfied with F3 data to find linear or nonlinear relationships between two sets of input and output data. Acoustic impedance is another seismic attribute that is widely used to estimate porosity distribution of reservoir rocks. Given the acoustic impedance attribute in the inversion procedure, the simplest method to derive the appropriate relationship between porosity and AI is to cross

(a)

(b)

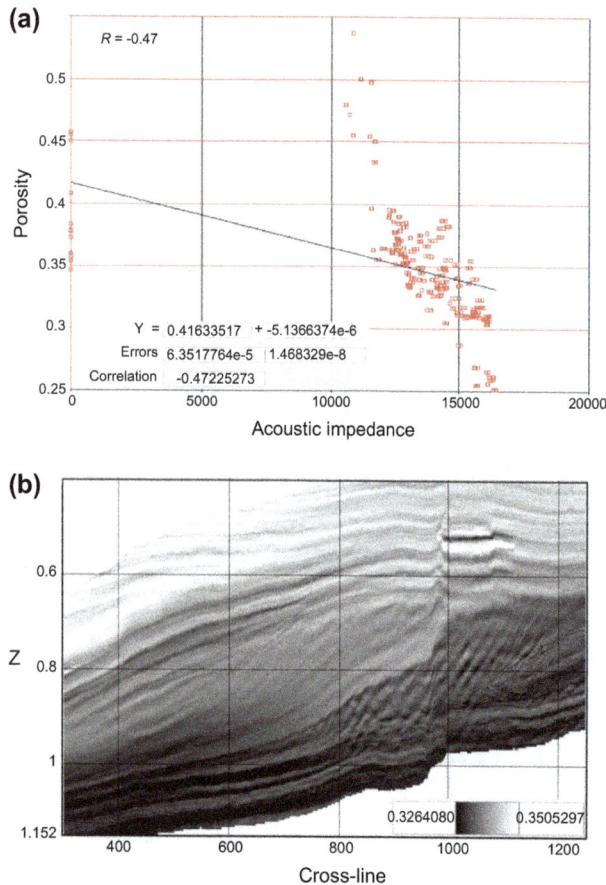

Fig. 3 **a** Cross plot between the target log (porosity) and the seismic attribute (AI); **b** Distribution of porosity estimated using the relationship established between porosity and AI

plot between two datasets. Assuming a linear relationship between porosity and AI, a straight line may be fitted by regression in Fig. 3. In Fig. 3a, the target log property (porosity) is plotted against AI attribute using OpendTect software. The cross correlation and the mean-squared prediction error are 47 % and 6.35e−6, respectively. From the linear regression fit, the distribution of porosity in the reservoir in inline-228 is estimated (Fig. 3b). In this research, the candidate attributes such as energy, envelope, spectral decomposition, and similarity are used to predict spatial distribution of porosity in the North Sea reservoir using an experimental model which is constructed using an algebraic technique, and then the results are compared with conventional method of cross-plot analysis (Fig. 3a).

5 Pseudo-forward modeling

The development of a mathematical model that is able to predict petrophysical properties should be performed based on the physical concepts. These equations (the so-called forward model) are often formulated using fundamental

seismic factors of the Earth such as wave velocity, density, etc. Unlike the conventional procedure, this work intends to extract an empirical model that is weakly supported by experimental data. Although it is possible to introduce a model from seismic data, the terms in the equation are empirical and any functional connection to physical concepts is not entirely justified. Therefore, because of empirical nature of the proposed model, the developed model is called pseudo-forward equation (PFE), in this work. As described before, some seismic attributes were chosen for prediction of spatial distribution of porosity. This paper has designed various mathematical structures based on the aforementioned attributes, but they have shown different degrees of accuracy. Implicitly, it has been assumed that multi-attribute functions are more valid than single-attribute ones over the target zone. Improvements have focused on accuracy enhancement, shorter equations, and improved representation of the sand and shale regions. Finally, in all these situations, various mixture models were developed that exhibit varied behavior in contrast to sand and shale layers. The ultimate empirical function is a single-attribute equation based on the similarity attribute. Similarity is a form of "coherency" that expresses how much two or more trace segments look alike. The coherency attribute is a measure of lateral changes in acoustic impedance caused by variations in structure, stratigraphy, lithology, porosity, and fluid content. The first coherency algorithm based on correlation was proposed by Bahorich and Farmer (1995) and then it was completed by Marfurt et al. (1998). For attributes expressing coherence, it is generally true that they are suitable for the indication of sudden changes between neighboring channel sections. They are extremely good for the detection of faults, fractured zones, or boundaries related to lithological changes. There are several types of coherence attributes. The best known one is the so-called coherency cube, while recently developed methods are semblance-type procedures, eigen structure or variance-based coherence, and coherence based on the calculation of least squares (Eichkitz et al. 2012). The so-called similarity attribute characteristic for the coherence, used by this research, is a simple one and can be calculated quickly. Its value between two channel sections could be given by (OpendTect dGB Plugins User Documentation 2012)

$$\text{sim}(x, y) = 1 - \frac{\sqrt{\sum_{i=1}^{N} (x_i - y_i)^2}}{\sqrt{\sum_{i=1}^{N} x_i^2} + \sqrt{\sum_{i=1}^{N} y_i^2}}, \tag{1}$$

where $\text{sim}(x, y)$ is the value of similarity between x and y vectors containing N number of data. N could be defined by a time gate. The numerator is the Euclidean distance in the N dimension of vectors x and y, and the denominator is the sum of the vectors' lengths. In various research (Santosh et al. 2013), the similarity attribute map is applied to enhance the fault structures and clear salt edges. To implement the PFE on F3 block, the similarity attribute of three wells (F06-1, F02-1, F03-2) besides the porosity values from density logs are used to construct the structure of PFE and well F03-4 is selected to evaluate the performance of PFE. The PFE model is introduced as

$$Y = a + bs\ln(s) + \frac{c}{\ln(s)}, \tag{2}$$

where Y is denoted as porosity and s is the similarity attribute. The constants of a, b, and c are fundamental parameters that depend on behavior of porosity in the reservoir. Equation 2 is proposed as an empirical model which could fit approximately to the dataset in the F3 block. Note that a forward model has a physical concept to analyze in inversion modeling but because of empirical nature of the proposed model, in this work, Eq. (2) is called the pseudo-forward equation (PFE). According to the above, this study should solve the pseudo-forward equation and estimate the optimized constants for a reliable prediction and finally verify the fit between predicted and observed data. To solve the pseudo-forward equation, a linear algebraic approach is developed to invert the pseudo-forward equation.

6 Inverse modeling

This research is faced with the situation that a quantity (similarity) is measured at the surface of the Earth and the aim is to know porosity of the rocks beneath the place where we made the measurements. For each set of measurements (similarity), a PFE is presented which approximately relates it to the porosity. The PFE is a nonlinear function which needs optimal constants to predict the porosity distribution in a reservoir. The nature of these constants originates the nature of the reservoir. Inverse theory is a method to infer the unknown physical property (porosity) from measurements (similarity). To solve the PFE first, it is represented in the form of an operator equation:

$$\mathbf{d} = \mathbf{Gm}, \tag{3}$$

where $\mathbf{d}$ is the vector of predictions (porosity), $\mathbf{m}$ is the vector of the unknown parameters of the model (a, b, c), and $\mathbf{G}$ is the theoretical function or the linear operator which makes it possible to calculate $\mathbf{d}$ (porosity) from an earth model defined by the $\mathbf{m}$ parameters (a, b, c). $\mathbf{G}$ is the theory that predicts the porosity distribution in a reservoir from the model parameters $\mathbf{m}$. This theory is based on

seismic attributes. Mathematically, $\mathbf{Gm}$ is a functional, a rule that unambiguously assigns a single real number to an element of a vector space. Now let us introduce the nomenclature of Eq. (3) more accurately. In these notes, vectors will be denoted by bold lowercase letters, and matrices will be denoted by bold uppercase letters. Suppose there exist N measurements (similarity) in a field then there are N values for the corresponding porosity data and we are trying to determine the values of three model parameters (a, b, c). Our nomenclature for data and model parameters will be

Data: $\mathbf{d} = [d_1, d_2, d_3, \ldots, d_N]^{\mathrm{T}}$;

$\quad d_i = (\text{Porosity value})_i \quad i = 1, 2, 3, \ldots, N$

Model parameters: $\mathbf{m} = [a, b, c]^{\mathrm{T}}$;

$\quad (a, b, c) = \text{constants of PFE},$ $\hspace{2em} (4)$

where $\mathbf{d}$ and $\mathbf{m}$ are N and three-dimensional column vectors, respectively, and T denotes transpose.

The model, or relationship between $\mathbf{d}$ and $\mathbf{m}$, could implement in elements of $\mathbf{G}$ matrix. Then, the equation of PFE can be written as

$$d_1 = a + bs_1\ln(s_1) + \frac{c}{\ln(s_1)}$$

$$d_2 = a + bs_2\ln(s_2) + \frac{c}{\ln(s_2)}$$

$$\vdots$$

$$d_N = a + bs_N\ln(s_N) + \frac{c}{\ln(s_N)} \tag{5}$$

Getting the PFE equation set up in matrix notation is essential before we can invert the system. Hence, the above statements are written as

$$\begin{bmatrix} d_1 \\ d_2 \\ \vdots \\ d_N \end{bmatrix} = \begin{bmatrix} 1 & s_1\ln(s_1) & 1/\ln(s_1) \\ 1 & s_2\ln(s_2) & 1/\ln(s_2) \\ \vdots & \vdots & \vdots \\ 1 & s_N\ln(s_N) & 1/\ln(s_N) \end{bmatrix} \begin{bmatrix} a \\ b \\ c \end{bmatrix}. \tag{6}$$

Then $\mathbf{d}$ and $\mathbf{m}$ are $N \times 1$ and 3×1 column vectors, respectively, and $\mathbf{G}$ is an $N \times 3$ matrix with constant coefficients. The logical next step is to invert Eq. (6) for an estimate of the model parameters $\mathbf{m}^{\mathrm{est}}$ as

$$\mathbf{m}^{\mathrm{est}} = (\mathbf{G})^{-1}\mathbf{d}. \tag{7}$$

This inverse problem reverses the process of predicting the values of porosities. It tries to invert the operator $\mathbf{G}$ to get an estimate of the model. A most common vector concerned is the data error or misfit vector which plays an essential role in the development of inverse methods (Menke 1989). If $\mathbf{d}^{\mathrm{pre}}$ is calculated by

$$\mathbf{d}^{\text{pre}} = \mathbf{G}\,\mathbf{m}^{\text{est}}. \tag{8}$$

The misfit vector (data error vector) will be provided by

Data error vector: $\mathbf{e} = \mathbf{d}^{\text{obs}} - \mathbf{d}^{\text{pre}}$. $\tag{9}$

The dimension of the error vector $\mathbf{e}$ is $N \times 1$. The total misfit E between observed ($\mathbf{d}^{\text{obs}}$) and predicted data ($\mathbf{d}^{\text{pre}}$) is considered as

$$E = \mathbf{e}^{\text{T}}\mathbf{e} = \begin{bmatrix} e_1 & e_2 & \cdots & e_N \end{bmatrix} \begin{bmatrix} e_1 \\ e_2 \\ \vdots \\ e_N \end{bmatrix} = \sum_{i=1}^{N} e_i^2. \tag{10}$$

The term E is a way to quantify the misfit between predicted and observed data. The solutions which implemented based on the misfit vector give rise to least squares solutions (Menke 1989). In the next stage, the least square procedure will be used to find a best fit model of PFE to the F3 block dataset and the corresponding codes were written in the Matlab environment.

6.1 Least square solution

The solution of an inverse problem consists of giving the best solution for the model from the inversion of Eq. (5). This relation is valid when the number of the equations is equal to the number of parameters of the model. In this case, the $\mathbf{G}$ matrix will be a square matrix which could be invertible if the determinant of the matrix is different from

development of the PFE model, is applied to obtain the overall accuracy of the PFE model. To get better model parameters for PFE, this work divided the training and validation sets into two groups: sand dataset (gamma ray <70 API) and shale dataset (gamma ray >70 API); hence, the PFE is solved in the form of two different datasets:

$$\begin{aligned} \mathbf{d} &= \mathbf{Gm}; \quad \text{sand dataset} \\ (268 &\times 1)(268 \times 3)(3 \times 1) \\ \mathbf{d} &= \mathbf{Gm}; \quad \text{shale dataset} \\ (215 &\times 1)(215 \times 3)(3 \times 1) \end{aligned} \tag{11}$$

The least square procedure is to take the partial derivative of E with respect to each element in $\mathbf{m}$ and set the resulting equations to zero. This will produce a system of three equations that can be manipulated in such a way that, in general, leads to a solution for the three elements of $\mathbf{m}$.

In summary, the least squares solution for $\mathbf{m}$ is given by

$$\mathbf{m}_{\text{LS}} = [\mathbf{G}^{\text{T}}\mathbf{G}]^{-1}\mathbf{G}^{\text{T}}\mathbf{d}. \tag{12}$$

The $\mathbf{m}_{\text{LS}}$ above is the solution that minimizes E, the total misfit. It is noted that there exists $\mathbf{m}_{\text{LS}}$ when the matrix $\mathbf{G}^{\text{T}}\mathbf{G}$ has a mathematical inverse (Menke 1989). Mathematically, the $\mathbf{G}^{\text{T}}\mathbf{G}$ has an inverse when the determinant of the matrix is different from zero and it is not zero for both datasets (sand and shale). The Eq. (12) is calculated for both datasets as follows:

$$\mathbf{m}_{\substack{\text{LS} \\ \text{SAND}}} = \begin{bmatrix} 0.5435 & 2.3385 & 0.0276 \\ 2.3385 & 10.4406 & 0.1158 \\ 0.0276 & 0.1158 & 0.0014 \end{bmatrix} \begin{bmatrix} 1 & 1 & \cdots & 1 \\ -0.1819 & -17.50 & \cdots & -0.0897 \\ -4.3743 & -4.5963 & \cdots & -10.1013 \end{bmatrix} \begin{bmatrix} 0.3003 \\ 0.3008 \\ \vdots \\ 0.3070 \end{bmatrix} = \begin{bmatrix} 0.4253 \\ 0.4780 \\ 0.0080 \end{bmatrix}$$

$$(3 \times 3) \qquad\qquad\qquad (3 \times 268) \qquad\qquad (268 \times 1)$$

$$\mathbf{m}_{\substack{\text{LS} \\ \text{SHALE}}} = \begin{bmatrix} 2.1986 & 8.8751 & 0.1386 \\ 8.8751 & 36.3829 & 0.5509 \\ 0.1386 & 0.5509 & 0.0089 \end{bmatrix} \begin{bmatrix} 1 & 1 & \cdots & 1 \\ -0.0981 & -0.1028 & \cdots & -0.1964 \\ -9.1346 & -8.6643 & \cdots & -3.9539 \end{bmatrix} \begin{bmatrix} 0.3468 \\ 0.3514 \\ \vdots \\ 0.2996 \end{bmatrix} = \begin{bmatrix} 0.5211 \\ 0.7619 \\ 0.0115 \end{bmatrix}$$

$$(3 \times 3) \qquad\qquad\qquad (3 \times 215) \qquad\qquad (215 \times 1)$$

$$\tag{13}$$

zero. In order to implement the PFE model with sufficient generality, the available data are divided into three subsets. The first subset is the training set derived from three wells (F02-1, F06-1, F03-2), which is used to estimate the model coefficients. The second subset is the validation set derived from the same wells. This set of data which is not applied during the development of the PFE model is used to validate the model. The third subset is the testing set derived from testing well (F03-4); this well which is not used during the

We note that the least square solutions $[0.4253 \quad 0.4780 \quad 0.0080]^{\text{T}}_{\text{SAND}}$ and $[0.5211 \quad 0.7619 \quad 0.0115]^{\text{T}}_{\text{SHALE}}$ do not fit the data exactly and are ones that minimize the misfit vector. Now one can calculate the minimized E as follows:

$$E_{\text{SAND}} = \mathbf{e}^{\text{T}}\mathbf{e} = 0.0091$$

$$\tag{14}$$

$$E_{\text{SHALE}} = \mathbf{e}^{\text{T}}\mathbf{e} = 0.0616.$$

In the next section, the accuracy and qualification of the responses obtained from the least square method are addressed; however, Eq. (14) has provided a measurement for validating the results. Before considering the next section, Eq. (14) shows that the least square solution of PFE for the sand dataset has more validity than for the shale dataset. Nevertheless, this viewpoint may tend to obscure an important aspect of the inverse problems. Namely, the nature of the problem depends more on the relationship between the data and model parameters than on the data or model parameters themselves. Therefore, it is essential that the qualification of the PFE itself is investigated in the following.

7 Assessment of the quality of PFE model

The PFE model is solved by a least square technique for two different layers:

$$\text{Porosity}_{\text{SAND}} = 0.4253 + 0.4780 s \ln(s) + \frac{0.0080}{\ln(s)}$$
$$\text{Porosity}_{\text{SHALE}} = 0.5211 + 0.7619 s \ln(s) + \frac{0.0115}{\ln(s)}. \quad (15)$$

Equation (15) is expected to estimate approximately the porosity of sand and shale layers of a reservoir using the similarity attribute. But the prediction power of these proposed equations should be discussed. The predictive performance of solved PFEs on validation set is illustrated in Fig. 4.

Figure 4 indicates obviously that PFE fitted the sand dataset with acceptable accuracy but shale data could not satisfy the PFE to estimate porosity values. The correlation coefficient reflects a model's ability to predict the output. In statistics, it indicates how well data points fit a statistical

model—sometimes simply a line or curve. It is a statistic used in the context of statistical models whose main purpose is either the prediction of future outcomes or the testing of hypotheses, on the basis of other related information. It provides a measure of how well observed outcomes are replicated by the model, as the proportion of total variation of outcomes explained by the model. A correlation coefficient of 1 indicates that the regression line perfectly fits the data (Steel and Torrie 1960). The correlation coefficient of PFE-SAND validates well the predictive power of this experimental equation (0.938); therefore, based on this evidence it should estimate the sand data of well F03-4 (test well) acceptably. According to the result of Fig. 4b, the structure of PFE cannot reliably estimate porosity of shale sediments; therefore, this paper just studies the PFE-SAND equation, and the purpose of PFE is the PFE-SAND in the following. To evaluate the solved PFE, 250 points of well F03-4 in reservoir with gamma ray lower than 70 API are extracted. This well has not been applied in implementation of the structure of PFE. Figure 5 illustrates the response of PFE to the test dataset.

According to Fig. 5, there is a main issue in the solved PFE and it is obvious that this experimental equation could not be successful in the prediction of the porosity values in a sandstone zone. The authors believe that the main reason of this issue is related to the nature of the inversion process; it means that the inverse problem does not have a unique response and it is known as an ill-posed problem. This is because the noise in the measured data affects the quality of the PFE. Any errors (noise) in the data will be mapped into errors in the estimates of the model parameters. For this reason, a model covariance matrix $[\text{cov}\, m]$ needs to be defined by assuming that $[\text{cov}\, d] = I_N$, that is, all the data variances are equal to 1 and the covariances are all 0 (uncorrelated data errors) (Menke 1989).

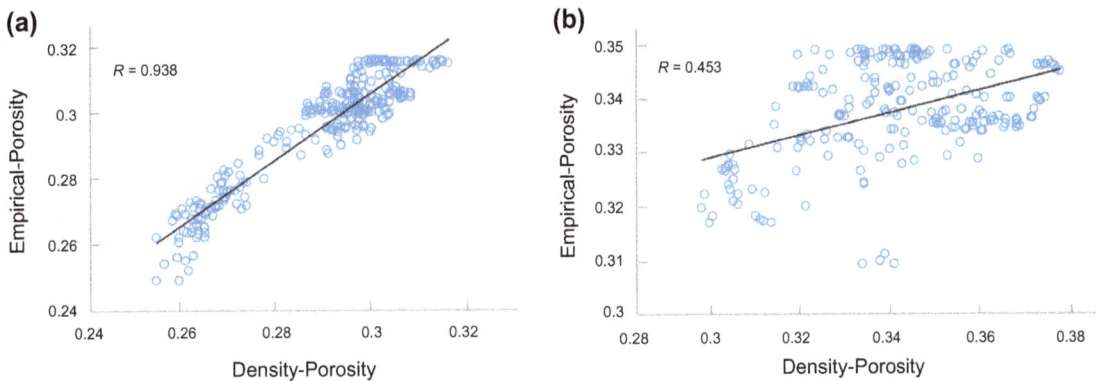

Fig. 4 a Cross plot between real and estimated porosity of validation set by PFE-sand model on sand dataset; **b** Cross plot between real and estimated porosity of validation set by PFE-shale model on shale dataset

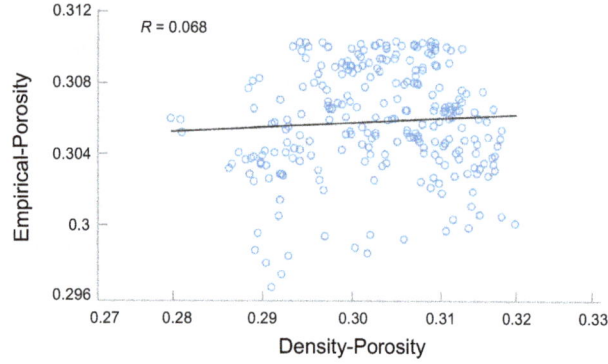

Fig. 5 Cross plot between real and estimated porosity of testing set by PFE on sand dataset

If

$$\mathbf{G}_g^{-1} = [\mathbf{G}^T\mathbf{G}]^{-1}\mathbf{G}^T, \tag{16}$$

then

$$\begin{aligned}[\text{cov}\,\mathbf{m}] &= \mathbf{G}_g^{-1}[\text{cov}\,\mathbf{d}][\mathbf{G}_g^{-1}]^T \\ &= \mathbf{G}_g^{-1}[\mathbf{G}_g^{-1}]^T\end{aligned}. \tag{17}$$

The Eq. (17) is very helpful for getting a sense of the basic stability of PFE. The mean of the stability is what the expected noise is in the solution. In fact, [cov m] is a function of the forward problem as expressed in **G**, and not a function of the actual data. Therefore, it could show the capability of the PFE equation to accept the noise in the data and it is not necessary to know [cov d] basically (Menke 1989). This research just wants to have a quick look at stability; hence, it is assumed that

$$[\text{cov}\,\mathbf{d}] = \mathbf{I}. \tag{18}$$

The diagonal terms of [cov m] are the variances of model parameters and the off-diagonal terms are the covariances. The [1, 1] entry in [cov m] is σ_a^2, the variance for a. Correspondingly, the standard deviation of the error for a is σ_a. Therefore, the perfect model should have a diagonal of zero in [cov m] matrix. Then the solution of PFE is, however, essentially meaningless if the diagonal entries of the corresponding covariance matrix are close to zero. To see this, consider the covariance matrix [cov m] for PFE:

$$\begin{aligned}[\text{cov}\,\mathbf{m}] &= \mathbf{G}_g^{-1}[\mathbf{G}_g^{-1}]^T = [\mathbf{G}^T\mathbf{G}]^{-1}\mathbf{G}^T[[\mathbf{G}^T\mathbf{G}]^{-1}\mathbf{G}^T]^T \\ &= \begin{bmatrix} 0.5435 & 2.3385 & 0.0276 \\ 2.3385 & 10.4406 & 0.1158 \\ 0.0276 & 0.1158 & 0.0014. \end{bmatrix}\end{aligned} \tag{19}$$

The above covariance matrix is a measure of how uncorrelated noise with unit variance in the data is mapped into uncertainties in the estimated model parameters. It means, in the inverse problem of PFE, every solution could be expressed as

$$\begin{aligned} a &= 0.4253; \quad \sigma_a^2 = 0.5435; \quad \sigma_a = 0.7372 \\ &\rightarrow a = 0.4253 \pm 0.7372 \\ b &= 0.4780; \quad \sigma_b^2 = 10.4406; \quad \sigma_b = 3.2312 \\ &\rightarrow b = 0.4780 \pm 3.2312 \\ c &= 0.0080; \quad \sigma_c^2 = 0.0014; \quad \sigma_c = 0.0374 \\ &\rightarrow c = 0.0080 \pm 0.0374. \end{aligned} \tag{20}$$

According to the above equations, these are very large variances for a, b, and c, which indicate that the solution, while fitting the data using the least square solution, is very unstable, or sensitive to noise in the data. Therefore, the solved PFE for sand dataset [Eq. (15)] could not be relied upon for prediction of porosity distribution as it is observed in Fig. 5. This work tries to improve the stability of PFE using the Tikhonov approach in the next part.

8 Improving the stability of PFE model using the Tikhonov approach

In 1902, Jacques Hadamard indicated the notion of a well-posed problem. A well-posed problem in the sense of Hadamard is a problem that fulfills the following three conditions:

(1) The solution exists.
(2) The solution is unique.
(3) The solution depends continuously on the problem data.

If any of these conditions is not realized, the problem becomes ill-posed. Note that both the first and second conditions deal with the feasibility of the problem, and the last condition relates to the possible implementation of a stable numerical procedure for its resolution. The solution of a problem is always based on some data, typically obtained from experimentation. If the solution does not depend "smoothly" on the problem data, small variations on the data could create huge variations on the solutions, resulting in strong instability which is not acceptable. When solving ill-posed problems, the concept of regularization immediately appears. Regularization is used to well-pose a problem that is ill-posed. Historically, the so-called Tikhonov regularization is one of the oldest and most well-known techniques for stabilization (Wiener 1942). To apply Tikhonov regularization for optimizing the PFE problem, the following minimization problem should be considered (Tikhonov and Arsenin 1977):

$$\min \|\mathbf{d} - \mathbf{Gm}\|^2 + \varepsilon^2 \|\mathbf{m}\|^2, \tag{21}$$

where ε^2 is a parameter that controls the influence of the regularization term. Using the above statement the following Tikhonov solution is proved:

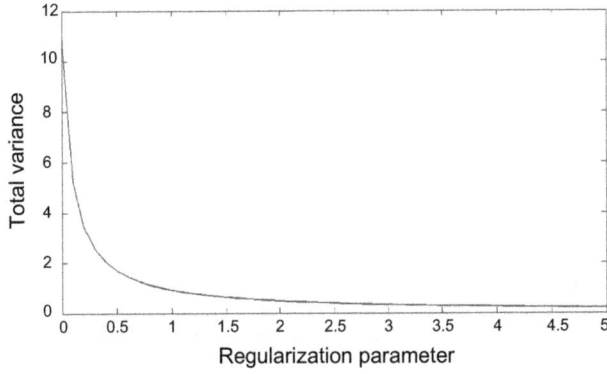

Fig. 6 Plot of total variance versus Tikhonov regularization parameter

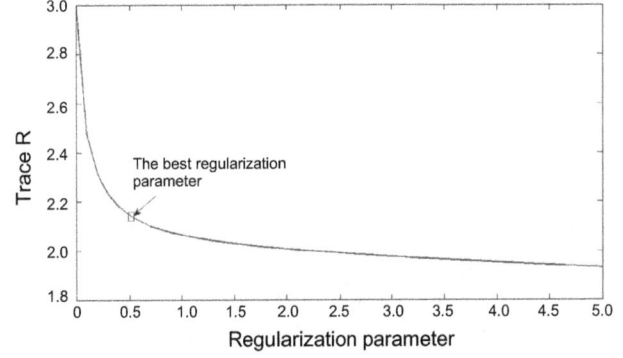

Fig. 7 Plot of trace (R) versus Tikhonov regularization parameter; the determined regularization parameter is the best balance between stability and resolution

$$\mathbf{m}_{\text{Tikh}} = \left[\mathbf{G}^{\text{T}}\mathbf{G} + \varepsilon^2 \mathbf{I}\right]^{-1}\mathbf{G}^{\text{T}}\mathbf{d}. \tag{22}$$

Care must be taken as it depends on the parameter ε^2; the choice of this parameter highly influences the estimated **m**. In practice, parameter ε^2 is determined by trial and error, with the attendant trade-off between resolution and stability. There are several heuristic ways to proceed in order to select ε^2 (Wabha 1990; Hansen 1992; Hilgendorf 1997), but the criterion described below is based on a balance between total variance of PFE and model resolution. It is a convenient graphical tool for displaying the trade-off between the size of a regularized solution and its fit to the given data, as the regularization parameter varies. In this research, to have a better solution for the PFE model, various total variances derived from different ε^2 are illustrated in a graph. The total variance is defined as the trace of the model covariance matrix, given by

Total variance = trace [cov **m**]

$$= \sigma^2 \left\{ \text{trace}\left[\mathbf{G}^{\text{T}}\mathbf{G} + \varepsilon^2 \mathbf{I}\right]^{-1} \right\}. \tag{23}$$

In the next step, we considered a plot of the total variance from Eq. (23) as a function of ε^2.

The total variance decreases, as expected, when the regularization parameter is increased. For $\varepsilon^2 = 0$, the total variance of PFE is maximum and finally the PFE does not have sufficient stability; it means this point is the least square solution as discussed before. According to Fig. 6, the best regularization parameter should be selected for minimum total variance. But just using this graph, it is hard to choose the most appropriate value for ε^2, because in order to select the best regularization parameter, it is important to achieve an acceptable balance between stability and accuracy of the solution by tuning carefully the regularization parameter. For this reason, the plots for total variance and trace (R) could help to choose the best one. The model resolution matrix is given by

$$R = \left[\mathbf{G}^{\text{T}}\mathbf{G} + \varepsilon^2 \mathbf{I}\right]^{-1}\mathbf{G}^{\text{T}}\mathbf{G}. \tag{24}$$

The model resolution matrix R measures the ability of the inverse operator to uniquely determine the estimated model parameters.

Figure 7 indicates that for $\varepsilon^2 = 0$, the PFE constants are determined perfectly. Comparing the plots of total variance and the trace of the model resolution matrix shows that as ε^2 increases, stability improves (total variance decreases) while resolution degrades. This is an inevitable trade-off. It seems that the most suitable value of the regularizing parameter ε^2 is determined by selecting one intermediate point on the corner of the trace (R) and total variance plots ($\varepsilon^2 = 0.5$). Such a point, indicated with a rectangle point in Fig. 7, is supposed to provide, in terms of accuracy and regularity, the value of the parameter corresponding to the most balanced perturbed solution of the inverse problem. As stated in the previous sections, the optimized solution of PFE inverse problem could be calculated as

$$\mathbf{m}_{\text{Tikh}} = \left[\mathbf{G}^{\text{T}}\mathbf{G} + 0.5\,\mathbf{I}\right]^{-1}\mathbf{G}^{\text{T}}\mathbf{d} = \begin{bmatrix} 0.3075 \\ -0.0211 \\ 0.0020 \end{bmatrix} \tag{25}$$

$$[\text{cov } \mathbf{m}]^{\text{Tikh}} = \left[\mathbf{G}^{\text{T}}\mathbf{G} + 0.5\,\mathbf{I}\right]^{-1}\mathbf{G}^{\text{T}}\left[\left[\mathbf{G}^{\text{T}}\mathbf{G} + 0.5\,\mathbf{I}\right]^{-1}\mathbf{G}^{\text{T}}\right]^{\text{T}}$$
$$= \begin{bmatrix} 0.0301 & 0.0518 & 0.0021 \\ 0.0518 & 0.2493 & 0.0024 \\ 0.0021 & 0.0024 & 0.0002 \end{bmatrix}$$

$$\tag{26}$$

$a_{\text{Tikh}} = 0.3075;\qquad \sigma_a^2 = 0.0301;\qquad \sigma_a = 0.1734$

$\qquad \rightarrow \quad a_{\text{Tikh}} = 0.3075 \pm 0.1734$

$b_{\text{Tikh}} = -0.0211;\qquad \sigma_b^2 = 0.2493;\qquad \sigma_b = 0.4992$

$\qquad \rightarrow \quad b_{\text{Tikh}} = -0.0211 \pm 0.4992$

$c_{\text{Tikh}} = 0.0020;\qquad \sigma_c^2 = 0.0002;\qquad \sigma_c = 0.0141$

$\qquad \rightarrow \quad c_{\text{Tikh}} = 0.0020 \pm 0.0141$

$$\tag{27}$$

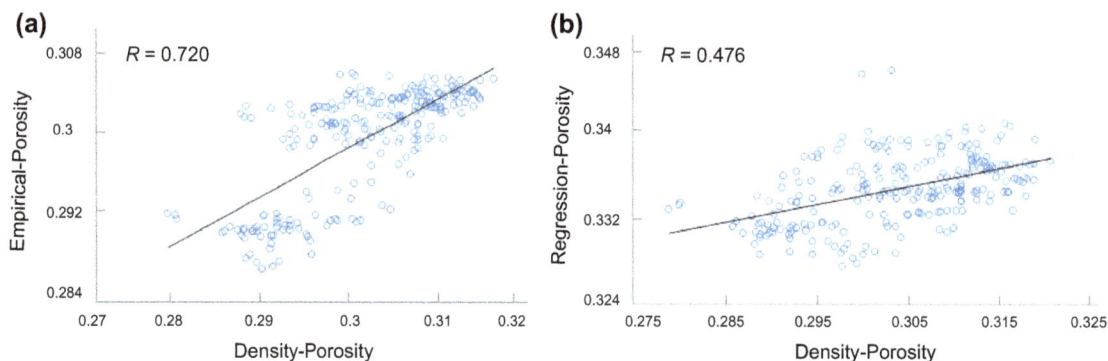

Fig. 8 **a** Cross plot between real and estimated porosity of testing set by PFE optimized by Tikhonov method in sand unit; **b** cross plot between real and estimated porosity of testing set by regression equation in sand unit

$$\text{Porosity}^{\text{Tikh}} = 0.3075 - 0.0211 s \ln(s) + \frac{0.0020}{\ln(s)}. \qquad (28)$$

Equation (27) presents the optimized constants and their variances. Variances are improved considerably by Tikhonov regularization compared with Eq. (20); in the following, to gain a better sense about prediction power of optimized PFE, the testing set is applied again in order to evaluate the new PFE [Eq. (28)].

The comparison between Figs. 5 and 8a indicates that Tikhonov regularization has increased the predictive performance of PFE and the optimized PFE could be an indicator that there is a considerable nonlinear relation between porosity values and similarity attribute. As the testing set is exactly the same, the cross correlation between actual porosity derived from the density log and the estimated one is around 47 % using a linear regression model (Fig. 8b), while the same increased to 72 % when the PFE model is used (Fig. 8a). Based on cross validation results, it seems that the developed PFE model could estimate porosity distribution of sand units of a reservoir with an acceptable quality. Although in shale units of reservoir, the results of the PFE model are not significant compared to the regression analysis.

According to the illustrated results in Fig. 9, it is found that the developed PFE model shows an inverse correlation faced with the shale units, while regression model could be adapted in shale sediments better than in sand ones. However, the PFE is not evaluated well with respect to the shale units, but it could be considered for sandstone reservoirs because it could obtain initial information about reservoirs. Also, further studies are needed absolutely to explore other aspects of this experimental equation. For example, this research was done with assumption of $[\text{cov } \mathbf{d}] = \mathbf{I}$ to obtain the model resolution matrix or the authors did not consider prior information in solving the PFE. Therefore, more research especially probabilistic approach should be applied to develop this model.

9 Reservoir characterization using PFE and regression analysis

In the following section, the authors attempt to apply the optimized PFE in various cross sections of the F3 block as an estimator and compare the illustrated outputs of the PFE with cross-plot analysis between AI and porosity.

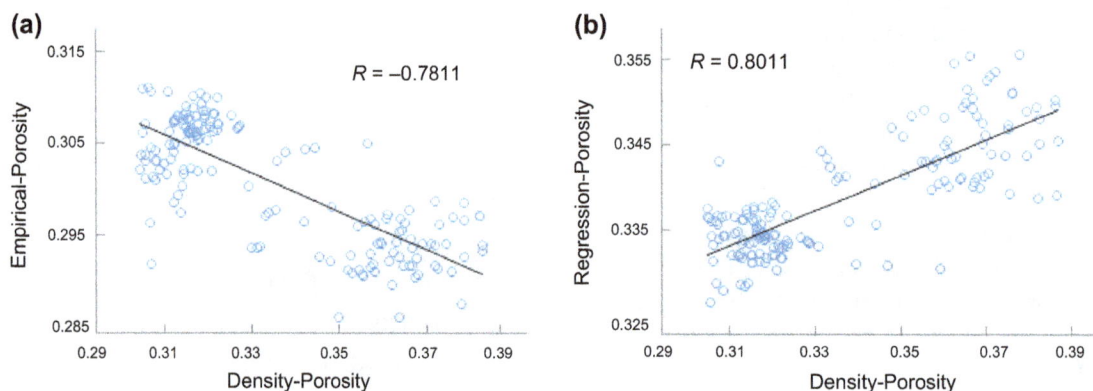

Fig. 9 **a** Cross plot between real and estimated porosity of testing set by PFE optimized by Tikhonov method in shale units; **b** cross plot between real and estimated porosity of testing set by regression equation in shale units

Two cross sections are illustrated in inlines 228 and 339 of F3 block which provide the porosity distributions of the reservoir using the optimized PFE model (Fig. 10a, c) and regression equation (Fig. 10b, d). These outputs provided in time range of 400–1150 ms indicate porosity distribution of only the upper package which is described in the geological setting part. In this package, three sedimentary units (units 1, 2, and 3) are identified. The boundary between the units is plotted in Figs. 11 and 12 provided by PFE model in inlines 244 and 442, respectively.

10 Discussion

In the previous sections, the PFE was introduced as a nonlinear mathematical model to have ability to estimate porosity. This model in a nonlinear mode is dependent on the similarity attribute. At first, this model was fitted on the implementing set using a least square solution and evaluated with a different well. The initial results showed that the PFE needed a different method to regularize. Then Tikhonov regularization method was employed and

optimized PFE could present the results relatively well in sand units, while when a conventional cross-plotting method between AI and porosity is used, the accuracy of results declined in sand units. In addition to address the cross correlation of the developed models, the illustration of the spatial distribution of porosity predicted by both models is interpreted in the target zone. In order to better differentiate the differences between PFE and cross-plot analysis results, the outputs of both models are presented in the same sections in Fig. 10. In Fig. 10c, d, it is evident that the unit 2 has a higher porosity value than other units (1 and 3) and does not exhibit any significant variations except close to a vertical discontinuity which is known as a gas chimney anomaly. The presence of gas chimneys has been interpreted as hydrocarbon leakage pathways, and mapping of such chimneys by neural network techniques has been established as an exploration tool. Wells drilled inside gas chimneys typically have higher pore fluid pressure, higher mud gas readings, higher mud gas wetness, more hydrocarbon shows, lower velocities, and higher temperatures than wells drilled outside gas chimneys (Løseth et al. 2008). Gas chimney and fault volumes

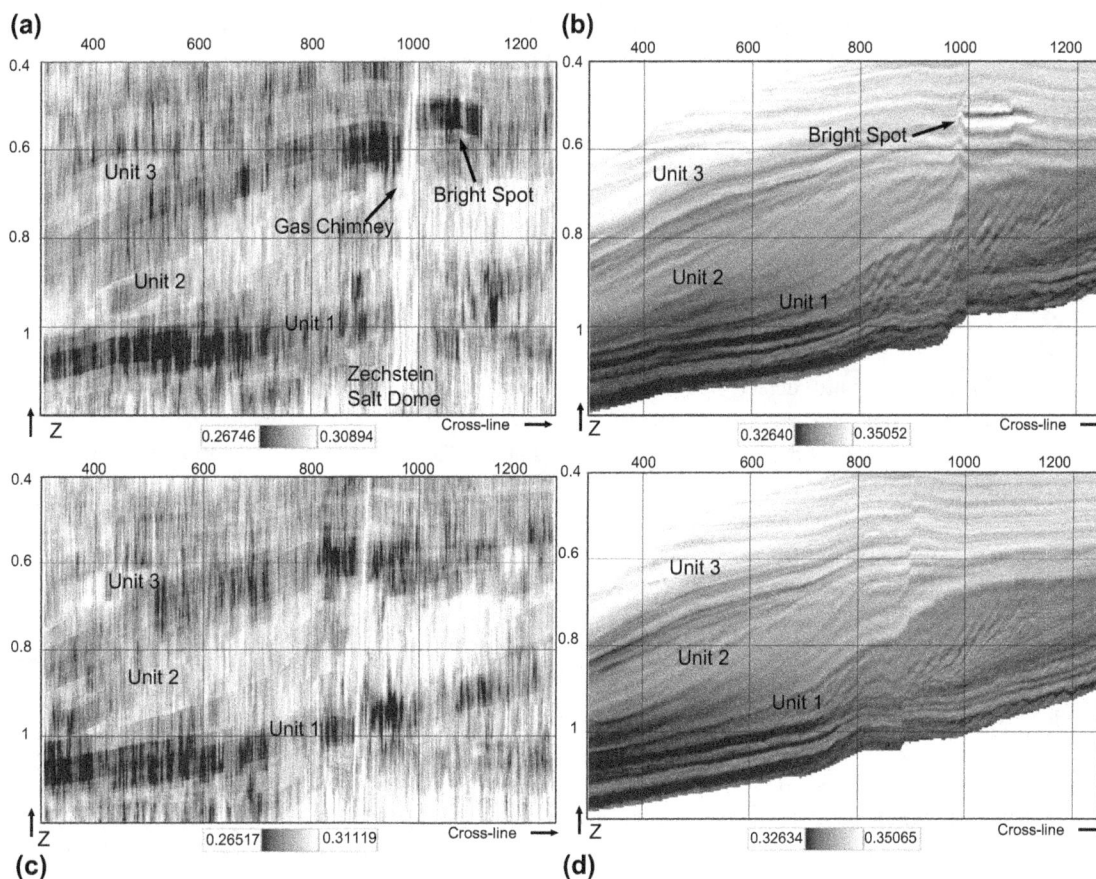

Fig. 10 a Distribution of porosity estimated by PFE in inline 228; b distribution of porosity estimated by linear regression in inline 228; c distribution of porosity estimated by PFE in inline 339; d distribution of porosity estimated by linear regression in inline 339

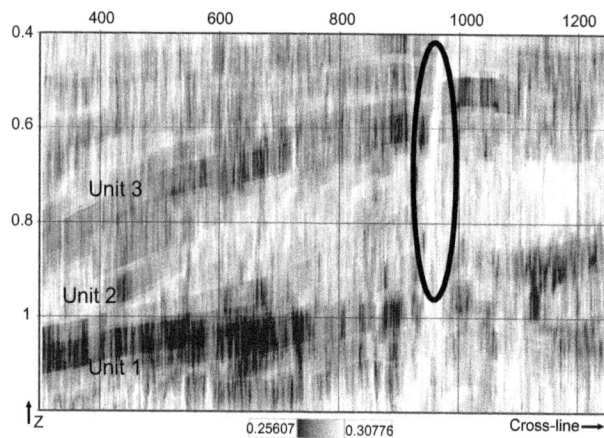

Fig. 11 Distribution of porosity estimated by PFE in inline 244

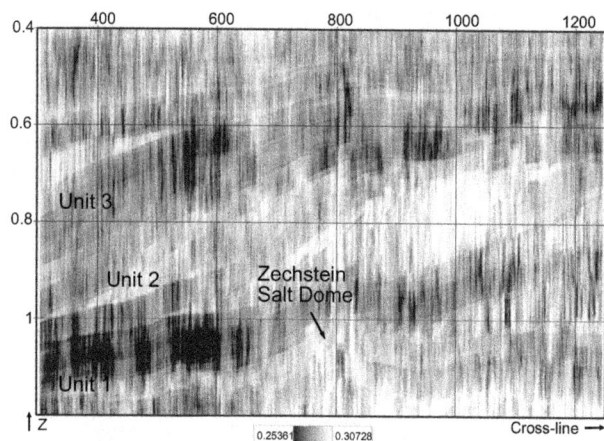

Fig. 12 Distribution of porosity estimated by PFE in inline 442

already defined in the PFE map, another anomaly could be found at about 530 ms in the map. In fact, F3 block contains a bright spot at about 530 ms possibly due to the presence of a gas pocket. Chopra and Marfurt (2007) demonstrate that reflections from gas-charged reservoir rocks showed much larger amplitudes than reflections from adjacent oil- or water-saturated zones. These are often known as bright spots. In the output of the PFE model (Fig. 10a), the bright spot is identified with a black arrow. In PFE results, multiple layers of shale and sand sediments are observed; however, it is proved that the PFE could not match the shale dataset (Figs. 11, 12). Because the PFE model was developed only using the dataset of sand sediments, this might allow it to tune more appropriately in the sand units than shale layers. However, despite the inaccuracy of the PFE model in the shale units, it seems that there is enough evidence of superiority in the results of PFE, and the observations suggest that the PFE model has performed well within the gas-bearing sand reservoir of the F3 block. Various seismic anomalies such as chimneys, faults, fractures, salt, bright spot, and sand bodies could be highlighted using the PFE technique that analyzes data with combinations of similarity attribute and PFE could present initial information about reservoir which is important for determination of optimum points for drilling operations.

11 Conclusions

This work provides a comparative analysis between a developed empirical model and conventional cross plotting to characterize a North Sea reservoir in term of porosity. The empirical model designed in a nonlinear mode has three unknown constants which were optimized using Tikhonov regularization based on the dataset taken from sand units. This method is different from the cross-plotting method, as it predicts porosity rather than acoustic impedance. This method did not succeed in matching the PFE on the shale dataset; therefore, this research concentrated on the behavior of the PFE on the sand dataset. A Tikhonov regularization parameter could improve the predictive power of the PFE and PFE validated with correlation coefficient equal to 0.72 for the testing set, while the same coefficient was only 0.47 for cross-plot analysis. But the developed models show a paradoxical behavior in shale units and it is evident that a single transform function such as PFE cannot be applied for estimation of petrophysical properties in various lithologies. The point that is significant in the seismic sections obtained by PFE is its capability in enhancing the gas chimneys. The reason for this behavior of PFE is that its intrinsic properties originate from the nature of similarity. The similarity attribute enhances the fault structures and salt edges. Unit 2, which

extracted from 3-D seismic data were rapidly becoming valuable tools for exploration and field development. In Fig. 10a, b, the PFE model could detect this anomaly which marks transition between the salt dome located in unit 1 (Zechstein) and near surface gas pockets in Fig. 12. The red polygon illustrated in Fig. 11 shows the possible areas of this occurrence. The reason for this behavior of optimized PFE is that its intrinsic properties originate from the nature of similarity. The similarity attribute enhances the fault structures and salt edges. Gas chimneys are a kind of fault structure whose similarity and amplitude attributes are usually used to detect these properties. In gas reservoirs, they identify the pathway of hydrocarbon migration and when the target of study is the determination of optimum drilling points, these structures could be an indicator to show the probable location of hydrocarbon accumulation. Also, in Figs. 10a and 12, a salt structure, the Zechstein salt dome, is identified by PFE in unit 1 and studying these structures is very important because they are traps for accumulation of hydrocarbon. In addition to the patterns is known as one of the main gas reservoirs of F3 block, shows higher porosity compared to the units 1 and 3 by PFE. According to the observations in the outputs of PFE, the ability to detect the geological structures such as faults (gas chimney), folds (salt dome), and bright spots besides porosity estimation of sandstone reservoirs could be a guideline to select the drilling points. The Tikhonov regularization approach showed that the bias represents a potentially significant component of the uncertainty in the results of calculations of inverse problem of PFE. Since the bias depends on something which is unknown for researchers, it will be necessary to use a priori information in order to estimate it. In this research, $[\text{cov}\,\mathbf{d}]$ is an important priori information assumed equal to the identity matrix while it could be considered with more accuracy. In the future work, determining $[\text{cov}\,\mathbf{d}]$ should be investigated and a probabilistic technique is proposed to apply to develop the PFE model with further predictive power. On the other hand, the physical base of PFE should be studied; if the physical relationship between porosity and similarity is investigated, probably the structure of PFE could be optimized.

References

Aminzadeh F, Groot PD. Neural networks and other soft computing techniques with applications in the oil industry. Amsterdam: EAGE; 2006.

Anderson JK. Limitations of seismic inversion for porosity and pore fluid: lessons from chalk reservoir characterization exploration. Denver: Society of Exploration Geophysicists, SEG Annual Meeting; 1996. p. 309–12.

Artun E, Mohaghegh S. Intelligent seismic inversion workflow for high-resolution reservoir characterization. Comput Geosci. 2011;37:143–57.

Baddari K, Aifa TA, Djarfour N, Ferahtia J. Application of a radial basis function artificial neural network to seismic data inversion. Comput Geosci. 2009;35:2338–44.

Bahorich M, Farmer S. The coherence cube. Lead Edge. 1995;14:1053–8.

Banchs RE, Michelena RJ. From 3D seismic attributes to pseudo-well-log volumes using neural networks: practical considerations. Lead Edge. 2002;21(10):996–1001.

Bhatt A, Helle HB. Committee neural networks for porosity and permeability prediction from well logs. Geophys Prospect. 2002;50:645–60.

Calderon JE, Castagna J. Porosity and lithologic estimation using rock physics and multi-attribute transforms in Balcon Field, Colombia. Lead Edge. 2007;26(2):142–50.

Chopra S, Marfurt KJ. Seismic attributes for prospect identification and reservoir characterization. Tulsa: Society of Exploration Geophysicists; 2007. p. 456.

Daniel PH, James SS, John AQ. Use of multiattribute transforms to predict log properties from seismic data. Geophysics. 2001;66(1):220–36.

Eichkitz CG, Amtmann J, Schreilechner MG. Enhanced coherence attribute imaging by structurally oriented filtering. First Break. 2012;30(3):75–81.

Gregersen U. Sequence stratigraphic analysis of Upper Cenozoic deposits in the North Sea based on conventional and 3-D seismic data and well-logs. Ph.D. thesis, University of Aarhus; 1997.

Hadamard J. Sur les problèmes aux dérivées partielles et leur signification physique. Bull Univ Princet. 1902;13(49–52):28.

Hansen PC. Analysis of discrete ill-posed problems by means of the L-curve. SIAM Rev. 1992;34:561–80.

Helle HB, Bhatt A, Ursin B. Porosity and permeability prediction from wireline logs using artificial neural networks: a North Sea case study. Geophys Prospect. 2001;49:431–44.

Hilgendorf A. Linear and nonlinear models for inversion of electrical conductivity profiles in field soils from EM38 measurements. M.Sc. Thesis, Institute of Mining and Technology; 1997.

Joel DW, et al. Interpreter's corner—seismic reservoir characterization of a U.S. Midcontinent fluvial system using rock physics, poststack seismic attributes and neural networks. Lead Edge. 2002;21:428–36.

Kevin PD, Curtis AL. Genetic-algorithm/neural-network approach to seismic attribute selection for well-log prediction. Geophysics. 2004;69(1):212–21.

Leiphart DJ, Hart BS. Comparison of linear regression and a probabilistic neural network to predict porosity from 3-D seismic attributes in Lower Brushy Canyon channeled sandstones, southeast New Mexico. Geophysics. 2001;66(5):1349–58.

Leite EP, Vidal AC. 3D porosity prediction from seismic inversion and neural networks. Comput Geosci. 2011;37:1174–80.

Lim JS. Reservoir properties determination using fuzzy logic and neural networks from well data in offshore Korea. J Petrol Sci Eng. 2005;49:182–92.

Liu ZP, Liu JQ. Seismic-controlled nonlinear extrapolation of well parameters using neural networks. Geophysics. 1998;63(6):2035–41.

Luthi SM. Geological well logs: their use in reservoir modeling. New York: Springer; 2001.

Løseth H, Wensaas L, Arntsen B, Gading M. Gas chimneys and other hydrocarbon leakage anomalies interpreted on seismic data. Oslo: International geological congress; 2008.

Marfurt KJ, Kirlin RJ, Farmer SL, Bahorich MS. 3-D seismic attributes using a semblance based coherency algorithm. Geophysics. 1998;63:1150–65.

Menke W. Geophysical data analysis: discrete inverse theory (Revised Edition). Waltham: Academic Press; 1989.

OpendTectdGB Plugins User Documentation version 4.2. dGB Earth Sciences. Copyright © 2002–2011. http://opendtect.org/rel/doc/User/dgb/index.htm. Accessed 25 Aug 2012.

Pramanik AG, et al. Estimation of effective porosity using geostatistics and multiattribute transforms: a case study. Geophysics. 2004;69(2):352–72.

Russell B, Hampson D, Schuelke J, Quirein J. Multiattribute seismic analysis. Lead Edge. 1997;16:1439–43.

Santosh D, Aditi B, Poonam K, Priyanka S, Rao PH, Hasan SZ, Harinarayana T. An integrated approach for faults and fractures delineation with dip and curvature attributes. In: 10th biennial international conference and exposition on petroleum geophysics, Kochi, 2013. p. 23–5.

Satinder C, Kurt JM. Emerging and future trends in seismic attributes. Lead Edge. 2008; 27:298–318.

Schultz PS, Ronen S, Hattori M, Corbett C. Seismic-guided estimation of log properties (Part 1: a data-driven interpretation methodology). Lead Edge. 1994;13:305–10.

Singh V, Painuly PK, Srivastava AK, Tiwary DN, Chandra M. Neural networks and their applications in lithostratigraphic interpretation of seismic data for reservoir characterization. In: 19th world petroleum congress, Madrid, 2007; 1244–60.

Sørensen JC, Gregersen U, Breiner M, Michelsen O. High-frequency sequence stratigraphy of Upper Cenozoic deposits in the central and southeastern North Sea areas. Mar Pet Geol. 1997;14:99–123.

Steeghs P, Overeem I, Tigrek S. Seismic volume attribute analysis of the Cenozoic succession in the L08 block (Southern North Sea). Glob Planet Change. 2000;27:245–62.

Steel RGD, Torrie JH. Principles and procedures of statistics with special reference to the biological sciences. New York: McGraw Hill; 1960. p. 287.

Tiab D, Donaldson EC. Petrophysics—theory and practice of measuring reservoir rock and fluid transport properties. 2nd ed. Melbourne: Elsevier; 2004.

Tikhonov AN, Arsenin VY. Solution of Ill-posed problems. New York: Winston-Wiley; 1977.

Tigrek S. 3D seismic interpretation and attribute analysis of the L08 block, Southern North Sea Basin. M.S. thesis, Delft University of Technology; 1988.

Tonn R. Neural network seismic reservoir characterization in a heavy oil reservoir. Lead Edge. 2002;21(3):309–12.

Wabha G. Spline models for observation data. Philadelphia: SIAM; 1990. p. 59.

Walls JD, et al. Seismic reservoir characterization of a U.S. Midcontinent fluvial system using rock physics, poststack seismic attributes, and neural networks, Society of Exploration Geophysicists, SEG Annual Meeting, Calgary, Alberta, 2000. p. 428–36.

Wiener N. Extrapolation, interpolation, and smoothing of stationary time series. Cambridge: MIT Press; 1942.

14

Numerical simulation of high-resolution azimuthal resistivity laterolog response in fractured reservoirs

Shao-Gui Deng[1] · Li Li[1] · Zhi-Qiang Li[2] · Xu-Quan He[3] · Yi-Ren Fan[1]

Abstract The high-resolution azimuthal resistivity laterolog response in a fractured formation was numerically simulated using a three-dimensional finite element method. Simulation results show that the azimuthal resistivity is determined by fracture dipping as well as dipping direction, while the amplitude differences between deep and shallow laterolog resistivities are mainly controlled by the former. A linear relationship exists between the corrected apparent conductivities and fracture aperture. With the same fracture aperture, the deep and shallow laterolog resistivities present small values with negative separations for low-angle fractures, while azimuthal resistivities have large variations with positive separations for high-angle fractures that intersect the borehole. For dipping fractures, the variation of the azimuthal resistivity becomes larger when the fracture aperture increases. In addition, for high-angle fractures far from the borehole, a negative separation between the deep and shallow resistivities exists when fracture aperture is large as well as high resistivity contrast exists between bedrock and fracture fluid. The decreasing amplitude of dual laterolog resistivity can indicate the aperture of low-angle fractures, and the variation of the deep azimuthal resistivity can give information of the aperture of high-angle fractures and their position relative to the borehole.

Keywords High-resolution azimuthal resistivity laterolog · Fractured reservoir · Fracture dipping angle · Fracture aperture · Fracture dipping direction

1 Introduction

Fracture is the smallest and the most complex structure in the crust. It can not only increase the pore space and permeability, but also control the formation, distribution, and capacity of oil and gas in place (Jiang et al. 2004; Zeng et al. 2007; Zhang et al. 2009; Weng et al. 2011; Nie et al. 2012; Kuchuk and Biryukov 2014; Reynolds et al. 2014; Yao et al. 2013; Zhao et al. 2014). Fracture identification and quantitative characterization are keys to effective exploration of fractured reservoirs (Bourbiaux 2010; Sun et al. 2011). Dual laterolog is widely used to study fracture development and fracture porosity because of good current focusing and detectability of fractured formation (Sibbit and Faivre 1985; Li et al. 1996; Deng and Li 2009; Noroozi et al. 2010; Le et al. 2011; Ja'fari et al. 2012; Deng et al. 2013). However, dual laterolog cannot accurately reflect complex heterogeneity and anisotropy in fractured reservoirs. Formation MicroScanner Image (FMI) figures can provide visual displays of sidewall geological characteristics of fractures, caves, etc. But shallow investigation depth limits its further application (Shen et al. 2009; Dershowitz et al. 2010; Sausse et al. 2010; Yang et al. 2011; Deng et al. 2012; Moinfar et al. 2010; Yun et al. 2013). Azimuthal resistivity imager (ARI) and high-resolution azimuthal

✉ Shao-Gui Deng
dengshg@upc.edu.cn

[1] School of Geosciences, China University of Petroleum, Qingdao 266580, Shandong, China

[2] China Research Institute of Radiowave Propagation, Xinxiang 453003, Henan, China

[3] China Petroleum Southwest Oil and Gas Field Branch, Chengdu 610051, Sichuan, China

Edited by Jie Hao

laterolog sonde (HALS) were earlier proposed to study three-dimensional distribution of resistivity surrounding the borehole. However, the logging response mechanism is not well understood and application examples about fractures are rare (Faivre 1993; Davies et al. 1994; Smits et al. 1995; Yang and Tao 1999; Karim et al. 2013; Olsen et al. 2014). This study aims to implement numerical simulation of the high-resolution azimuthal resistivity laterolog (HARL), in order to combine the radial detection of dual laterolog and azimuthal detection around the borehole, and then corresponding logging response characteristics and identification method of fractures are investigated to aid fractured reservoir evaluation.

2 Three-dimensional finite element model of fractured reservoirs

2.1 Fundamental theory

HARL can provide two measurement modes, high-resolution dual laterolog mode and azimuthal resistivity

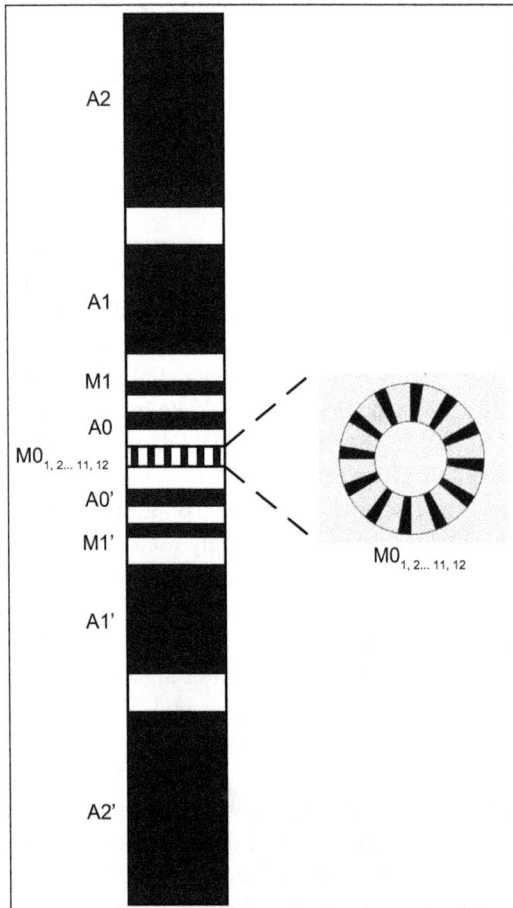

Fig. 1 High-resolution azimuthal resistivity laterolog

measurement mode. As shown in Fig. 1, differing from the conventional dual laterolog, the main current electrode of high-resolution dual laterolog mode is divided into A0 and A0′, and then an electrical potential guiding electrode (M0) is added between A0 and A0′. According to the potential difference of three electrical potential guiding electrodes, the instrument constantly adjusts the focusing voltage. This configuration not only improves the vertical resolution, but also significantly reduces the instrument length. Azimuthal resistivity measurement mode is achieved by the electric potential difference between 12 azimuthal electrodes disposed in the electrical potential guiding ring (M0). The angle of a single azimuthal electrode is 10°. The angle between the center axes of two adjacent azimuthal electrodes is 30°. Therefore, the HARL can simultaneously measure high-resolution dual laterolog and azimuthal resistivity and then obtain two images of shallow and deep investigation depth, respectively.

High-resolution dual laterolog mode sets the electrical potential equal between the average value of 12 azimuthal electrodes and the two electrical potential guiding electrodes (M1, M1′), then the expression formulas of high-resolution dual laterolog mode are obtained as follows:

$$\text{HARLd} = K_d \frac{U_{d(M1)}}{I_{d(A0)} + I_{d(A0')}} \tag{1}$$

$$\text{HARLs} = K_s \frac{U_{s(M1)}}{I_{s(A0)} + I_{s(A0')}}. \tag{2}$$

Based on the potential difference between the main current electrode and the 12 azimuthal electrodes, the expression formulas of azimuthal resistivity measurement mode can be obtained as follows:

$$\text{HARLd}(i) = K_d \frac{U_{d(M1)}}{I_{d(A0)} + I_{d(A0')}} \frac{\sum_{i=1}^{12} \Delta U_{d(M0_i A0)}}{12 \Delta U_{d(M0_i A0)}} \tag{3}$$
$$i = 1, 2, \ldots, 12$$

$$\text{HARLs}(i) = K_s \frac{U_{s(M1)}}{I_{s(A0)} + I_{s(A0')}} \frac{\sum_{i=1}^{12} \Delta U_{s(M0_i A0)}}{12 \Delta U_{s(M0_i A0)}} \tag{4}$$
$$i = 1, 2, \ldots, 12,$$

where K_d and K_s are coefficients of instrument in deep and shallow laterologs; $U_{d(M1)}$ and $U_{s(M1)}$ are potential values of electrical potential guiding electrode (M1) in deep and shallow laterologs; $I_{d(A0)}$, $I_{s(A0)}$, $I_{d(A0')}$, and $I_{s(A0')}$ are current values of main current electrodes (A0, A0′) in deep and shallow laterologs; $\Delta U_{d(M0_i A0)}$ and $\Delta U_{s(M0_i A0)}$ are potential difference between main current electrodes (A0, A0′) and the azimuthal electrode (M0$_i$) in deep and shallow laterologs, respectively.

2.2 Calculation principle

Assuming that a fractured reservoir consists of fractures and bedrock and that the fractures exist in the form of fracture groups, we use the plane model of parallel fractures (Zeng et al. 2007). As shown in Fig. 2, the conductivities of bedrock and fracture fluid are σ_b and σ_f and the aperture and dipping angle of each fracture are h and Ω, respectively. The response of HARL is a comprehensive effect of all media, including fractures and bedrock within the detection range. Assuming that the potential distribution generated by HARL around the borehole is U, the potential gradient of bedrock can be expressed as ∇U, and the potential gradient in any fracture plane can be decomposed into a normal component (E_{bn}) and a tangential component (E_{bt}), which are given by

$$E_{bn} = e_n(-\nabla U \cdot e_n) \tag{5}$$

$$E_{bt} = (-\nabla U \cdot e_t + e_n(\nabla U \cdot e_n) \cdot e_t) \cdot e_t, \tag{6}$$

where e_n is the normal direction of the fracture plane and e_t is the tangential direction of the fracture plane. Because the normal component of the current and the tangential

component of the potential in both sides of the fracture plane are continuous, the current density (J_f) of fracture part can be decomposed into normal (J_{fn}) and tangential (J_{ft}) components which are given by

$$J_{fn} = \sigma_b e_n(-\nabla U \cdot e_n) = \sigma_b(U_x \sin \Omega - U_z \cos \Omega)e_n \tag{7}$$

$$J_{ft} = \sigma_f[-\nabla U - (U_x \sin \Omega - U_z \cos \Omega)e_n]. \tag{8}$$

The formula of using the three-dimensional finite element method to simulate the logging response is given below:

$$\Phi = \frac{1}{2} \iiint_V J \cdot E \mathrm{d}x\mathrm{d}y\mathrm{d}z - \sum_E U_E I_E, \tag{9}$$

where Φ is the energy functional, V is the area of the three-dimensional space minus the electrode system, U_E is the potential of all electrodes, and I_E is the supply current. According to Eqs. (5)–(9), the functional equations of HARL in fractured formations can be written as

$$\Phi = \Phi_b + \Phi_f - \sum_E U_E I_E \tag{10}$$

$$\Phi_f = \frac{1}{2} \iiint_{V_f} [(\sigma_b^2/\sigma_f - \sigma_f)(-U_x \sin \Omega + U_z \cos \Omega)^2 + \sigma_f(\nabla U)^2]\mathrm{d}V_f \tag{11}$$

$$\Phi_b = \frac{1}{2} \iiint_{V_b} \sigma_b(\nabla U)^2 \mathrm{d}V_b, \tag{12}$$

where the integration function of Φ_b is only in bedrock and the integration function of Φ_f is only in fractures. Generally, $\phi_f << 1$ and $\sigma_f >> \sigma_b$ are established for fracture porosity and fracture fluid conductivity in fractured reservoir, respectively. So Eq. (12) can be written as

$$\Phi = \frac{1}{2} \iiint_V [(\sigma_b + \sigma_f \phi_f \cos^2 \Omega)(U_x)^2 + (\sigma_b + \sigma_f \phi_f)](U_y)^2 + (\sigma_b + \sigma_f \phi_f \sin^2 \Omega)(U_z)^2 + 2\sigma_f \phi_f \sin \Omega \cos \Omega \, U_x U_z]\mathrm{d}x\mathrm{d}y\mathrm{d}z - \sum_E U_E I_E, \tag{13}$$

where ϕ_f is the fracture porosity. According to Eq. (13) and the specific boundary conditions, we can use an improved frontal solver to rapidly calculate the response of HARL (Zhang 1984).

3 HARL response of fracture intersecting borehole

3.1 Calculation condition

The borehole diameter is 8 inches. Mud resistivity is 1 Ω m. Fracture fluid resistivity is 1 Ω m. The numerical simulation

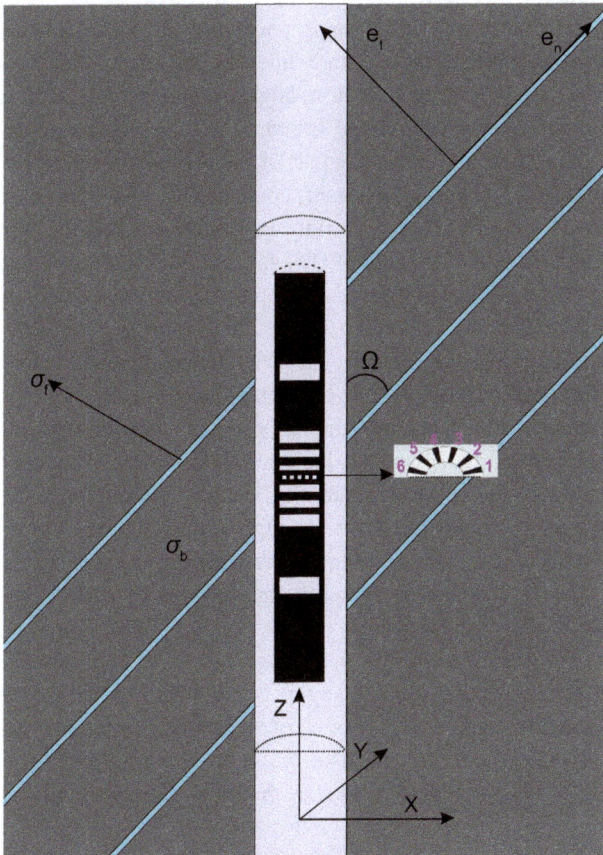

Fig. 2 Fractured reservoir model

model is axisymmetric. So we only observe six azimuthal resistivity curves, and the other six azimuthal resistivity curves are symmetrical about the symmetry plane. As shown in Fig. 2, the positive direction of the x-axis is set as the initial direction of the zero degree angle, and then the measured angle of the 1st–6th azimuthal resistivity successively increases. Because the response characteristics of deep and shallow azimuthal resistivities are similar, except in special circumstances, the article only shows deep azimuthal resistivity.

3.2 Relationship of HARL response and fracture dipping angle

The relationship of HARL response and different fracture dipping angles is shown in Fig. 3a, b, in which the

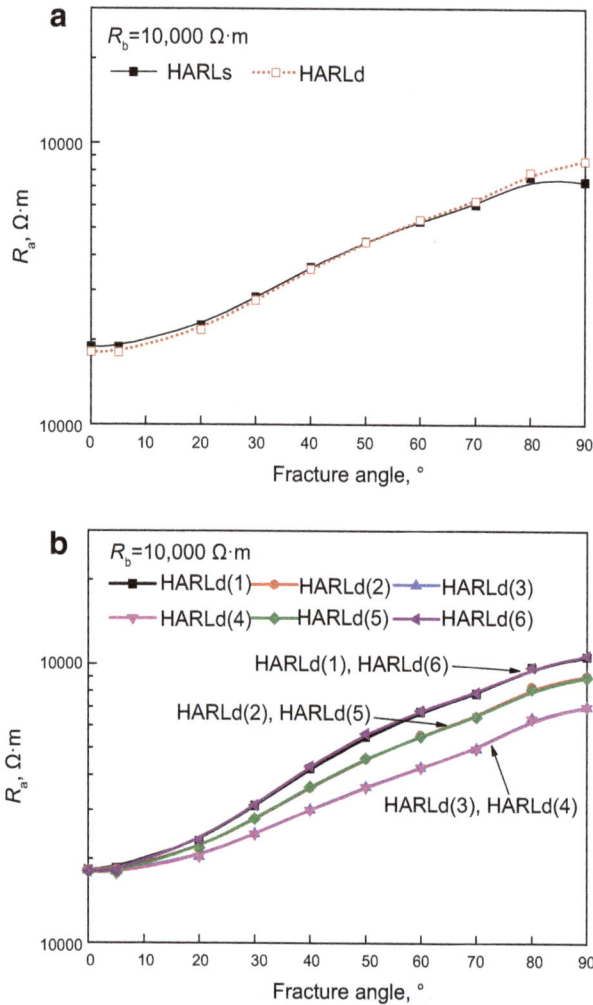

Fig. 3 Relationship of HARL response and different fracture dipping angles. **a** high-resolution dual laterolog curves, **b** deep azimuthal resistivity curves

fracture aperture is 50 μm and the bedrock resistivity is 10,000 Ω m. Negative separations between the deep and shallow resistivity occur in low-angle fractures, while positive separations occur in high-angle fractures. As the fracture dipping angle increases, the apparent resistivity and the variation of azimuthal resistivity increase. The 3rd and 4th azimuthal resistivity become the lowest when the two azimuthal electrodes are close to the fracture strike direction. The 1st and 6th azimuthal resistivity become the highest when the two azimuthal electrodes are close to the fracture dipping direction. The results indicate that high-resolution dual laterolog curves can reflect the fracture dipping angle, and azimuthal resistivity curves contain information of the fracture dipping direction.

3.3 Relationship of HARL response and fracture aperture

The relationship of the apparent resistivity in high-resolution dual laterolog mode and different fracture aperture is shown in Fig. 4a. As the fracture aperture increases, the apparent resistivity decreases. The decrease rate is larger for horizontal fractures than for vertical fractures. So the apparent resistivity of low-angle fractures is obviously lower than that of high-angle ones with the same fracture aperture. When the fracture aperture is large, there are negative separations for low-angle fractures and positive separations for high-angle fractures. The corrected apparent conductivity means the part of the bedrock conductivity minus the logging apparent conductivity. As shown in Fig. 4b, the bedrock resistivity is 3000 Ω m, and the dipping angles are respectively 0°, 45°, and 90°. The corrected apparent conductivities of deep and shallow laterologs are almost linearly related to the fracture aperture. Apparent conductivities increase as the fracture aperture increases. The conductivity of the horizontal fracture is large; in terms of dual laterolog, the response is stronger for low-angle fractures than for high-angle fractures. The deep conductivity is larger than the shallow conductivity for low-angle fractures, while for high-angle fractures the results are opposite. The bigger the fracture aperture is, the larger the separation between the deep and shallow conductivities becomes.

The relationship of azimuthal resistivity curves and different fracture dipping angles is shown in Fig. 4c, d. The azimuthal resistivity curves of horizontal fractures are completely overlapped. The variation of azimuthal resistivity becomes larger as the fracture aperture and the

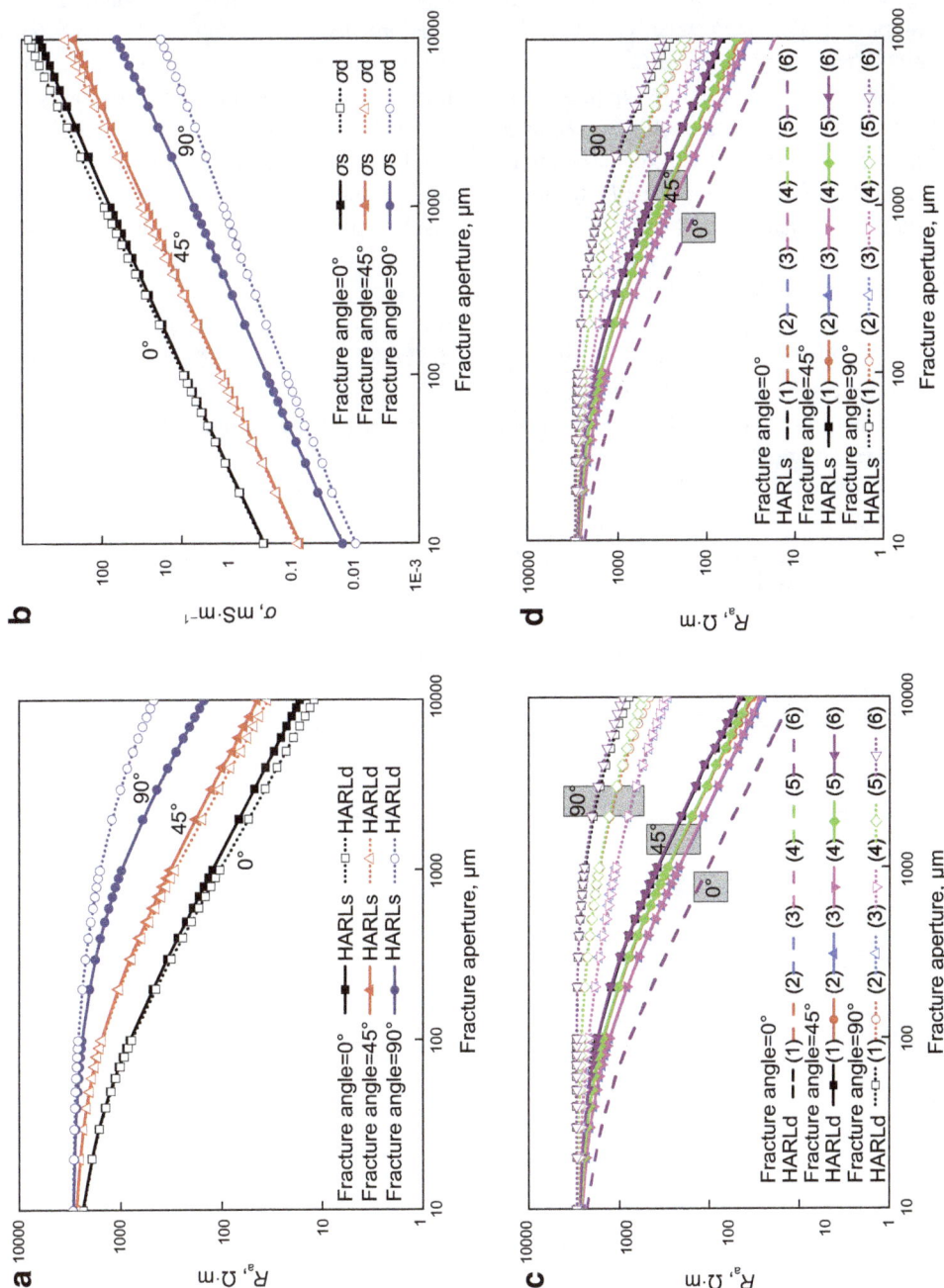

Fig. 4 Relationship of HARL response and different fracture aperture. **a** High-resolution dual laterolog curves, **b** corrected conductivities of high-resolution dual laterolog curves, **c** deep azimuthal resistivity curves, **d** shallow azimuthal resistivity curves

fracture dipping angle increase, which reaches maximum as the fracture dipping angle reaches 90°.

3.4 Azimuthal resistivity imaging of fractures

The azimuthal resistivity imaging of fractures with different dipping angles is shown in Fig. 5a. From top to bottom, fracture dipping angles are respectively 0°, 45°, and 75°; the bedrock resistivity is 3000 Ω m; and the fracture aperture is 200 μm. As the fracture dipping angle increases, the bending degree of sine curves becomes bigger and the apparent aperture of fractures becomes larger. When the aperture of fractures is small, the identification of fractures, especially high-angle fractures, becomes challenging, as shown in Fig. 5b, in which the aperture of fractures reduces to 5 μm with the same other conditions as in Fig. 5a.

When using the azimuthal resistivity to identify fractures, low-angle fractures can be identified by the decreased amplitude of apparent resistivity, and high-angle fractures can be identified by the variation of azimuthal resistivity. Assuming that the resistivity ratio of bedrock and fracture fluid is 3000:1, for low-angle fractures, the smallest detectable aperture can be less than 10 μm when we set 10 % (the dotted line in Fig. 6a) as an identification threshold in terms of the ratio of the resistivity of the fracture and the bedrock (the formula in Fig. 6a). For high-angle fractures, the smallest detectable aperture can be larger than 10 μm in terms of the ratio of the maximum variation of the

azimuthal resistivity and the laterolog resistivity (the formula in Fig. 6b). For medium-angle fractures, the smallest detectable aperture is also larger than 10 μm considering both resistivity amplitude difference and azimuthal variation, as the formulas and the dotted lines shown in Fig. 6c, d. When the resistivity ratio increases to 10,000:1, according to all the solid lines shown in Fig. 6, the detectable aperture becomes smaller. When the fracture aperture is large enough, the maximum variation no longer significantly increases and even decreases for medium-angle and high-angle fractures, as shown in Fig. 6b, d. The spatial resolution of the FMI-HD tool is 0.2 in (5.08 mm), representing the button size of each electrode. But the high-resolution electrodes are sensitive enough to identify fluid-filled fractures less than 10 μm in width.

4 HARL response of crossing fractures

A fracture group usually exists in the form of crossing fractures. This article only discusses the situation of two crossing fractures, as shown in Fig. 7, in which the bedrock resistivity is 10,000 Ω m and the fracture aperture is 50 μm. Assuming that one arbitrary dipping angle fracture respectively crosses with one horizontal fracture and one vertical fracture, by comparing their responses (solid lines) with the response of parallel fractures (dotted lines), the apparent resistivity of crossing fractures increases more slowly, as shown in

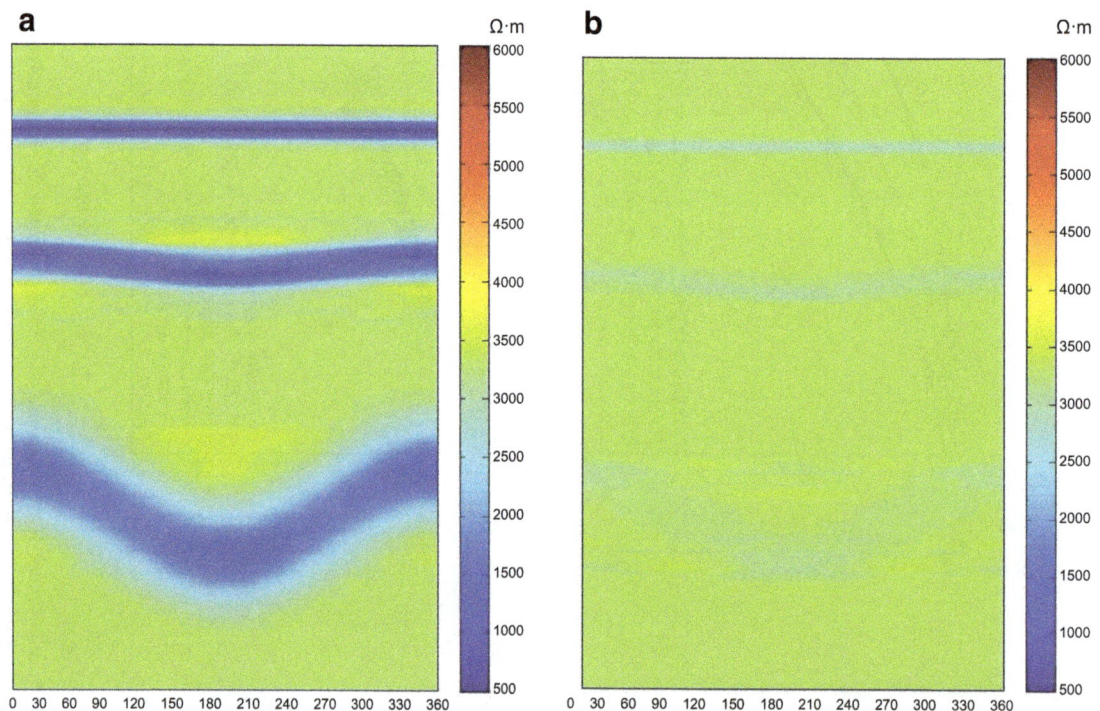

Fig. 5 Deep azimuthal resistivity imaging of fractures with different dipping angles. **a** Fracture aperture 200 μm, **b** fracture aperture 5 μm

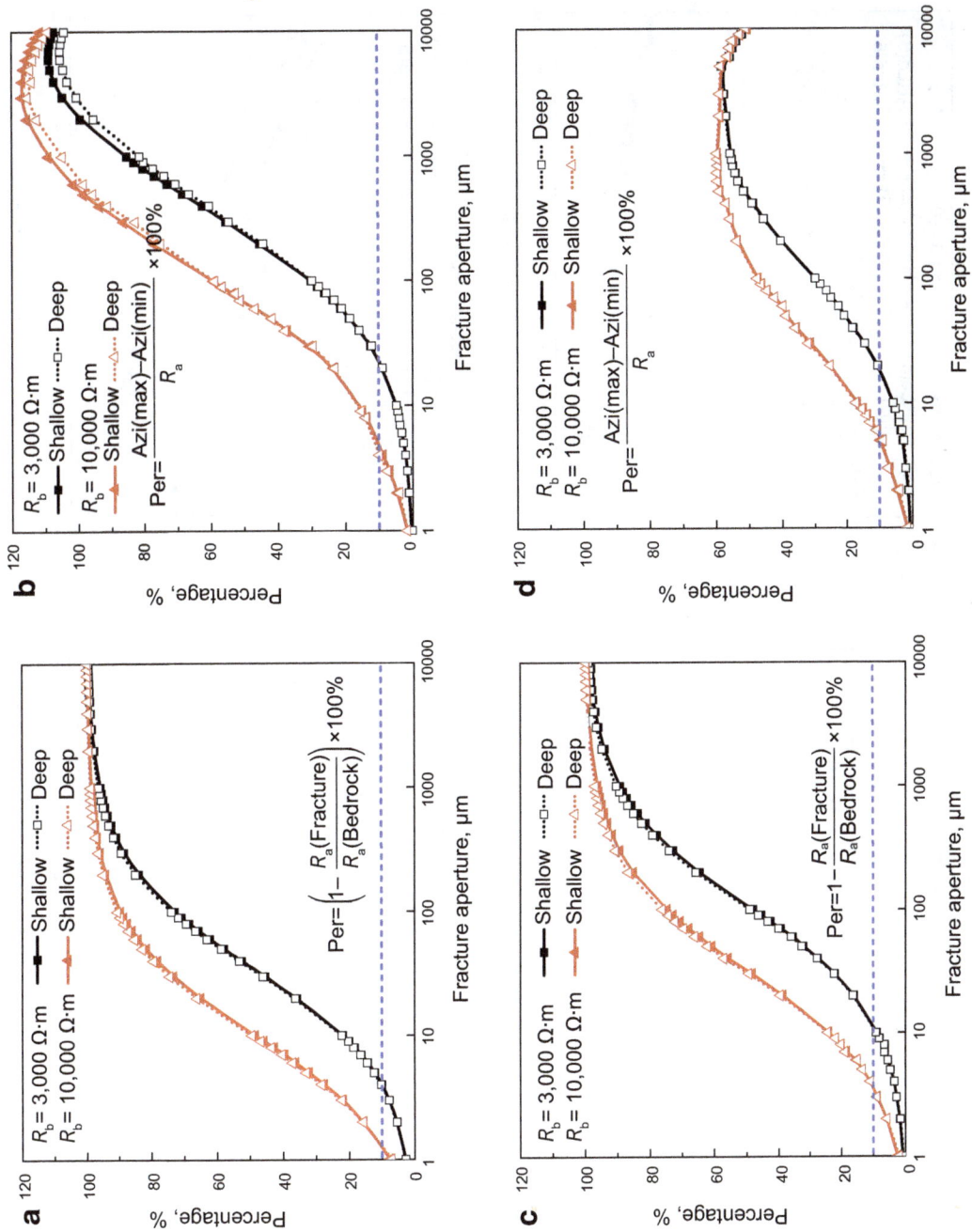

Fig. 6 Determine the smallest detectable fracture aperture. **a** The resistivity reduction ratio for horizontal fractures, **b** the maximum variation of azimuthal resistivity for vertical fractures, **c** the resistivity reduction ratio for 45° fractures, **d** the maximum variation of azimuthal resistivity for 45° fractures

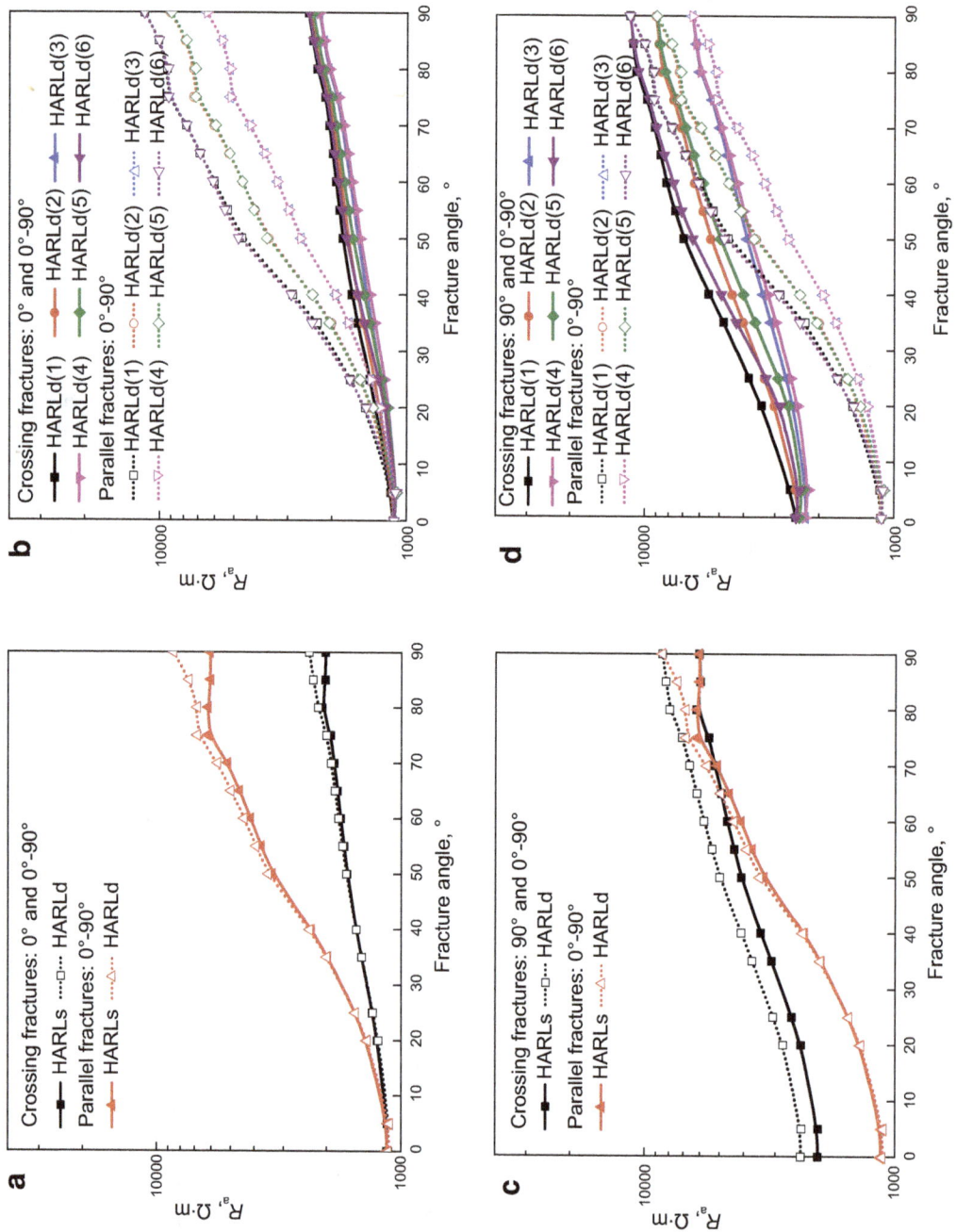

Fig. 7 HARL response of crossing fractures. **a** High-resolution dual laterolog curves, **b** deep azimuthal resistivity curves, **c** high-resolution dual laterolog curves, **d** deep azimuthal resistivity curves

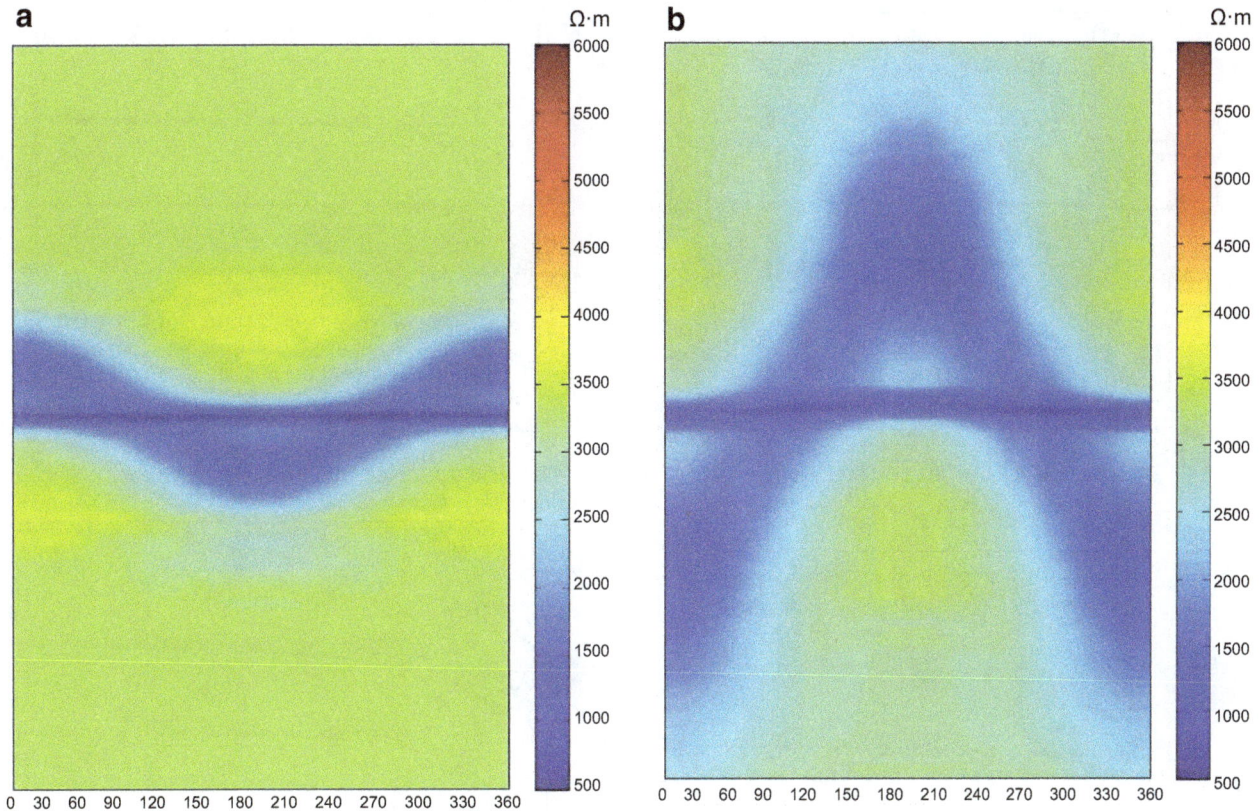

Fig. 8 Azimuthal resistivity imaging of crossing fractures. **a** Deep azimuthal resistivity imaging of 0° and 75° crossing fractures, **b** shallow azimuthal resistivity imaging of 120° and 170° crossing fractures

Fig. 7a, c. In comparison, the variation of azimuthal resistivity of crossing fractures that include horizontal fracture is weaker and those that include vertical fracture gradually increase as the dipping angle increases, as shown in Fig. 7b, d.

Figure 8a shows a deep azimuthal resistivity imaging of crossing fractures with one horizontal and one 75° high-angle fracture, and Fig. 8b shows a shallow azimuthal resistivity imaging of crossing fractures with one 120° and one 170° fracture, in which the bedrock resistivity is 3000 Ω m, and the fracture aperture is 200 μm. The results show that, in crossing fractures, the apparent resistivity is mainly affected by the low-angle fracture, while the variation of azimuthal resistivity is mainly affected by the high-angle fracture. The combination type of crossing fractures can be clearly displayed.

5 HARL response of high-angle fracture beside the borehole

Fracture and structure prediction beside the borehole is very important, but it is challenging for conventional logging. The HARL response is utilized to study vertical fractures beside the borehole, and the apparent resistivities

are shown in Fig. 9a, c, and e, in which the fracture apertures are 50, 200, and 1000 μm, and the bedrock resistivities are 10,000 and 3000 Ω m, respectively. When the vertical fracture intersects the borehole, which means that the distance from the fracture to the center of borehole is less than 0.1 m, the apparent resistivity is significantly small and a positive separation occurs between the deep and shallow resistivities. As the vertical fracture is gradually far away from the borehole, the positive separation gradually disappears. For the vertical fracture far from the borehole, if the fracture aperture and the resistivity ratio are large enough, negative separation occurs. This is because the impact of deep laterolog is greater than that of the shallow one, resulting in faster decrease of the deep lateral resistivity.

The azimuthal resistivity response is shown in Fig. 9b, d, and f. When the vertical fracture intersects the borehole, the azimuthal electrode closest to the fracture plane has the minimal resistivity, and the variation of azimuthal resistivity is obvious. When the vertical fracture is near the sidewall, the variation is the largest. As the vertical fracture is gradually far away from the borehole, the variation gradually weakens. As the distance between the 1st azimuthal electrode and vertical fracture is the nearest, it has the smallest resistivity, and the 6th

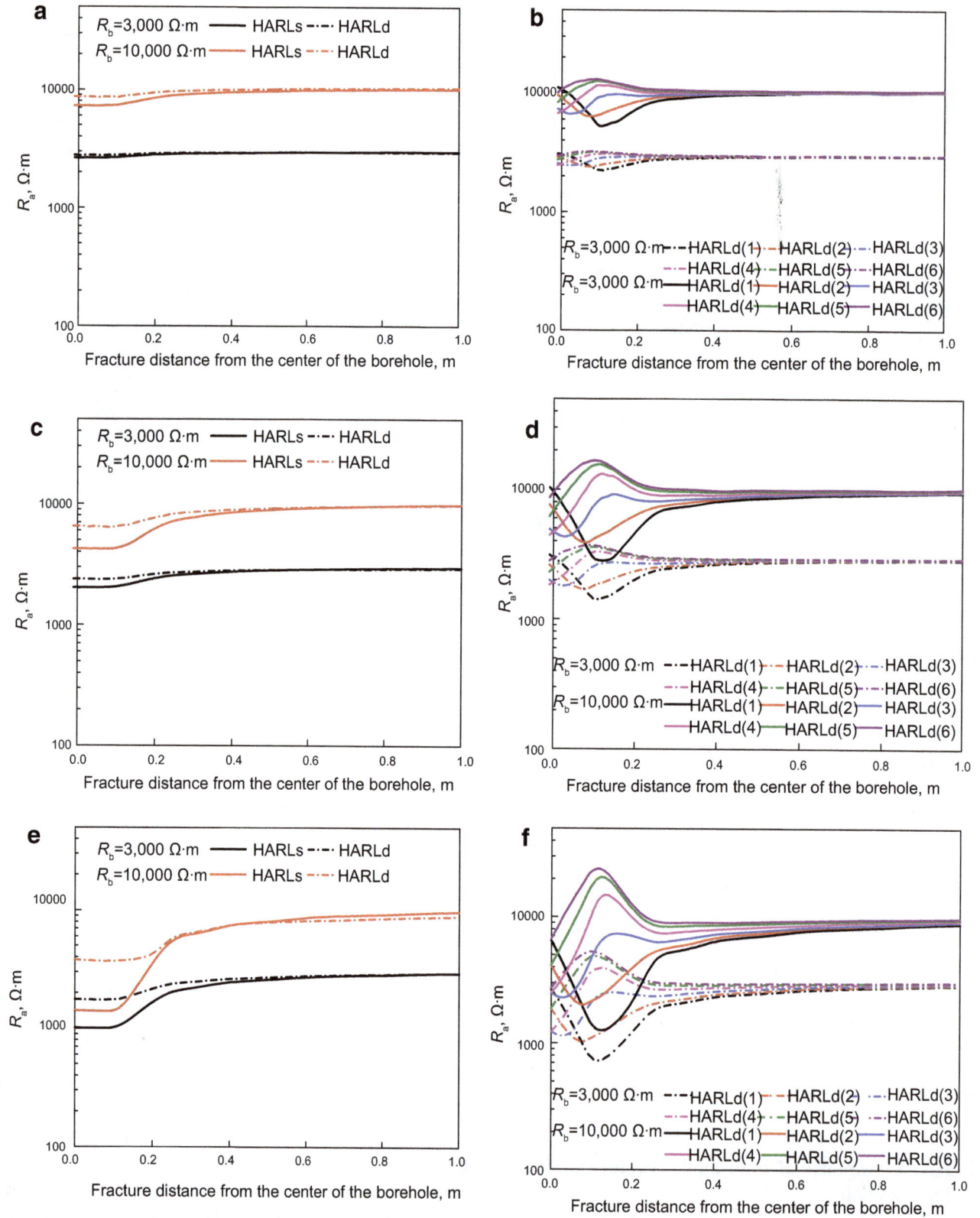

Fig. 9 HARL response of vertical fracture with different distances from the center of borehole. **a** High-resolution dual laterolog curves of a 50 μm fracture, **b** deep azimuthal resistivity curves of a 50 μm fracture, **c** high-resolution dual laterolog curves of a 200 μm fracture, **d** deep azimuthal resistivity curves of a 200 μm fracture, **e** high-resolution dual laterolog curves of a 1000 μm fracture, **f** deep azimuthal resistivity curves of a 1000 μm fracture

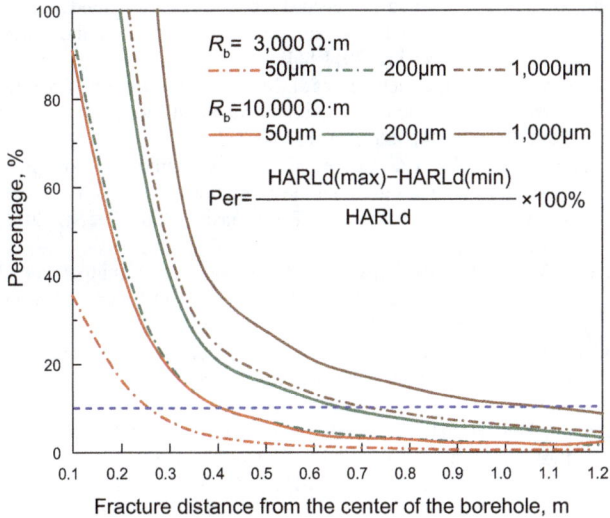

Fig. 10 Determination of the farthest detectable distance of a vertical fracture beside the borehole

electrode has the largest resistivity because of its farthest distance.

For vertical fractures beside the borehole, the farthest detectable distance is determined as the ratio of the maximum variation of deep azimuthal resistivity and the deep lateral resistivity that reaches 10 %. As shown in Fig. 10, when the bedrock resistivity is 3000 Ω m and the fracture apertures are 50, 200, and 1000 µm, the farthest detectable distances are 0.25, 0.40, and 0.7 m, respectively. When the bedrock resistivity is 10,000 Ω m, and the fracture apertures are 50, 200, and 1000 µm, the farthest detectable distances are 0.40, 0.67, and 1.1 m, respectively.

6 Conclusions

A good correlation exists between HARL and fracture aperture. A relatively low apparent resistivity and negative separations between deep and shallow resistivities present for low-angle fractures, while a relatively high azimuthal resistivity and positive separations exist for high-angle fractures.

Azimuthal resistivity imaging can be used to identify the presence of the fractures and their combination type. As the fracture dipping angle, aperture, or the resistivity ratio increases, the variation of azimuthal resistivity increases.

According to the positive or negative separations between the deep and shallow resistivities and the variation of azimuthal resistivity, high-angle fractures surrounding the borehole can be determined when the fracture aperture is relatively large and the resistivity ratio of bedrock and fracture medium is high.

Acknowledgments This research has been co-funded by the National Natural Science Foundation of China (41174099, 41474100), the Fundamental Research Funds for the Central Universities (14CX06077A), and National Major Science & Technology Projects of China (2011ZX05003, 2011ZX05009, 2011ZX05020, 2011ZX 05035).

References

Bourbiaux B. Fractured reservoir simulation: a challenging and rewarding issue. Oil & Gas Sci Technol-Rev IFP. 2010;65(2): 227–38.

Davies DH, Faivre O, Gounot MT, et al. Azimuthal resistivity imaging: a new generation laterolog. SPE Form Eval. 1994;9(03):165–74 SPE 24676.

Deng SG, Li ZQ. Simulation of array laterolog response of fracture in fractured reservoir. Earth Sci: J China Univ Geosci. 2009;34(5): 841–7 (in Chinese).

Deng SG, Mo XX, Lu CL, et al. Numerical simulation of the dual laterolog response to fractures and caves in fractured-cavernous formation. Pet Explor Dev. 2012;39(6):751–7 (in Chinese).

Deng SG, Wang Y, Hu YY, et al. Integrated petrophysical log characterization for tight carbonate reservoir effectiveness: a case study from the Longgang area, Sichuan Basin, China. Pet Sci. 2013;10(3):336–46.

Dershowitz WS, Cottrell MG, Lim DH, et al. A discrete fracture network approach for evaluation of hydraulic fracture stimulation of naturally fractured reservoirs. In: 44th U.S. rock mechanics symposium and 5th U.S.–Canada rock mechanics symposium, June 27–30, Salt Lake City, Utah (ARMA 10475); 2010.

Faivre O. Fracture evaluation from quantitative azimuthal resistivity. In: Annual technical conference and exhibition. SPE annual technical conference and exhibition, October 3–6, Houston, Texas (SPE 26434); 1993.

Ja'fari A, Kadkhodaie-Ilkhchi A, Sharghi Y, et al. Fracture density estimation from petrophysical log data using the adaptive neuro-fuzzy inference system. J Geophys Eng. 2012;9(1):105.

Jiang LZ, Gu JY, Guo BC. Characteristics and mechanism of low permeability clastic reservoir in Chinese petroliferous basin. Acta Sedimentol Sin. 2004;22(1):13–8 (in Chinese).

Karim SM, Murugesu TS and Li B. The utilisation of high-definition borehole images to determine fracture properties and their relative age for fractured basement reservoirs. In: International petroleum technology conference, March 26–28, Beijing, China (IPTC 17144); 2013.

Kuchuk F, Biryukov D. Pressure-transient behavior of continuously and discretely fractured reservoirs. SPE Reserv Eval Eng. 2014; 17(1):82–97 (SPE 158096).

Le F, Zhou ZQ, Maurer HM, et al. 3-D study of resistivity tool responses in formations with vertical fractures or horizontal transverse isotropy. In: SPWLA 52nd annual logging symposium, May 14–18, Colorado Springs, Colorado; 2011.

Li SJ, Xiao CW, Wang HM, et al. Mathematical model of dual laterolog response to fracture and quantitative interpretation of fracture porosity. Acta Geophysi Sin. 1996;39(6):845–52 (in Chinese).

Moinfar A, Varavei A, Sepehrnoori K, et al. Development of a novel and computationally-efficient discrete-fracture model to study IOR processes in naturally fractured reservoirs. In: SPE improved oil recovery symposium, April 14–18, Tulsa, Oklahoma, USA (SPE 154246); 2012.

Nie RS, Meng YF, Jia YL, et al. Dual porosity and dual permeability modeling of horizontal well in naturally fractured reservoir. Transp Porous Media. 2012;92(1):213–35 (SPE 11242).

Noroozi MM, Moradi B and Bashiri G. Effects of fracture properties on numerical simulation of a naturally fractured reservoir. In: Trinidad and tobago energy resources conference, June 27–30, Port of Spain, Trinidad (SPE 132838); 2010.

Olsen TN, Germinario MP, Reinmiller R, et al. Horizontal lateral image analysis applied to fracture stage optimization in Eastern Barnett Shale, Tarrant and Dallas Counties, Texas. In: SPE/AAPG/SEG unconventional resources technology conference, August 25–27, Denver, Colorado, USA (SPE 1922149); 2014.

Reynolds MM, Bachman RC and Peters WE. A comparison of the effectiveness of various fracture fluid systems used in multistage fractured horizontal wells: Montney Formation unconventional gas. In: SPE hydraulic fracturing technology conference, February, 4–6, The Woodlands, Texas, USA (SPE 168632); 2014.

Sausse J, Dezayes C, Dorbath L, et al. 3D model of fracture zones at Soultz-sous-Forêts based on geological data, image logs, induced microseismicity and vertical seismic profiles. CR Geosci. 2010;342(7): 531–45.

Shen J, Su B, Guo N. Anisotropic characteristics of electrical responses of fractured reservoir with multiple sets of fractures. Pet Sci. 2009;6(2):127–38.

Sibbit AM and Faivre O. The dual laterolog response in fractured rocks. In: SPWLA 26th annual logging symposium, June 17–20, Dallas, Texas; 1985.

Smits JW, Benimeli D, Dubourg I, et al. High resolution from a new laterolog with azimuthal imaging. In: Society of petroleum engineers annual technical conference. 1995, p. 563–76 (SPE 30584).

Sun SZ, Zhou XY, Yang HJ, et al. Fractured reservoir modeling by discrete fracture network and seismic modeling in the Tarim Basin, China. Pet Sci. 2011;8(4):433–45.

Weng X, Kresse O, Cohen CE, et al. Modeling of hydraulic-fracture-network propagation in a naturally fractured formation. Soc Pet Eng. 2011;26(04):368–80 (SPE 140253).

Yang HJ, Sun SD, Cai LL, et al. A new method of formation evaluation for fractured and caved carbonate reservoirs: a case study from the Lundong area, Tarim Basin, China. Pet Sci. 2011; 8(4):446–54.

Yang W and Tao G. Forward and inversion of azimuthal lateral resistivity logs. In: SEG annual meeting, 31 October–5 November, Houston, Texas (SEG 1999-0120); 1999.

Yao J, Sun H, Fan DY, et al. Numerical simulation of gas transport mechanisms in tight shale gas reservoirs. Pet Sci. 2013;10(4): 528–37.

Yun Y, Hou HJ, Hui H, et al. Understanding reservoir properties through high-definition microresistivity images in horizontal shale oil wells drilled with OBM in China. In: SPE middle east oil and gas show and conference, March 10–13, Manama, Bahrain (SPE 164360); 2013.

Zeng LB, Li YG, Wang ZG, et al. Type and sequence of fractures in the second member of Xujiahe Formation at the south of western Sichuan Depression. Earth Sci: J China Univ Geosci. 2007;32(2): 194–200 (in Chinese).

Zhang GJ. Electrolog (I). Beijing: Petroleum Industry Press; 1984. p. 1–38 (in Chinese).

Zhang Z, Tong HM, Bao ZD. Development characteristics and quantitative prediction of reservoir fractures in the Chaoyanggou oil field. Min Sci Technol (China). 2009;19(3):373–9.

Zhao XM, Liu L, Hu JL, et al. The tectonic fracture modeling of an ultra-low permeability sandstone reservoir based on an outcrop analogy: a case study in the Wangyao Oilfield of Ordos Basin, China. Pet Sci. 2014;11(3):363–75.

15

Genetic mechanisms of secondary pore development zones of Es_4^x in the north zone of the Minfeng Sag in the Dongying Depression, East China

Yan-Zhong Wang[1] · Ying-Chang Cao[1] · Shao-Min Zhang[1] · Fu-Lai Li[1] · Fan-Chao Meng[1]

Abstract The genetic mechanisms of the secondary pore development zones in the lower part of the fourth member of the Shahejie Formation (Es_4^x) were studied based on core observations, petrographic analysis, fluid inclusion analysis, and petrophysical measurements along with knowledge of the tectonic evolution history, organic matter thermal evolution, and hydrocarbon accumulation history. Two secondary pore development zones exist in Es_4^x, the depths of which range from 4200 to 4500 m and from 4700 to 4900 m, respectively. The reservoirs in these zones mainly consist of conglomerate in the middle fan braided channels of nearshore subaqueous fans, and the secondary pores in these reservoirs primarily originated from the dissolution of feldspars and carbonate cements. The reservoirs experienced "alkaline–acidic–alkaline–acidic–weak acidic", "normal pressure–overpressure–normal pressure", and "formation temperature increasing–decreasing–increasing" diagenetic environments. The diagenetic evolution sequences were "compaction/gypsum cementation/halite cementation/pyrite cementation/siderite cementation–feldspar dissolution/quartz overgrowth–carbonate cementation/quartz dissolution/feldspar overgrowth–carbonate dissolution/feldspar dissolution/quartz overgrowth–pyrite cementation and asphalt filling". Many secondary pores (fewer than the number of primary pores) were formed by feldspar dissolution during early acidic geochemical systems with organic acid when the burial depth of the reservoirs was relatively shallow. Subsequently, the pore spaces were slightly changed because of protection from early hydrocarbon charging and fluid overpressure during deep burial. Finally, the present secondary pore development zones were formed when many primary pores were filled by asphalt and pyrite from oil cracking in deeply buried paleoreservoirs.

Keywords Secondary pore development zone · Genetic mechanism · Diagenetic evolution sequences · Secondary pores · Dongying depression

1 Introduction

As the degree of hydrocarbon exploration in middle-shallow formations continues to improve, deeply buried formations are gradually becoming important targets for hydrocarbon exploration (Hu et al., 2013; Sun et al. 2013, 2015). Studies of anomalously high porosity and permeability in deeply buried sandstone reservoirs by Bloch et al. (2002) showed that deep formations can still develop abnormally high porosity zones that can form oil and gas fields with commercial value (Bloch et al. 2002). Understanding the genesis of deep, abnormally high porosity zones (AHPZs) is important for precisely predicting deeply buried high-quality reservoirs. This issue has been studied many times. Although the controlling factors of the formation of deep, abnormally high porosity zones vary in different regions, most geologists generally believe that mineral dissolution, shallow fluid overpressure, early hydrocarbon charging, and grain rims are the main factors that control the development of deep AHPZs (Bloch et al. 2002; Ehrenberg 1993; Warren and Pulham 2001; Meng et al. 2011, 2010; Taylor et al. 2010; Wilkinson and

✉ Yan-Zhong Wang
wangyanzhong1980@163.com

[1] School of Geosciences, China University of Petroleum, Qingdao 266580, Shandong, China

Edited by Jie Hao

Haszeldine 2011; Cao et al. 2014; Yuan et al. 2015; Ajdukiewicz and Larese 2012; Wang et al. 2014; Jiang et al. 2009).

Secondary pore development zones (SPDZ) are a typical type of abnormally high porosity zone. Since Schmidt and McDonald (1977) proposed the theory that secondary pores in clastic reservoirs can form during diagenetic processes (Schmidt and McDonald 1977), scientists worldwide have made significant progress in identifying the distinguishing features, genetic mechanisms, distribution, and geological significance of secondary pores (Schmidt and McDonald 1979; Giles and De Boer 1990; Osborne and Swarbrick 1999; Higgs 2007; Zhu et al. 2006; Zhang 2007; Zeng 2001; Yuan and Wang 2001; Liu and Zhu 2006; Wang et al. 1995; Wang and Zhao 2001; Ma et al. 2005; Zhang et al. 2008; Dutton and Loucks 2010; Taylor et al. 2015). Currently, SPDZs in shallow layers, which correspond to the mature stage of organic matter evolution, are considered to be generated from the dissolution of aluminum silicates and carbonate minerals by organic acids and CO_2 from organic matter evolution (Zeng 2001; Yuan and Wang 2001; Liu and Zhu 2006; Wang et al. 1995; Zhang et al. 2014). However, different viewpoints exist regarding the generation of deep SPDZs. Some authors believe that these features form by mineral dissolution in low porosity reservoirs at deep burial depths (Yuan and Wang 2001). Other authors, however, believe that the large amounts of secondary pores that form in these shallow formations are effectively preserved during deep burial, while primary pores are destroyed, which results in the formation of deep SPDZs (Wang et al., 2001; Ma et al. 2005; Zhang et al. 2008; Bjørlykke and Jahren 2012; Bjørlykke 2014).

Petroleum exploration in the deeply buried nearshore subaqueous fans in the Es_4^x sub-member in the Minfeng Sag, Dongying Depression, has greatly improved in recent years. For example, the daily oil production from depths of 4316.6–4343 m in the Fengshen1 well in Es_4^x is 81.7 t, and the daily gas production is 118,336 m^3. The reservoirs in the nearshore subaqueous fans, which are closely related to gypsum layers and the source rocks, experienced a complex burial evolution, including tectonic subsidence—uplift—subsidence and an alternating acidic-alkaline diagenetic environment. The current reservoir space mainly consists of secondary pores, and primary pores have been mostly destroyed. Exploration of the Dongying Depression shows that four sets of source rocks exist in the Es_4^x, Es_4^s, Es_3^x, and Es_3^z sub-members and that the total thickness of these source rocks exceeds 2000 m.

The reservoirs in the Es_4^x sub-member could potentially form abundant secondary pores during shallow burial because the organic matter in Es_4^x's source rocks could supply organic acids in a shallow open system. However, all the deep SPDZs in the Paleogene sandstones in the

Dongying Depression are simply interpreted as having formed from deep burial mineral dissolution (Yuan and Wang 2001; Zhu et al. 2007). Some questions require further studies, such as whether these deep secondary pore development zones were originally shallow secondary pore development zones that were preserved effectively during deep burial and how these shallow secondary pore development zones were preserved during deep burial. These unresolved problems produce great difficulties and risks for the exploration and development of hydrocarbons in these reservoirs. For instance, the drilling of the Fengshen2 and Fengshen3 wells close to the Fengshen1 well was not successful (the former were dry wells and the latter only produced 2.64×10^4 m^3 of gas daily) (Zhong et al. 2004).

The characteristics and genetic mechanisms of the SPDZs in Es_4^x in the Minfeng Sag were systematically studied based on a combination of core observations, thin section identification, SEM observations, fluid inclusion analysis, core X-ray analysis, vitrinite reflectance tests, and core properties analysis, along with additional knowledge regarding the tectonic evolution history, thermal evolution of organic matter, and history of hydrocarbon accumulation in the study area. The results of this study show that the deep SPDZs in the study area experienced the formation of significant secondary pores at shallow depth and the occupation of many primary pores by massive asphalt and pyrite from oil cracking in deeply buried paleo-reservoirs under high temperature. This achievement is significant for re-evaluating the genetic mechanisms and distribution of deep high-quality reservoirs and the deployment of hydrocarbon exploration in deep formations.

2 Geological background

The Dongying Depression is a sub-tectonic unit that lies in the southeastern part of the Jiyang sub-basin of the Bohai Bay Basin in East China. This unit is a Meso-Cenozoic half graben rift-downwarped lacustrine basin, which developed on Paleozoic bedrock paleotopography (Yuan and Wang 2001). The Dongying Depression, which lies east of the Qingtuozi Salient, south of the Luxi Uplift and Guangrao Salient, west of the Linfanjia and Gaoqing Salients, and north of the Chenjiazhuang-Binxian Salient, covers an area of 5850 km^2 with an east–west axis of 90 km and a north–south axis of 65 km. Additionally, the depression is generally NE-trending. In profile view, the Dongying Depression is a half graben with a faulted northern margin and a gentle southern margin. In plan view, this depression is further subdivided into secondary structural units, such as a northern steep slope zone, a middle uplift, the Lijin, Minfeng, Niuzhuang, and Boxing sags, and a southern gentle slope zone (Zhang et al. 2006) (Fig. 1).

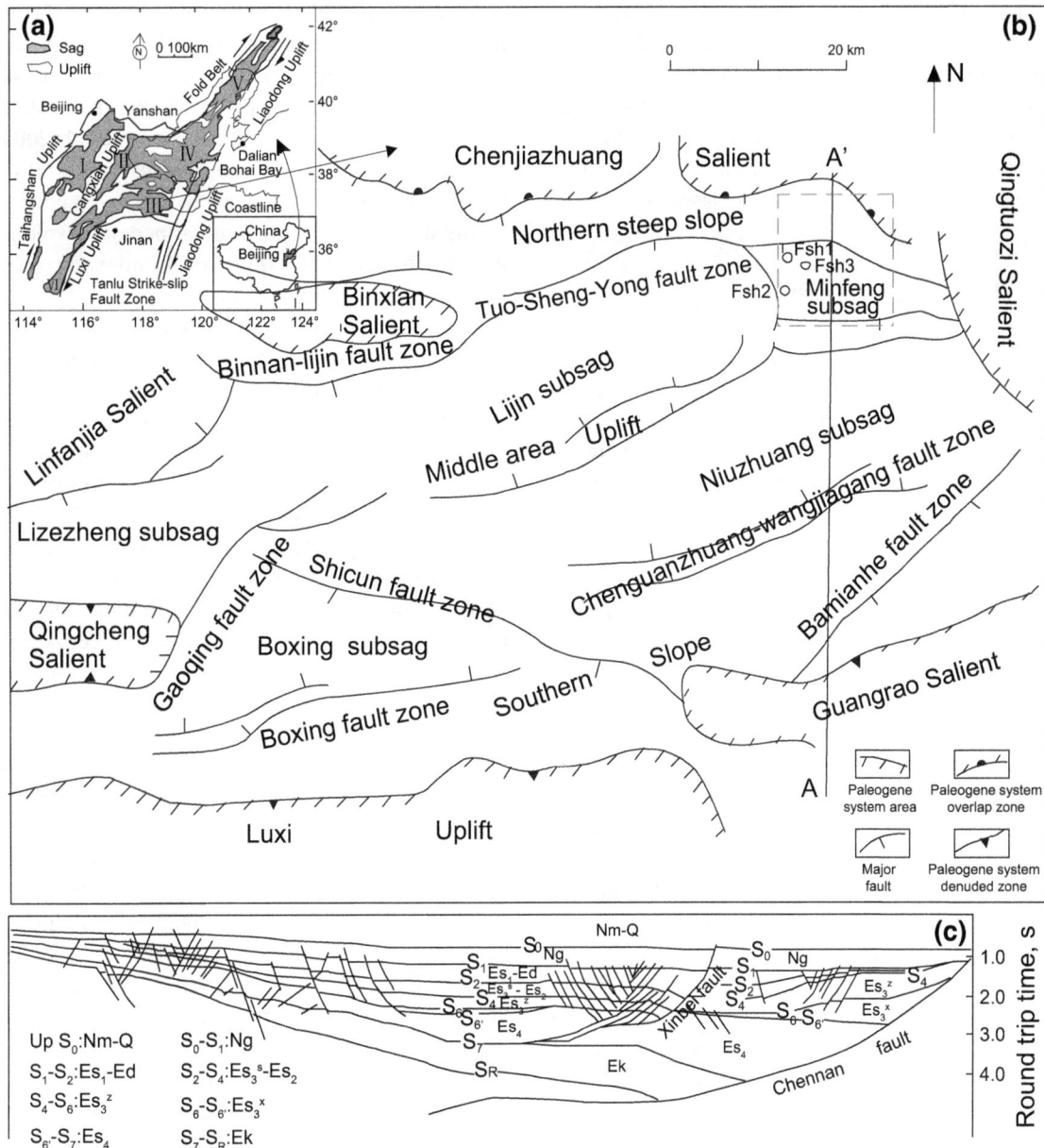

Fig. 1 a Tectonic setting of the Dongying Depression in the Jiyang Sub-basin (III) of the Bohai Bay Basin. Other sub-basins in the Bohai Bay Basin in East China include the Jizhong Sub-basin (I), Huanghua Sub-basin (II), Bozhong Sub-basin (IV), Liaohe Sub-basin (V), and Dongpu Sub-basin (VI) (according to Liu et al. 2012). **b** Structural map of the Dongying Depression with well locations. The area in the green line is the study area, which is located in the northern zone of the Minfeng Sag of the Dongying Depression. **c** Section of A–A′ in map (**b**). S_0, S_1, S_2, S_4, S_6, $S_{6'}$, S_7, and S_R are major seismic reflection boundaries. *Nm* Neocene Minghuazhen Formation; *Q* Quaternary Period; *Ng* Neocene Guantao Formation; *Ed* Paleogene Dongying Formation; *Es₁* The first member of the Paleogene Shahejie Formation (Es); *Es₂* The second member of Es; Es_3^s The upper part of the third member of Es; Es_3^z The middle part of the third member of Es; Es_3^x The lower part of the third member of Es; *Es₄* The fourth member of Es; *Ek* Paleogene Kongdian Formation (Liu et al. 2012)

The Minfeng Sag lies in the northeastern area of the Dongying Depression, north of the Chenjiazhuang Salient, south of the middle uplift, east of the Qingtuozi Salient, and west of the Lijin Sag (Fig. 1). During the depositional period of the Es_4^x sub-member, the Dongying Depression in this early rift stage was characterized by an arid climate, a

small lake-water area, and high salinity. The northern steep slope zone of the Minfeng Sag is a structural belt in the steep slope zone that is controlled by the Chennan boundary fault, which is located near the subsidence center with deeper water. Terrigenous clastic sediments were transported by seasonal floods to the deep lake, leading to the deposition of the nearshore subaqueous fans in the downthrown side of the Chennan fault. These fans were distributed close to lacustrine source rocks (Sui et al. 2010). During flood stagnation, the water evaporated rapidly, and thick gypsum and halite were deposited. In vertical profile view, the strata show a sedimentary assemblage of interbedded gypsum and clastic rocks.

3 Methodology and database

This study used cast thin sections and SEM observations to analyze the characteristics of the reservoir spaces based on effective reservoir porosity cutoffs. Porosity data were combined with porosity cutoffs to determine the distribution of SPDZs. The sedimentary characteristics, secondary pore features, and diagenetic evolution sequences of the reservoirs in these SPDZs were studied based on the identification of these zones. The genetic mechanism and evolutionary model of the secondary pore development zones, the evolution of the diagenetic environment, and the reservoir reconstruction process were discussed.

The database that was used in the study includes approximately 250-m cores from eight wells, porosity data from 172 samples, 70 thin sections, 30 cast thin sections, six SEM samples, and 80 fluid inclusion samples. The core samples were provided by the Geological Scientific Research Institute of the Sinopec Shengli Oilfield Company. The porosity was tested by a 3020-62 helium porosity analyzer at the Exploration and Development Research Institute of the Sinopec Zhongyuan Oilfield Company. The SEM samples were examined with a Quanta200 SEM with an EDAX energy dispersive X-ray spectrometer at the Geological Scientific Research Institute of the Sinopec Shengli Oilfield Company. The thin sections were made by the CNPC Key Laboratory of Oil and Gas Reservoirs at the China University of Petroleum and were examined by the authors with an Axio Scope A1 APOL digital polarizing microscope, which was produced by the German company Zeiss. The fluid inclusions were analyzed using a THMSG600 conventional inclusion temperature measurement system, which was produced by the British Company Linkam.

4 Results

4.1 Distribution of SPDZs

The term "secondary pore development zone" has been generally applied by scientists around the world but has not been defined precisely and scientifically. Most geologists

suggest that an SPDZ is the depth interval for which the real porosity evolution curve is higher than the normal porosity evolution curve and the percent of secondary pores is over 50 % (Liu and Zhu 2006; Zhu et al. 2007; Shi et al. 2005; Zhong et al. 2003; Zhang et al. 2003; Zheng and Wu 1996). However, the porosity values of reservoirs in secondary pore development zones and the methods that are used to determine the normal porosity evolution curve have not been clearly explained. Additionally, the methods that are used to determine the real porosity evolution curve have not yet been unified. For example, one method that is used to determine this curve is to fit the functional relationship between the porosity and depth (Liu and Zhu 2006; Zhu et al. 2007; Zhong et al. 2003; Zhang et al. 2003). Another method uses the porosity envelope curve of the porosity-depth profile as the real porosity evolution curve (Yuan and Wang 2001; Zhu et al. 2010; Liu et al. 2010). In this paper, an SPDZ is first defined as a zone where high porosity reservoirs with more than 50 % secondary pores develop. This definition includes three meanings: (1) the percent of secondary pores is greater than 50 %; (2) the porosity of the reservoirs is higher than the effective reservoir porosity cutoff because the absolute content of secondary pores is high; (3) high porosity reservoirs concentrate to form belts at a particular depth interval in the porosity-depth profile, with the porosity envelope curve bulging towards higher porosities.

The effective reservoir porosity cutoff is the basis for determining SPDZs. Only when the effective porosity cutoff is known, can the development of high porosity reservoirs be confirmed and the distribution of secondary pore development zones determined. Based on collections and arrangements of a large number of porosity and permeability data and the interpretation of oil, gas, water layers, and dry layers, the quantitative functional relationship between the effective porosity cutoff and burial depth from the Es_3^z to Es_4^x sub-members in the northern zone of the Minfeng Sag was obtained using the oil test method, which was developed by Wang (2010):

$$\varphi_{\text{cutoff}} = -8.1623 \ln(H) + 73.765 \quad R^2 = 0.8833$$

φ_{cutoff}: porosity cutoff, %; H: burial depth, m.

Using point-counting methods, a quantitative analysis of various pores in cast thin sections shows that the percentage of secondary pores in the reservoirs in the nearshore subaqueous fans in the Es_4^x sub-member is greater than 50 % (Fig. 2). Overlapping the φ_{cutoff} curve and the porosity envelope curve in the porosity-depth profile shows that two secondary pore development zones, whose depths range from 4200 to 4500 m and from 4700 to 4900 m, developed in the Es_4^x sub-member in the northern zone of the Minfeng Sag.

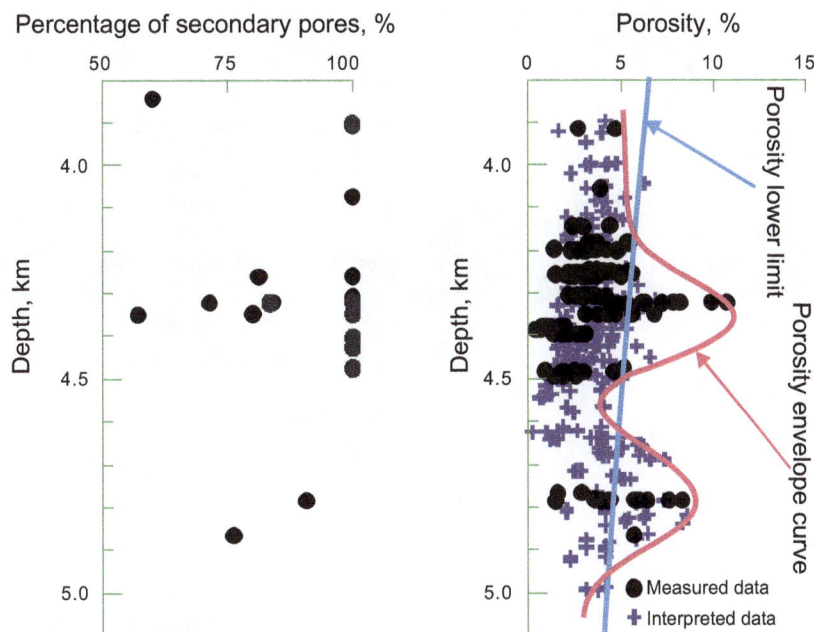

Fig. 2 By overlapping the effective reservoir porosity cutoff in the porosity-depth profile, the relationship between the porosity envelope curve and the porosity lower limit shows two secondary pore development zones, which range from 4200 to 4500 m and from 4700 to 4900 m, in Es_4^x in the northern zone of the Minfeng Sag

4.2 Reservoir characteristics in secondary pore development zones

4.2.1 Sedimentary characteristics

Multi-phase nearshore subaqueous fans developed in the Es_4^x sub-member in the northern zone of the Minfeng Sag. The nearshore subaqueous fans can be subdivided into inner fan, middle fan, and outer fan sub-facies according to the sedimentary features and hydrodynamic conditions. The inner fan sub-facies are dominated by major channels that are mainly filled with thick matrix-supported conglomerates and lack normal lacustrine mudstones between multi-phase fans. The poorly sorted conglomerates have a high proportion of matrix, with sub-angular grains floating among them, and scoured bases can be identified, which indicates proximal and rapid accumulation. The middle fans are dominated by braided channels and inter-distributaries. The lithology of the braided channels mainly consists of massive gravel sandstones and superimposed coarse sandstones with scoured bases, which have grain-supporting characteristics, medium-poor sorting, moderate thickness, and low matrix content. Normally graded bedding, scouring structure, and intensely contemporaneous deformation structures are present. The inter-distributaries are typical turbidites, which contain thin and fine-grained sediments with high matrix content. Lacustrine mudstones are generally deposited among multi-stage middle fans (Fig. 3). The lithology of the outer fan sub-facies includes

typical turbidites, which mainly consist of dark-gray mudstones with thin interbedded sandstones and pebbly sandstones (Fig. 3).

The relationship between different micro-facies in the nearshore subaqueous fans and their physical properties were identified based on sedimentary analyses of individual wells. This relationship shows that the high porosity reservoirs in the SPDZs are located in the center of thick sand beds that were deposited in braided channels in the middle fan, in contrast to the reservoirs in the inner and outer fans or the inter-distributaries in the middle fan. The low porosity reservoirs that correspond to the SPDZs include thin sand bodies in the outer fan, inter-distributaries in the middle fan, marginal reservoirs in thick sand beds in the middle fan, and thick conglomerates in the inner fan (Fig. 3).

4.2.2 Secondary pores

The secondary pores in the SPDZ reservoirs in the Es_4^x sub-member include pores that formed from the dissolution of feldspars and acid extrusive rock debris, detrital quartz grains and quartz overgrowths, carbonate and pyrite cements and compacted cracks in feldspars and other brittle grains. These secondary pores originated from the dissolution of feldspars and rock debris. They mainly occur as intra-granular pores and grain boundary pores (Appendix Fig. 7A, B, C). The quartz grains and quartz overgrowths usually dissolved along the boundaries, which formed irregular pores (Appendix Fig. 7D). Euhedral ankerite

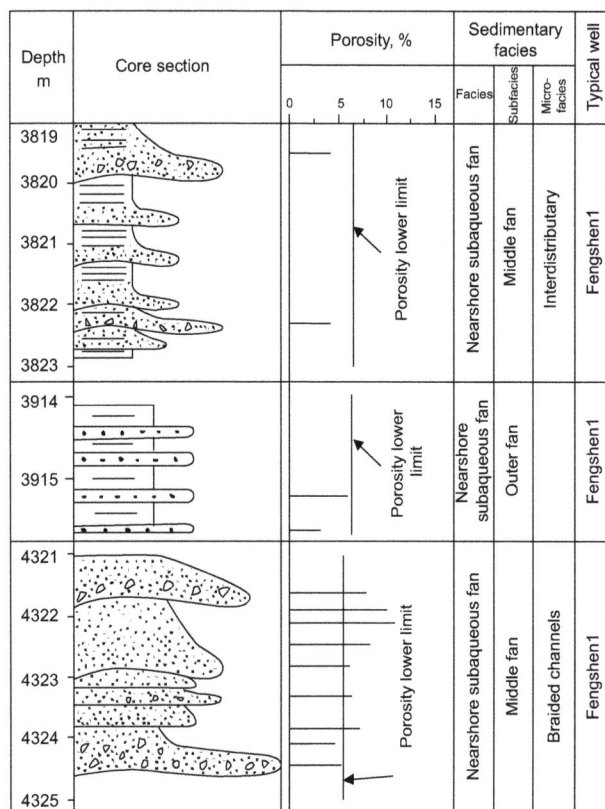

Fig. 3 Sedimentary characteristics and physical properties of the reservoirs in the Es_4^x sub-member in the northern zone of the Minfeng Sag. The effective reservoirs are the conglomerate in the central part of the positive sedimentary cycle of braided channels in the middle fan of the nearshore subaqueous fans

Table 1 Types and percentages of secondary pores in Es_4^x's secondary pore development zones in the northern zone of the Minfeng Sag

Secondary pore development zone Types of secondary pores	4200–4500 m	4700–4900 m
Feldspar-dissolved pores, %	25–70/48.6	62.5–72.9/ 67.7
Carbonate-dissolved pores, %	12.7–69.0/ 40.4	10.4–15.6/ 13.0
Rock debris-dissolved pores, %	0–6.33/1.58	0–6.25/3.13
Quartz-dissolved pores, %	0–12.5/4.7	6.25–15.6/ 10.9
Pyrite-dissolved pores, %	0–12.5/3.13	0–6.25/3.13
Cracks, %	0–6.33/1.58	0–4.17/2.08

Note: "25-70/48.6" means "Minimum–Maximum/Average"

The diagenetic evolution sequences of the reservoirs in the SPDZs were established based on an analysis of the types and features of the diagenesis, including the texture of authigenic minerals, the metasomatism-crosscutting relationship, the dissolution-filling relationship, and the homogenization temperatures of fluid inclusions.

The siderite cements are mainly granular and lumpy and are products of early diagenesis (Appendix Fig. 8A). The halite is completely crystalline (Appendix Fig. 8B), and the anhydrite was replaced by dolomite or ankerite, which suggests that the halite and gypsum were early cements, with gypsum turning into anhydrite after dehydration at high temperatures. Quartz overgrowths were replaced by ankerite and pyrite, which demonstrates that the quartz overgrowths formed earlier than the ankerite and pyrite (Appendix Fig. 8D, E). Multi-stage quartz overgrowths can be identified in thin sections. The homogenization temperature (Th) of the aqueous inclusions in the early stage of quartz overgrowths in Es_4^x is only 115 °C (Table 2). The combination of Th with burial and thermal history of the Fengshen8 well suggests the precipitation of the quartz cements at 42 Ma. The homogenization temperature of the late stage quartz overgrowths reached 155–160 °C (Table 2), which suggests that the quartz overgrowths occurred later. Ankerite was identified in the secondary pores in the feldspar grains (Appendix Fig. 8F) and carbonate cements (Appendix Fig. 7E, F), which indicates that the reservoir experienced two stages of acidic dissolution: an early stage of feldspar dissolution and a late stage of carbonate cement dissolution.

In an acidic geochemical environment, SiO_2 (aq) that is released from feldspar dissolution can precipitate in the form of quartz overgrowths. In this study, the homogenization temperatures of the oil inclusions in the quartz overgrowths and the fillings of the feldspar-dissolved pores

cement dissolved along its boundaries, while dolomite cement dissolved to form secondary pores in cements (Appendix Fig. 7E, F). Pyrite-dissolved pores are mostly found within cements (Appendix Fig. 7G). Compacted cracks in feldspars, which generally cut through the grains, are wide at one end and narrow at the other side of the grains (or are irregular) (Appendix Fig. 7H). Quantitative data regarding the amounts of different types of secondary pores in cast thin sections show that feldspar-dissolved pores and carbonate-dissolved pores dominate in the reservoirs in the SPDZs in the Es_4^x sub-member at depths from 4200 to 4500 m and from 4700 to 4900 m, followed by quartz-dissolved pores, a few acid extrusive rock debris-dissolved pores, pyrite-dissolved pores, and compacted cracks (Table 1).

4.2.3 Diagenetic evolution sequence

The diagenesis processes that occurred in the Es_4^x reservoirs in the SPDZs include the multi-stage dissolution of minerals (e.g., feldspar, carbonate, and quartz), multi-stage cementation (e.g., carbonate, silica, anhydrite, pyrite, and asphalt), and complex replacement (Appendices 1, 2, 3).

Table 2 Homogenization temperatures of fluid inclusions from reservoirs in Es_4^x in the northern zone of the Minfeng Sag

Well number	Depth, m	Horizon	Host minerals	Types	Inclusion number	Average homogenization temperature, °C
Feng8	4397.5	Es_4^x	Quartz overgrowth	Brine	6	115
Feng8	4200.7	Es_4^x	Quartz cement	Brine	5	124.1
Fengshen3	3785.6	Es_4^x	Quartz cement	Brine	8	133.9
Feng8*	4055.35	Es_4^x	Quartz overgrowth	Brine	3	143.3
Fengshen3	4867	Es_4^x	Quartz overgrowth	Brine	1	155
Fengshen3	4785.7	Es_4^x	Quartz overgrowth	Brine	1	155
Fengshen10	4260.6	Es_4^x	Quartz overgrowth	Brine	1	155.5
Fengshen3	4785.7	Es_4^x	Quartz overgrowth	Brine	1	160
Feng8*	4055.35	Es_4^x	Quartz overgrowth	Oil	3	99.1
Feng8*	4055.35	Es_4^x	Quartz overgrowth	Oil	5	112.6
Feng8*	4201.1	Es_4^x	Fillings of feldspar-dissolved pores	Oil	5	88.7
Feng8*	4055.35	Es_4^x	Fillings of feldspar-dissolved pores	Oil	9	91.9
Fengshen1*	4321.6	Es_4^x	Fillings of feldspar-dissolved pores	Oil	4	98.8
Feng8*	4201.1	Es_4^x	Fillings of feldspar-dissolved pores	Oil	4	108.1
Fengshen1*	4348.8	Es_4^x	Fillings of feldspar-dissolved pores	Oil	2	108.7
Feng8*	4055.35	Es_4^x	Fillings of feldspar-dissolved pores	Oil	4	109.6

* Means data from the Geological Scientific Research Institute of the Sinopec Shengli Oilfield Company

(such as SiO_2) mainly range from 88 °C to 110 °C, and the homogenization temperatures of paragenetic aqueous inclusions are about 115 °C, which suggest that the oil and aqueous inclusions formed simultaneously. This observation means that feldspar dissolution and early quartz overgrowth cementation occurred roughly during the same early period. Both carbonate cementation and quartz dissolution occur in an alkaline environment, so they may have formed during the same period. The replacement of feldspar overgrowths by ankerite (Appendix Fig. 8G) suggests that the ankerite formed later than the feldspar overgrowths, whereas the ankerite and feldspar overgrowths both formed in an alkaline environment, which indicates that they are probably products from the same period. Strong asphalt cementation is typical in the reservoirs in the SPDZs of Es_4^x. Many primary pores and various secondary pores (from the dissolution of feldspars, ankerite, quartz grains, and quartz overgrowths) are largely filled by asphalt (Appendix Fig. 8H, 3A, B, C, D, E, F). These textures suggest that asphalt formed very late, and many primary and secondary pores existed in the reservoirs before being filled with asphalt. For example, the thin section porosity of the asphalt in the reservoir at a depth of 4323.3 m in the Fengshen1 well is approximately 10 %, and the porosity that was filled by asphalt may be 22 % according to the relationship between the thin section porosity and core porosity. Song et al. (2009a) suggested that crude oil in the deep buried reservoirs in the Fengshen1 well started to crack into gas and asphalt during the late depositional period of the Minghuazhen Formation. Li et al. (2010a, b) proposed that the asphalt in the Fengshen1 well was a product of oil pyrolysis when temperatures exceeded 160 °C. Additionally, Li et al. (2010a, b) found that secondary pores in the feldspar grains were filled with asphalt, and tension fractures that are associated with the secondary pores were produced by overpressure from oil cracking (Appendix Fig. 8H). The above analysis shows that the asphalt formed relatively late. Before being filled with asphalt, the reservoirs should have had high porosity, which indicates that the porosities were well preserved during deep burial. Pyrite cements developed extensively in the northern zone of the Minfeng Sag. Partial cloddy pyrites are products of early cementation (Appendix Fig. 9G). Because mostly pyrite cements replaced quartz overgrowths (Appendix Fig. 8E), feldspar overgrowths, and ankerite (Appendix Fig. 9H), they should have formed during a late diagenetic stage. The textures of both the pyrite cements and asphalt suggest that they formed during the same period. During paragenesis with asphalt, pyrite is considered to be a reaction product of hydrogen sulfide (H_2S) from crude oil cracking under high temperatures and Fe^{2+} in reservoir fluids.

According to these comprehensive analyses, the diagenetic evolution sequence of the reservoirs in the SPDZs of Es_4^x is as follows: compaction/gypsum cementation/halite cementation/pyrite cementation/siderite cementation → feldspar dissolution/quartz overgrowth → carbonate cementation/quartz dissolution/feldspar overgrowth → carbonate dissolution/feldspar dissolution/quartz overgrowth → pyrite cementation and asphalt filling.

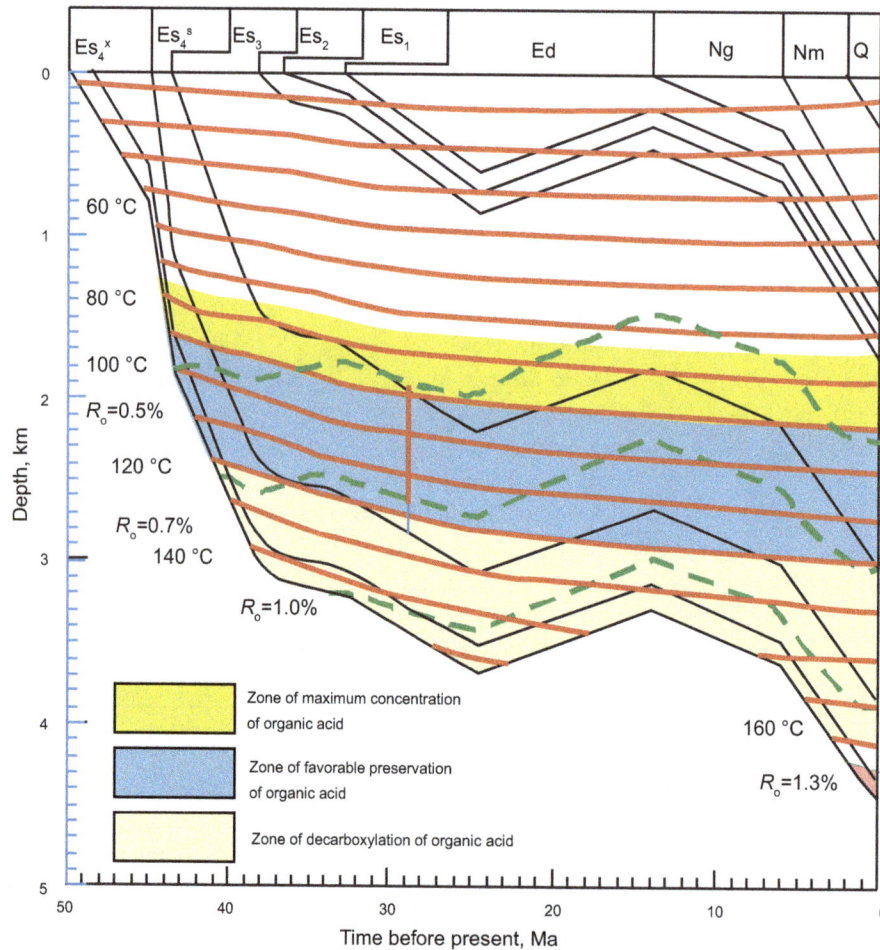

Fig. 4 The burial history and evolutionary history of organic matter from the Fengshen1 well (modified from Song et al. 2009a)

5 Discussion

Models of the diagenetic environment evolution and reservoir reconstruction were established using the burial evolution history of the Fengshen1 well based on research of the sedimentary characteristics, secondary pore features, and diagenetic evolution of the reservoirs in Es_4^x's SPDZs in the northern zone of the Minfeng Sag (Fig. 4). The genetic mechanism and evolutionary model of the SPDZs in Es_4^x are discussed from the perspective of the SPDZ at depths from 4200 m to 4500 m (Fig. 5, Fig. 6).

Gypsum-halite layers were deposited in the Es_4^x sub-member, and the thickness of the gypsum-halite layers in Es_4 is 1287.5 m in the Fengshen2 well and 267.7 m in the Fengshen1 well. Three sets of high-quality source rocks developed in Es_4 (Song et al. 2009a), and studies suggest that the gypsum-halite layer is contemporaneous with deep water source rocks. Gypsum precipitates under physico-chemical conditions with pH higher than 7.8 (Qiu and Jiang 2006), which suggests the development of an alkaline-reducing environment in the salt lake during the depositional period of the Es_4^x sub-member.

From the deposition of the Es_4^x sub-member to 44 Ma before the present (the end of the deposition of Es_4^s), the top boundary of the Es_4^x was buried to a depth shallower than 750 m at formation temperatures below 50 °C, and the bottom boundary was buried less than 1400 m at temperatures below 75 °C. The main diagenesis during this period was compaction, which led to the drainage of formation water. At this time, the salinity of the water in the pore spaces increased, which resulted in the early precipitation of gypsum and halite. Anaerobic bacteria broke down organic matter and SO_4^{2-} in the pore water, releasing organic acids, H_2S, CO_2, and other gases. Under these conditions, the Fe^{3+} in the sediments was reduced to Fe^{2+} and formed spherulitic pyrite and agglomerate siderite cement (Curtis 1978). Because the organic acids that formed during this period were mostly destroyed by bacteria, the formation water remained alkaline, and the formation exhibited normal fluid pressure. The conglomerate bodies of the nearshore subaqueous fans were similar to dome-shaped anticlines, with a flat bottom and convex top in a cross-section and was defined by Zhong et al. (2004) as "fan-anticlines" that formed by sedimentation. The inner

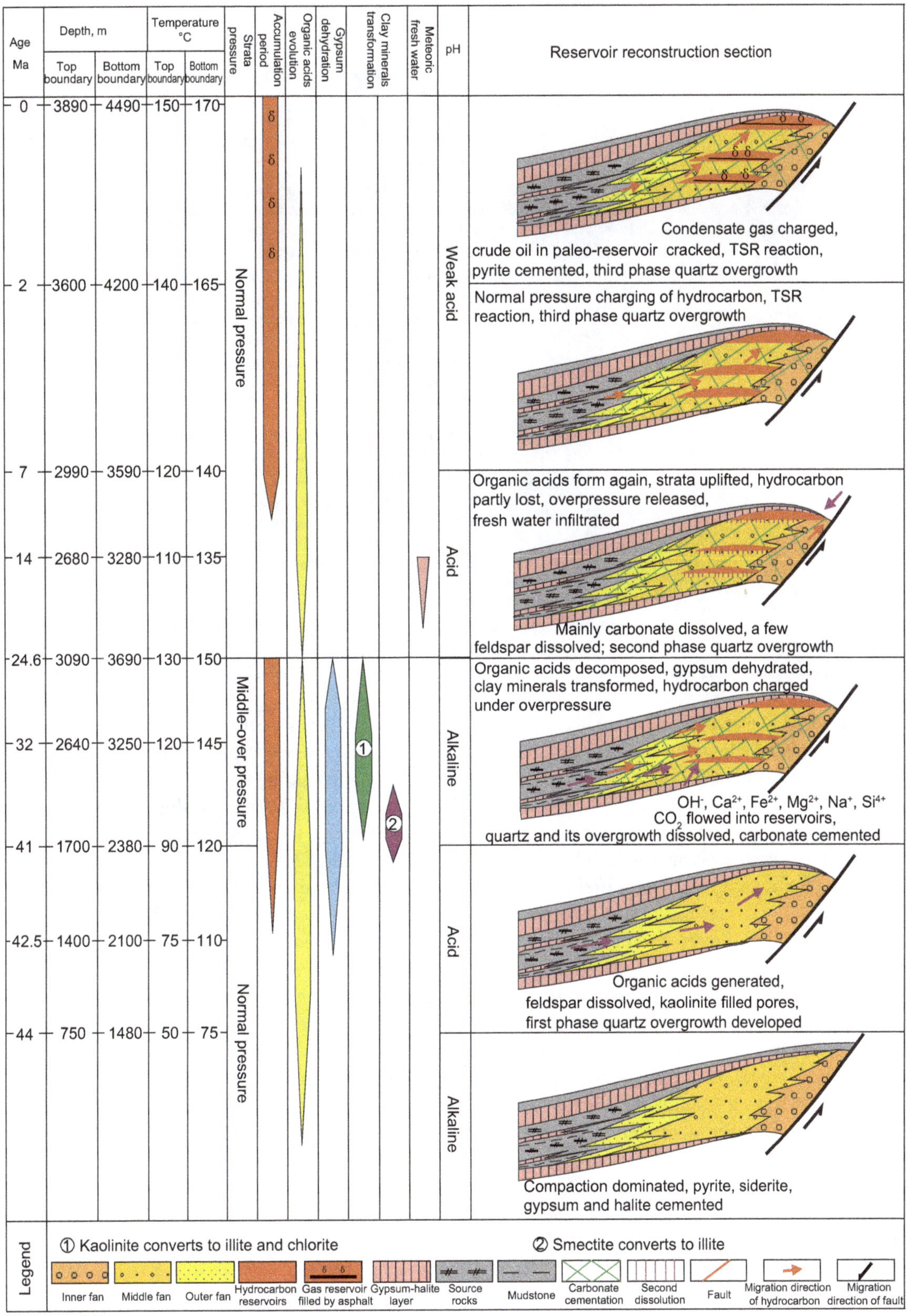

Fig. 5 Diagenetic environment evolution and reservoir reconstruction model of Es_4^x in the northern zone of the Minfeng Sag

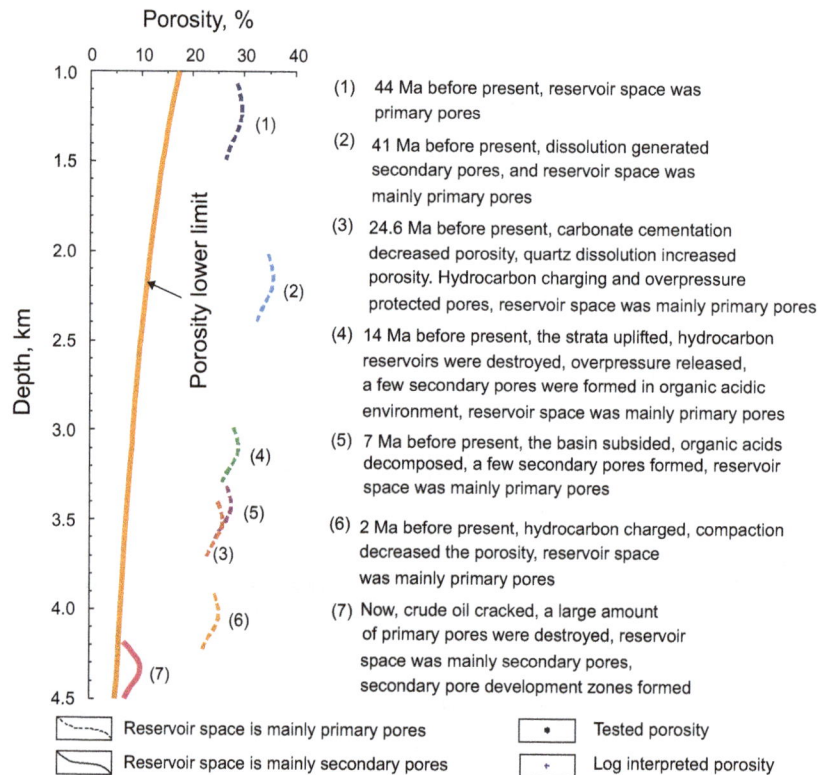

Fig. 6 Genetic mechanism and evolutionary model of Es_4^x's secondary pore development zones in the northern zone of the Minfeng Sag

fans of the nearshore subaqueous fans are mainly composed of matrix-supported conglomerates, whose resistance to compaction is weak. The middle fans mainly consist of pebbly sandstones and sandstones in braided channels, whose resistance to compaction is relatively strong. The high part of the conglomerate fans had an anticlinal attitude and formed dome traps as a result of differential compaction (Wang 2003). During this period, the reservoir spaces were dominated by primary pores after compaction and early cementation (Fig. 6).

From 44 to 41 Ma before the present (the early period of the deposition of Es_3^z), the top boundary of the strata was buried at 1700 m at temperatures of 90 °C, and the bottom boundary was buried at 2380 m at temperatures of 120 °C. According to Surdam's studies (Surdam et al. 1984, 1989), significant organic acid generation occurs during burial evolution. The temperature range of the maximum concentration of short-chain carboxylic acids is 75–90 °C (a peak of kerogen releasing oxygen-containing groups), and the optimum temperature for organic acid preservation is 80–120 °C. At lower temperatures, organic acids may be decomposed by bacteria. When the temperature rose to 120–160 °C, carboxylate anions were converted into hydrocarbons and CO_2 by thermal decarboxylation, raising the concentration of CO_2 in solution and reducing the concentration of organic acids. However, the presence of organic acids maintained the pH of the fluids at 5–6 during

this time. When the temperature was higher than 160 °C, the organic acids were completely converted to CO_2, and the pH of the solution during this time was mainly controlled by the concentration of CO_2. During this stage, the organic matter in Es_4^x had begun to mature and released a large quantity of organic acids. The temperature range during this stage was favorable for the preservation of organic acids, which caused the pH of the formation water to become acidic. Because of the shallow burial, the development of primary porosity, and good pore connectivity, the strata also exhibited the properties of an open hydrologic system with normal fluid pressure. In this environment, feldspar dissolved to form secondary pores. This resulted in the precipitation under appropriate conditions of authigenic kaolinite and first phase quartz overgrowths. As stated above, the time of the early feldspar dissolution as determined by the homogenization temperatures of aqueous inclusions in the quartz overgrowths was approximately dated to 42 Ma before the present, which is the same as the feldspar dissolution under an organic acid environment. A study by Wang (2010) suggested that gypsum began to convert to anhydrite through dehydration as the formation temperature exceeded 90 °C, and large-scale dehydration can be expected during 100–150 °C. Thus, at approximately 42 Ma, significant amounts of gypsum started to dehydrate as the bottom temperature reached 100 °C, with a portion of OH^- and Ca^{2+} dissolving in the water from dehydration of gypsum.

However, the concentration of organic acids in the strata during this period reached a maximum, which caused the formation water to remain acidic. Under conditions of acidic formation water, Ca^{2+} does not precipitate as carbonate. As the temperature reached 100 °C, smectite gradually transformed into illite through the middle state of the mixed-layer of illite/smectite (I/S) in alkaline potassium-rich solutions (Wang 2010). Although the temperature reached the conversion temperature of smectite to I/S, and although the formation water was rich in K^+ and Al^{3+} because of feldspar dissolution, smectite was not converted to I/S because of the presence of the acidic formation water. Although secondary pores were abundant during this stage, primary pores still dominated the reservoir spaces in the SPDZs in Es_4^x because of the short dissolution time (only 3 Ma) and the development of primary pores (Fig. 6). From 41 to 24.6 Ma before the present (late depositional period of the Dongying Formation), the top boundary of the strata was buried at 3090 m at temperatures of 130 °C, and the bottom boundary was buried at 3690 m at temperatures of 150 °C. The decarboxylation of organic acids began during this period, which formed CO_2 and hydrocarbons and significantly reduced the concentration of organic acids. Gypsum entered the large-scale dehydration stage, during which the presence of alkaline water controlled the pH of the formation water. In this alkaline environment, smectite quickly transformed into I/S, which released metal ions such as Ca^{2+}, Na^+, Fe^{2+}, Mg^{2+}, and Si^{4+}. In this alkaline environment, which was rich in K^+, Fe^{2+}, and Mg^{2+}, the early kaolinite was rapidly converted to illite and chlorite. Under alkaline conditions with Ca^{2+}, Fe^{2+}, and Mg^{2+}, detrital quartz and the early stage overgrowths dissolved to form secondary pores, and significant amounts of carbonate cements precipitated to fill primary pores and early feldspar pores. The reservoirs in Es_4^x exhibit three stages of hydrocarbon accumulation. The first stage occurred from the end of the depositional period of Es_2 (or the early depositional period of Es_3^z) to the late depositional period of the Dongying Formation, from approximately 38 Ma (or 41 Ma) to 24.6 Ma before the present. The second and third stages experienced continuous charging of hydrocarbons, which occurred from the mid-late depositional period of the Guantao Formation to the present (Song et al. 2009a, b). A study by Sui et al. (2010) showed that when the nearshore subaqueous fans in the northern zone of the Minfeng Sag were buried deeper than 3200 m, the inner fan subfacies acted as lateral seals for the hydrocarbon reservoirs. Therefore, the middle fan could have formed lithologic traps because of the lateral plugging of the inner fan and the normal lacustrine mudstone seal. Thus, when the organic matter became highly mature, massive amounts of oil and gas migrated into the top part of the "fan-anticlines" and middle fan lithologic traps at depths greater than 3200 m, which formed early hydrocarbon reservoirs. A study on the

evolution of formation pressure showed that this hydrocarbon charging period was accompanied by fluid overpressure, which protected the reservoir pores. The point contacts of grains and abundant primary pores can still be identified in the Fengshen1 well (Wang 2007) (Appendix Fig. 10A). Carbonate cementation was significantly inhibited in reservoirs with hydrocarbon charging, and the present reservoirs are characterized by a lower amount of euhedral carbonate cements (Appendix Fig. 10B). However, reservoirs without hydrocarbon charging were intensely filled by carbonate cements because of the long duration of the alkaline environment. A statistical analysis of the percentage of secondary pores after carbonate cementation showed that the reservoir spaces in the SPDZs were mainly primary pores, and the percentage of secondary pores was less than 15 % (Wang 2010) (Fig. 6).

The strata experienced uplift and then subsidence from 24.6 to 7 Ma before the present (the end of the depositional period of the Guantao Formation). During this period, the top boundary was uplifted to 2680 m and then subsided to 2990 m, while the bottom was uplifted to 3280 m and then subsided to 3590 m. The temperature of the top boundary fell to 110 °C and then increased to 120 °C, while the temperature of the bottom boundary fell to 135 °C and then increased to 140 °C. During this period, the organic matter stopped producing hydrocarbons, but the evolution of the organic matter still generated large amounts of organic acids, which were preserved at a favorable temperature and caused the formation water to be acidic. The tectonic movements of the Chennan Fault destroyed the initial hydrocarbon reservoirs and caused the loss of hydrocarbons and release of fluid overpressure. Meanwhile, meteoric freshwater penetrated deep formations using faults as conduits. The strata in Es_4^x were thought to have been buried relatively deeply during the uplift stage; thus, meteoric fresh water had a weak effect on reservoir reconstruction.

During this stage, the organic acids primarily reconstructed the reservoirs at the bottom of "paleo-reservoirs", which had been charged with fewer hydrocarbons, or in reservoirs where hydrocarbons had leaked. The organic acids dissolved carbonate cements and small amounts of feldspars, which caused second phase quartz overgrowths to develop in the reservoirs. This process occurred because these reservoirs were protected by overpressure and hydrocarbons, which made the early cementation relatively weak and led to higher porosity and good fluidity in these reservoirs. However, acidic fluids had difficulty in flowing into reservoirs with strong carbonate cementation, which acted as an obstacle for reservoir reconstruction and effective reservoir formation. The degree of reservoir reconstruction was limited during this stage; thus, the reservoir spaces in the secondary pore development zones were mainly primary pores (Fig. 6).

The strata subsided quickly from 7 to 2 Ma before the present (the end of the depositional period of the Minghuazhen Formation). The top boundary was buried at 3600 m, and the bottom was buried at 4200 m. The temperatures of the top and bottom boundaries were 140 and 165 °C, respectively. Organic matter reached the second hydrocarbon generation peak and produced large amounts of crude oil and associated gas. The center of the Minfeng Sag had already entered the condensate gas stage. Many organic acids began to decompose by thermal decarboxylation, which decreased the acidity of the formation water. Later, the reservoirs were mainly charged with oil and gas, which were further supplements to the "paleo-reservoirs". Li et al. (2010a, b) showed that normal pressure charging occurred during this period. When little formation water was flowing, the degree of hydrocarbon charging was limited. Thus, the oil charging during this period was only a supplement to early stage hydrocarbon reservoirs. According to Li et al. (2010a, b), the gas reservoirs in Es_4^x in the Fengshen1 well were mainly gas from oil cracking, which demonstrated that the contribution of later-stage charging to the present reservoirs was subordinate. During this period, the rocks had basically consolidated and the emplacement of hydrocarbons inhibited diagenesis, so compaction and cementation had little effect on reservoir reconstruction. At the same time, the formation temperature reached the threshold for the thermochemical sulfate reduction (TSR) reaction, and organic acids and H_2S were formed from hydrocarbon and anhydrite reactions. These acidic fluids could dissolve feldspars and carbonates, forming third phase quartz overgrowths with fluid inclusions with homogenization temperatures from 140 to 160 °C (Table 2). However, the formation was a relatively closed geochemical system during this period, and large amounts of water flow and material transport were impossible, so the amount of acidic fluids from the TSR reaction and the amount of secondary pores from the dissolution of feldspar and carbonate were small. Therefore, primary pores still dominated in the reservoirs in the SPDZs (Fig. 6).

From 2 Ma to the present, the top boundary of the strata was buried at 3890 m at temperatures of 150 °C, and the bottom boundary was buried at 4490 m at temperatures of 170 °C. During this stage, the organic acids had almost been completely decomposed, which caused the formation water to be weakly acidic with a pH of 6–7. Organic matter began generating condensate gas, and crude oil in the "paleo-reservoirs" started to crack and form large amounts of asphalt and H_2S. The TSR reactions also produced some H_2S. Later pyrite from the reaction between H_2S and Fe^{2+} in the reservoirs replaced the initial cements and filled the pores. Evidence of petroleum cracking in the "paleo-reservoirs" was discussed in detail as part of the diagenetic evolution sequence. The later cementation of asphalt and pyrite destroyed a large amount of primary porosity in the "paleo-reservoirs", which caused the proportion of secondary pores to exceed that of primary pores (Fig. 2, Fig. 6). According to the definition of the SPDZ, the SPDZs in Es_4^x in the northern zone of the Minfeng Sag formed when many primary pores were filled with asphalt and pyrite from oil cracking since 2 Ma.

6 Conclusions

(1) Secondary pore development zones can be defined in three ways: (1) the percent of secondary pores is greater than 50 %; (2) the porosity of the reservoirs is higher than the effective reservoir porosity cutoff because the absolute content of secondary pores is high; and (3) high porosity reservoirs concentrate to form belts at particular depth intervals in the porosity-depth profile, with the porosity envelope curve bulging toward higher porosities. Accordingly, two secondary pore development zones exist in Es_4^x in the northern zone of the Minfeng Sag, which range from 4200 to 4500 m and from 4700 to 4900 m.

(2) The secondary pore development zones in Es_4^x in the northern zone of the Minfeng Sag experienced the following processes. Significant numbers of secondary pores (although fewer than primary pores) formed from the dissolution of feldspar in an early organic acid environment. During this stage, the burial depth of the reservoirs was shallow. The pore spaces were slightly changed during strata subsidence—uplift—subsidence because of early hydrocarbon charging and overpressure protection. Finally, the present secondary pore development zones formed when many primary pores were filled by massive asphalt and pyrite from oil cracking in deeply buried paleo-reservoirs.

Acknowledgments The research is co-funded by National Natural Science Foundation of China (Grant No. 41102058, Grant No. U1262203, and Grant No. 41202075), the National Science and Technology Special Grant (Grant No. 2011ZX05006-003), the Fundamental Research Funds for the Central Universities (Grant No. 14CX02181A, Grant No. 15CX08001A, and Grant No. 15CX0 5007A), and Shandong Natural Science Foundation (Grant No. ZR2011DQ017). The authors thank the Geological Scientific Research Institute of Sinopec Shengli Oilfield Company, the Exploration and Development Research Institute of Sinopec Zhongyuan Oilfield Company, and the CNPC key laboratory of oil and gas reservoirs in China University of Petroleum for providing database and technological assistance.

Appendix 1

See Appendix Fig. 7.

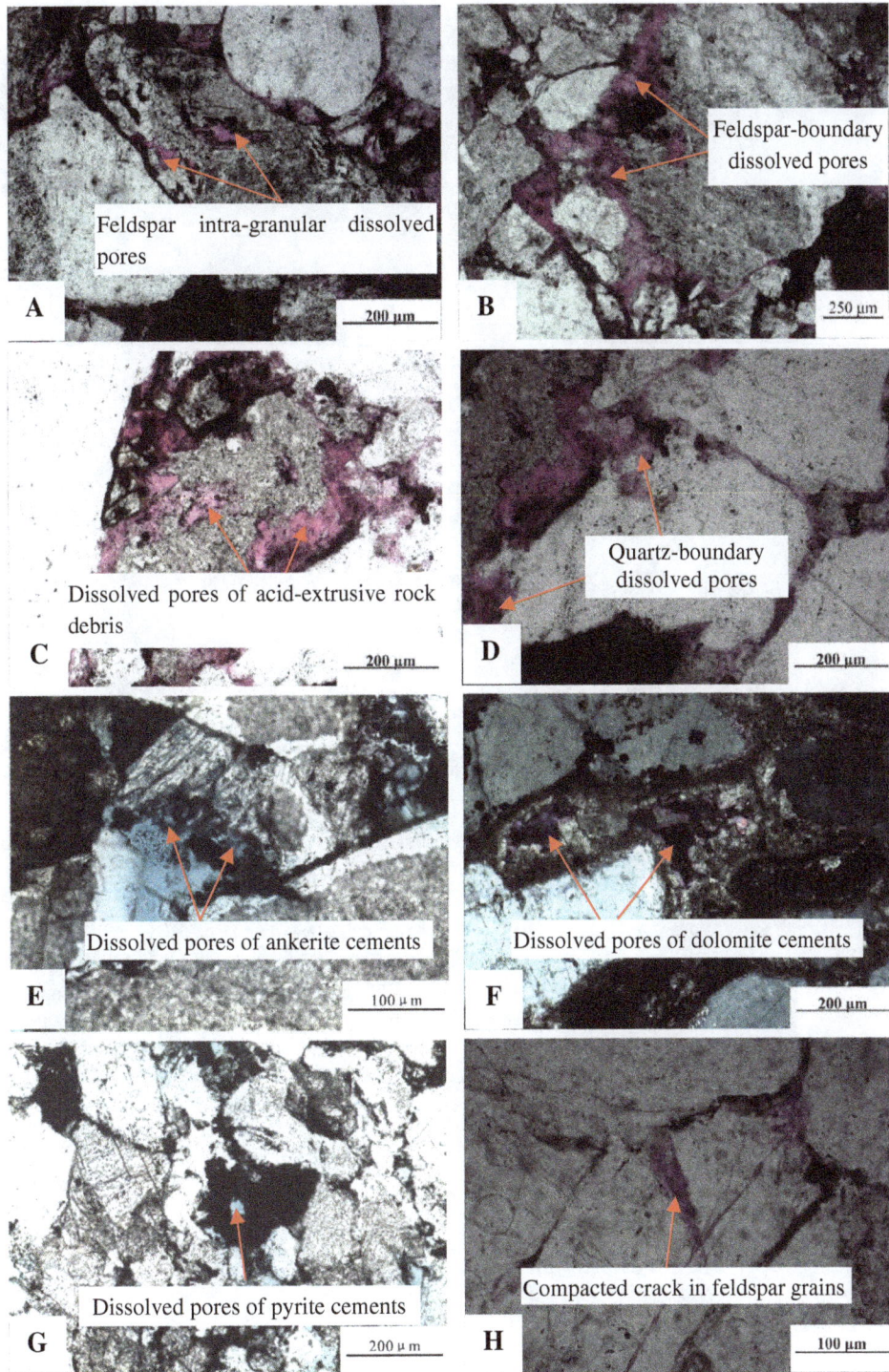

Fig. 7 **A** Fengshen1 well, 4321.9 m, intra-granular-dissolved pores in feldspar grains. **B** Fengshen1 well, 4321.9 m, dissolved pores along feldspar grain boundaries. **C** Fengshen1 well, 4321.9 m, dissolved pores in acid extrusive rock debris. **D** Fengshen1 well, 4321.9 m, dissolved pores along quartz grain boundaries. **E** Fengshen1 well, 4323.3 m, dissolved pores in ankerite cement. **F** Fengshen4 well, 4476.15 m, dissolved pores in dolomite cement. **G** Fengshen1 well, 4348.25 m, dissolved pores in pyrite cement. **H** Fengshen3 well, 4867 m, compacted cracks in feldspar grains

Appendix 2

See Appendix Fig. 8.

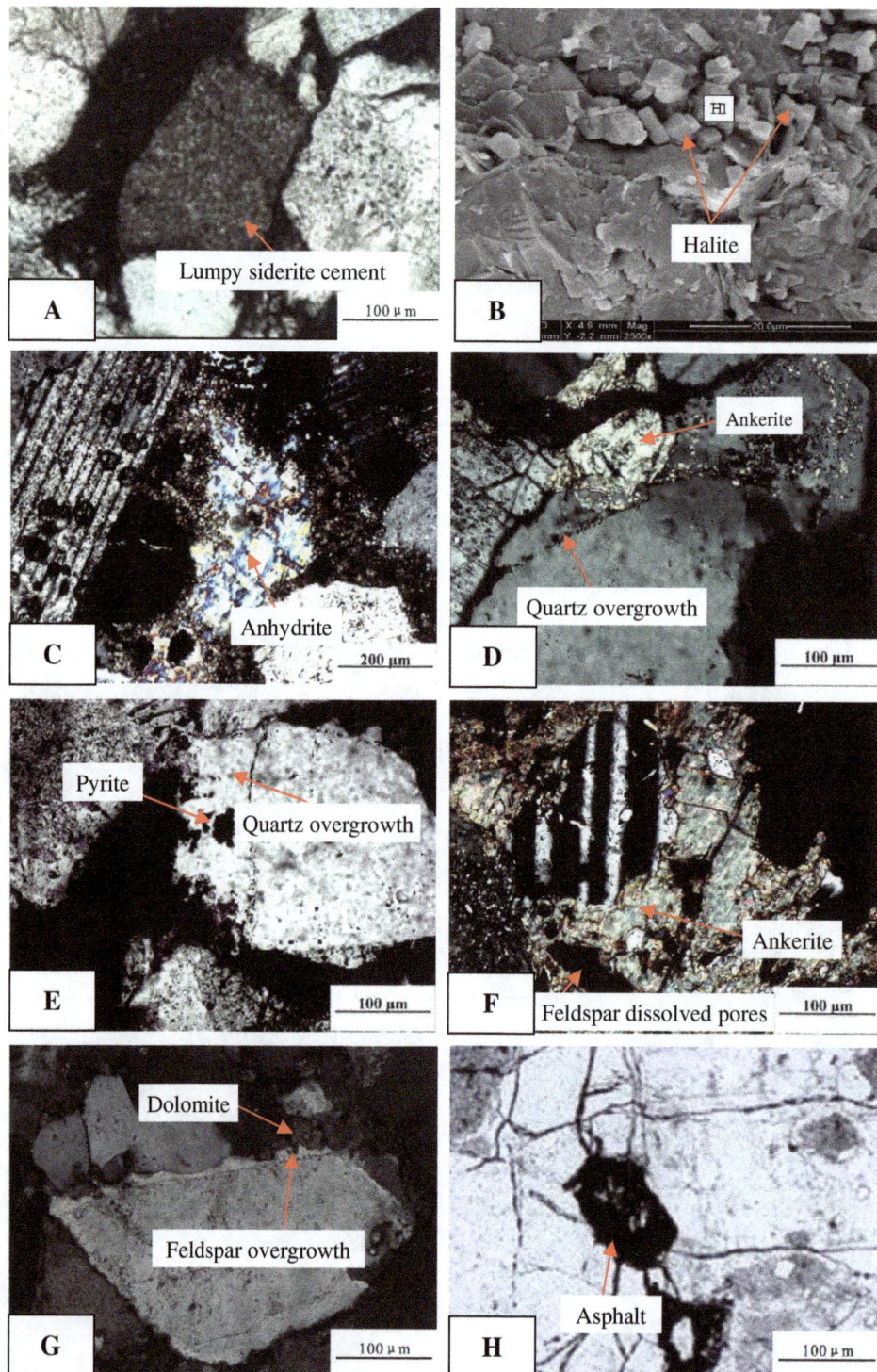

Fig. 8 **A** Fengshen1 well, 4323.3 m, lumpy siderite cement. **B** Feng8 well, 4397.15 m, halite-filled pores (SEM). **C** Feng8 well, 4397.15 m, anhydrite that replaced ankerite. **D** Fengshen3 well, 4867 m, ankerite that replaced quartz overgrowths. **E** Fengshen3 well, 4867 m, pyrite that replaced quartz overgrowths. **F** Fengshen3 well, 4867 m, ankerite filling parts of the feldspar-dissolved pores. **G** Fengshen1 well, 4350 m, dolomite that replaced feldspar overgrowths. **H** Fengshen1 well, 4321.9 m, asphalt filling intra-granular-dissolved pores in feldspar grains (Li et al. 2010a, b)

Appendix 3

See Appendix Fig. 9.

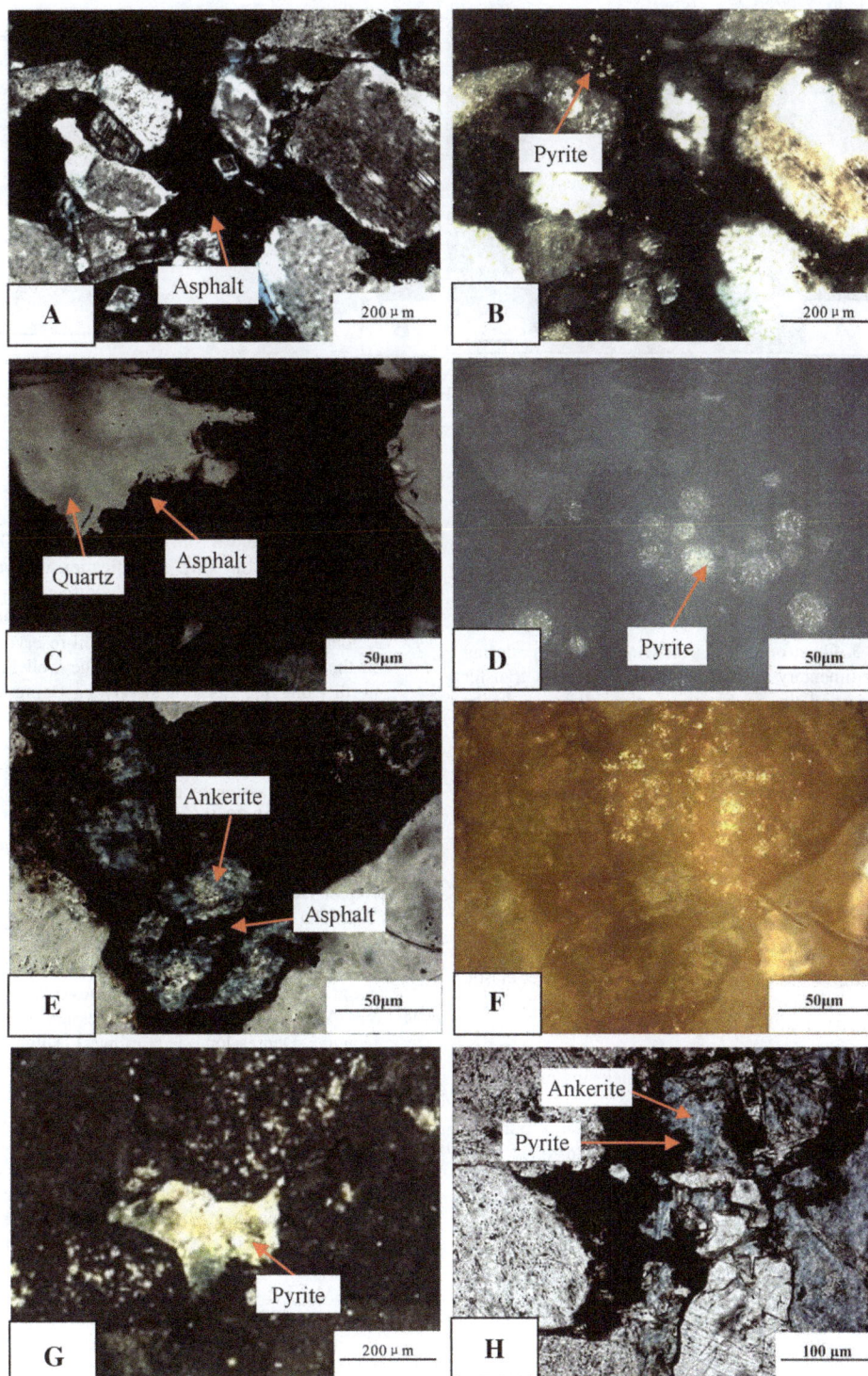

Fig. 9 A Fengshen1 well, 4323.3 m, asphalt and star-like pyrite filling intergranular pores. **B** Fengshen1 well, 4323.3 m, reflected light, the same visual area as A. **C** Fengshen1 well, 4321.9 m, asphalt and pyrite filling quartz-dissolved pores. **D** Fengshen1 well, 4321.9 m, reflected light, the same visual area as C. **E** Fengshen1 well, 4321.9 m, asphalt filling dissolved pores in ankerite cement. **F** Fengshen1 well, 4321.9 m, reflected light, the same visual area as E. **G** Fengshen1 well, 4323.3 m, reflected light, pyrite cement. **H** Fengshen3 well, 4867 m, pyrite that replaced ankerite

Appendix 4

See Appendix Fig. 10.

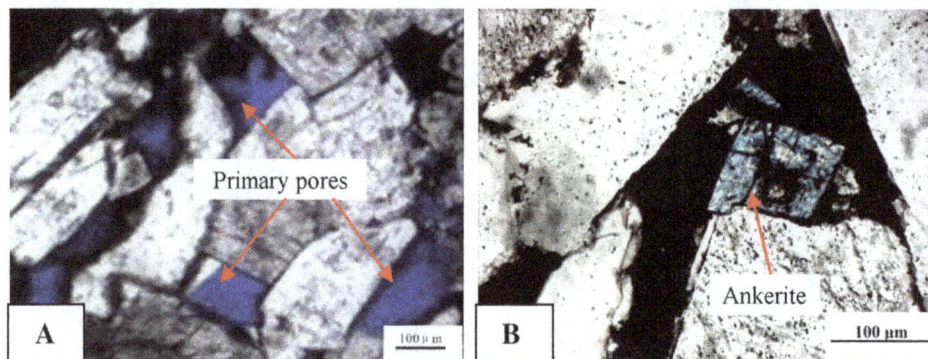

Fig. 10 A Fengshen1 well, 4322.5 m, primary pores (Wang 2010). **B** Fengshen1 well, 4321.9 m, complete crystal ankerite filling intergranular pores

References

Ajdukiewicz JM, Larese RE. How clay grain coats inhibit quartz cement and preserve porosity in deeply buried sandstones: observations and experiments. AAPG Bulletin. 2012; 96(11):2091–119.

Bjørlykke K, Jahren J. Open or closed geochemical systems during diagenesis in sedimentary basins: constraints on mass transfer during diagenesis and the prediction of porosity in sandstone and carbonate reservoirs. AAPG Bulletin. 2012;96(12): 2193–214.

Bjørlykke K. Relationships between depositional environments, burial history and rock properties: some principal aspects of diagenetic process in sedimentary basins. Sed Geol. 2014;301: 1–14.

Bloch S, Lander RH, Bonnell L. Anomalously high porosity and permeability in deeply buried sandstone reservoirs: origin and predictability. AAPG Bulletin. 2002;86:301–28.

Cao YC, Yuan GH, Li XY, et al. Characteristics and origin of abnormally high porosity zones in buried Paleogene clastic reservoirs in the Shengtuo area, Dongying Sag, East China. Pet Sci. 2014;11:346–62.

Curtis CD. Possible links between sandstone diagenesis and depth-related geochemical reactions occurring in enclosing mudstones. J Geolog Soc. 1978;135:107–17.

Dutton SP, Loucks RG. Diagenetic controls on evolution of porosity and permeability in Lower Tertiary Wilcox sandstones from shallow to ultradeep (200–6700 m) burial, Gulf of Mexico Basin, U.S.A. Mar Pet Geol. 2010;27(1):69–81.

Ehrenberg SN. Preservation of anomalously high porosity in deeply buried sandstones by grain-coating chlorite: examples from the Norwegian Continental Shelf. AAPG Bull. 1993;77:1260–86.

Giles MR, De Boer RB. Origin and significance of redistributional secondary porosity. Mar Pet Geol. 1990;7:378–97.

Higgs KH, Zwingmann H, Reyes AR, et al. Diagenesis, porosity evolution, and petroleum emplacement in tight gas reservoirs, Taranaki Basin, New Zealand. J Sediment Res. 2007;77: 1003–25.

Hu WR, Bao JW, Hu B. Trend and progress in global oil and gas exploration. Pet Explor Dev. 2013;40(4):409–13 (**in Chinese**).

Jiang ZX, Qiu LW, Chen GJ. Alkaline diagenesis and its genetic mechanism in the Triassic coal measure strata in the Western Sichuan Foreland Basin, China. Pet Sci. 2009;6(4):354–65.

Li CQ, Chen HH, Liu HM. Identification of hydrocarbon charging events by using micro-beam fluorescence spectra of petroleum inclusions. Earth Sci. 2010a;35(4):657–62 (**in Chinese**).

Li YJ, Song GQ, Li WT, et al. A fossil oil-reservoir and the gas origin in the Lower Sha-4 Member of the well Fengshen-1 area, the north Dongying Zone of the Jiyang Depression. Oil Gas Geol. 2010b;31(2):173–9 (**in Chinese**).

Liu H, Jiang ZX, Zhang RF, et al. Gravels in the Daxing conglomerate and their effect on reservoirs in the Oligocene Langgu Depression of the Bohai Bay Basin, North China. Mar Pet Geol. 2012;29:192–203.

Liu W, Zhu XM. Distribution and genesis of secondary pores in Tertiary clastic reservoir in southwestern Qaidam Basin. Pet Explor Dev. 2006;33(3):315–8 (**in Chinese**).

Liu YX, Zhu M, Zhang SM. Diagenesis and pore evolution of reservoir of the Member4 of Lower Cretaceous Quantou Formation in Sanzhao Sag, northern Songliao Basin. J Palaeogeogr. 2010;12(4):480–7 (**in Chinese**).

Ma WM, Wang XL, Ren LY. Overpressure and secondary pores in Dongpu Depression. J Northwest Univ (Nat Sci Edition). 2005;35(3):325–30 (**in Chinese**).

Meng YL, Liang HW, Meng FJ, et al. Distribution and genesis of the anomalously high porosity zones in the middle-shallow horizons of the Songliao Basin. Pet Sci. 2010;7(3):302–10.

Meng YL, Liu WH, Meng FJ, et al. Distribution and origin of anomalously high porosity zones of the Xujiaweizi Fault Depression in Songliao Basin. J Palaeogeogr. 2011;13(1):75–84 (**in Chinese**).

Osborne MJ, Swarbrick RE. Diagenesis in North Sea HPHT clastic reservoirs—consequences for porosity and overpressure prediction. Mar Pet Geol. 1999;16:337–53.

Qiu LW, Jiang ZX. Alkaline diagenesis of Terrigenous clastic rocks. Beijing: Geological Publishing Press; 2006. p. 22 (**in Chinese**).

Schmidt V, McDonald DA. Role of secondary porosity in sandstone diagenesis. AAPG Bull. 1977;61:1390–1.

Schmidt V, McDonald D. A texture and recognition of secondary porosity in sandstones. Special Publication. Soc Econ Paleontol Mineral, 1979;26:209–225.

Shi ZF, Zhang ZC, Ye SD. The mechanism of secondary pores in the reservoir of Funing Formation in Gaoyou Depression of Subei Basin. Acta Sedimentol Sin. 2005;23(3):429–36 (in Chinese).

Song GQ, Jiang YL, Liu H, et al. Pooling history of cracked gas in middle-deep reservoirs in Lijin-Minfeng areas of the Dongying Sag. Nat Gas Ind. 2009a;29(4):14–7 (in Chinese).

Song GQ, Jin Q, Wang L. Study on kinetics for generating natural gas of Shahejie Formation in deep-buried sags of Dongying Depression. Acta Pet Sinica. 2009b;33(5):672–7 (in Chinese).

Sui FG, Cao YC, Liu HM. Physical properties evolution and hydrocarbon accumulation of Paleogene nearshore subaqueous fan in the eastern north margin of the Dongying Depression. Acta Geol Sinica. 2010;84(2):246–56 (in Chinese).

Sun LD, Fang CL, Li F, et al. Innovations and challenges of sedimentology in oil and gas exploration and development. Pet Explor Dev. 2015;42(2):129–36 (in Chinese).

Sun LD, Zou CN, Zhu RK, et al. Formation, distribution and potential of deep hydrocarbon resources in China. Pet Explor Dev. 2013;40(6):641–9 (in Chinese).

Surdam RC, Boese SW, Crossey GJ. The chemistry of secondary porosity. AAPG Memoir. 1984;37:127–51.

Surdam RC, Crossey LJ, Hagen ES, et al. Organic-inorganic interactions and sandstone diagenesis. AAPG Bull. 1989;73:1–23.

Taylor TR, Giles MR, Hathon LA, et al. Sandstone diagenesis and reservoir quality prediction: models, myths, and reality. AAPG Bull. 2010;94:1093–132.

Taylor TR, Kittridge MG, Winefield P, et al. Reservoir quality and rock properties modeling—Triassic and Jurassic sandstones, Greater Shearwater Area, UK Central North Sea. Mar Pet Geol. 2015;65:1–21.

Wang P, Zhao CL. An approach to the generating mechanism of secondary pores in Duqiaobai area of Dongpu Depression. Pet Explor Dev. 2001;28(4):44–6 (in Chinese).

Wang SP. Diagenesis researches of natural gas reservoir with gyprock and salt bed in Minfeng area of the lower part of Number4 of Shahejie Formation in Dongying Depression. Master's Thesis. China University of Petroleum. 2007 (in Chinese).

Wang YZ. Genetic mechanism and evolution model of secondary pore development zone of Paleogene in the north zone in Dongying Depression. Ph. D. Thesis. China University of Petroleum. 2010 (in Chinese).

Wang YZ, Cao YC, Ma BB, et al. Mechanism of diagenetic trap formation in nearshore subaqueous fans on steep rift lacustrine basin slopes—a case study from the Shahejie Formation on the north slope of the Minfeng Subsag, Bohai Basin, China. Pet Sci. 2014;11:481–94.

Wang ZG. Pooling model of steep slope structure and lithological zone in north Dongying Sag. Pet Explor Dev. 2003;20(4):10–2 (in Chinese).

Wang ZY, Wang FP, Lin XY. Pore types and origin of secondary pores of Triassic and Jurassic in north Tarim. Oil Gas Geol. 1995;16(3):203–10 (in Chinese).

Warren EA, Pulham AJ. Anomalous porosity and permeability preservation in deeply buried Tertiary and Mesozoic sandstones in the Cusiana field, Llanos Foothills, Colombia. J Sediment Res. 2001;71(1):2–14.

Wilkinson M, Haszeldine RS. Oil charge preserves exceptional porosity in deeply buried, overpressured sandstones: Central North Sea, UK. J Geolog Soc. 2011;168:1285–95.

Yuan GH, Gluyas J, Cao YC, et al. Diagenesis and reservoir quality evolution of the Eocene sandstones in the northern Dongying Sag, Bohai Bay Basin, East China. Mar Pet Geol. 2015;62:77–89.

Yuan J, Wang ZQ. Distribution and generation of deep reservoir secondary pores, Paleogene, Dongying Sag. J Mineral Pet. 2001;21(1):43–7 (in Chinese).

Zeng JH. Effect of fluid-rock interaction on porosity of reservoir rocks in Tertiary system, Dongying Sag. Acta Pet Sinica. 2001;22(4):39–43 (in Chinese).

Zhang Q, Zhong DK, Zhu XM. Pore evolution and genesis of secondary pores in Paleogene clastic reservoirs in Dongying Sag. Oil Gas Geol. 2003;24(3):281–5 (in Chinese).

Zhang Q, Zhu XM, Steel RJ, et al. Variation and mechanisms of clastic reservoir quality in the Paleogene Shahejie Formation of the Dongying Sag, Bohai Bay Basin, China. Pet Sci. 2014;11:200–10.

Zhang SW. "Water consumption" in diagenetic stage and its petroleum geological significance. Acta Sedimentol Sin. 2007;25(5):701–7 (in Chinese).

Zhang WC, Li H, Li HJ. Genesis and distribution of secondary porosity in the deep horizon of Gaoliu area, Nanpu Sag. Pet Explor Dev. 2008;35(3):308–12 (in Chinese).

Zhang YG, Xu WP, Wang GL. Hydrocarbon Reservoir Formation Aggregation in a Continental Rift Basin, East China. Beijing: Petroleum Industry Press; 2006. p. 127–31 (in Chinese).

Zheng MY, Wu RL. Diagenesis and pore zonation of sandstone reservoirs in Huanghua Depression. Oil Gas Geol. 1996;17(4):268–75 (in Chinese).

Zhong DK, Zhu XM, Zhang ZH. Origin of secondary porosity of Paleogene sandstone in the Dongying Sag. Pet Explor Dev. 2003;30(6):51–3 (in Chinese).

Zhong JH, Wang GZ, Gao XC. Sedimentary features, genesis and relation to hydrocarbon of fan-anticline in the north slope of the Dongying Sag. Chin J Geol (Scientia Geologica Sinica). 2004;3(4):625–36 (in Chinese).

Zhu SF, Zhu XM, Wang YB. Dissolution characteristics and pore evolution of Triassic reservoir in Ke-Bai area, northwestern margin of Junggar Basin. Acta Sediment Sin. 2010;28(3):548–54 (in Chinese).

Zhu XM, Mi LJ, Zhong DK. Paleogene diagenesis and its control on reservoir quality in Jiyang Depression. J Palaeogeogr. 2006;8(3):296–305 (in Chinese).

Zhu XM, Wang YG, Zhong DK. Pore types and secondary pore evolution of Paleogene reservoir in the Jiyang Sag. Acta Geol Sinica. 2007;81(2):197–204 (in Chinese).

16

Factors influencing physical property evolution in sandstone mechanical compaction: the evidence from diagenetic simulation experiments

Ke-Lai Xi[1,3] · Ying-Chang Cao[1] · Yan-Zhong Wang[1] · Qing-Qing Zhang[1] ·
Jie-Hua Jin[1] · Ru-Kai Zhu[2] · Shao-Min Zhang[1] · Jian Wang[1] · Tian Yang[1] ·
Liang-Hui Du[1]

Abstract In order to analyze the factors influencing sandstone mechanical compaction and its physical property evolution during compaction processes, simulation experiments on sandstone mechanical compaction were carried out with a self-designed diagenetic simulation system. The experimental materials were modern sediments from different sources, and the experiments were conducted under high temperature and high pressure. Results of the experiments show a binary function relation between primary porosity and mean size as well as sorting. With increasing overburden pressure during mechanical compaction, the evolution of porosity and permeability can be divided into rapid compaction at an early stage and slow compaction at a late stage, and the dividing pressure value of the two stages is about 12 MPa and the corresponding depth is about 600 m. In the slow compaction stage, there is a good exponential relationship between porosity and overburden pressure, while a good power function relationship exists between permeability and overburden pressure. There is also a good exponential relationship between porosity and permeability. The influence of particle size on sandstone mechanical compaction is mainly reflected in the slow compaction stage, and the influence of sorting is mainly reflected in the rapid compaction stage. Abnormally high pressure effectively inhibits sandstone mechanical compaction, and its control on sandstone mechanical compaction is stronger than that of particle size and sorting. The influence of burial time on sandstone mechanical compaction is mainly in the slow compaction stage, and the porosity reduction caused by compaction is mainly controlled by average particle size.

Keywords Primary porosity · Mechanical compaction · Unconsolidated sand · Diagenetic simulation experiment

1 Introduction

With increasing oil and gas exploration and the growing demand for oil and gas reserves, the oil and gas exploration targets of clastic rocks have turned to low porosity and permeability reservoirs, even to tight sandstone reservoirs, and they have gradually become the main source of increasing reserves and production of oil and gas (Wang and Tian 2003; Dai et al. 2012; Hart 2006; Zou et al. 2013; Tobin et al. 2010; Jia et al. 2012; Worden et al. 2000; Bloch et al. 2002; Wang et al. 2010; Zhang et al. 2011). Mechanical compaction, as one of the major destructive aspects of diagenesis, can cause dramatic changes in sandstone pore structure and distribution. It results in a substantial reduction of pore space, which is the main factor that damages reservoirs and forms low porosity, low permeability reservoirs (Zhu et al. 2008; Chester et al. 2004; Zhu et al. 2013; Lv and Liu 2009; Maast et al. 2011; Taylor et al. 2010; Ajdukiewicz et al. 2010; Zhang et al. 2008; Wang et al. 2015; Zhang et al. 2014; Liu et al. 2014). Previous studies have suggested that mechanical

✉ Ke-Lai Xi
kelai06016202@163.com

✉ Ying-Chang Cao
cyc8391680@163.com

[1] School of Geosciences, China University of Petroleum, Qingdao, Shandong 266580, China

[2] Research Institute of Petroleum Exploration and Development, CNPC, Beijing 100083, China

[3] Department of Geosciences, University of Oslo, Oslo 0316, Norway

Edited by Jie Hao

compaction is influenced by a combination of various factors such as burial depth, sediment composition, particle size, sorting, abnormally high pressure, and burial time (Liu et al. 2007; Aplin et al. 2006).

However, current studies about various factors influencing compaction mainly focus on simple and qualitative description, and little work has been done on quantitative analysis based on simulation experiments, leading to the vague understanding of the evolution of physical properties and influencing mechanisms of various factors during mechanical compaction processes. This directly restricts the accurate characterization of the formation of low porosity and permeability sandstone reservoirs and their densification processes. Therefore, carrying out sandstone mechanical compaction simulation experiments and understanding the evolution of physical properties and the influencing mechanisms of various factors during mechanical compaction have not only an important theoretical significance for diagenesis, but also an important practical significance in physical property prediction of low porosity and permeability reservoirs.

2 Experimental facility and experimental procedures

The experiment was carried out in the Diagenetic Simulation Laboratory of China University of Petroleum, using a self-designed diagenetic simulation experiment system. This facility consists of two modules: the porosity and permeability testing module and the diagenetic simulation module. It measures sandstone porosity and permeability, simulating the high temperature and pressure conditions of subsurface strata, and monitoring physical property changes of sandstones in the course of diagenesis in real time. The sandstone mechanical compaction simulator is mainly composed of a constant current–constant voltage (CCCV) pump, intermediate container group, displacement sensors, axial compression control pump, core holder (with a built-in heating device), back-pressure booster pump, fluid-receiving scale, and automatic control system (Fig. 1). The upper temperature limit of the system is 300 °C, and the pressure limit is 80 MPa.

During sandstone mechanical compaction simulation experiments, the sample is in the core holder with movable pistons covering both ends. The axial compression pump is used to simulate overburden pressure and the micro back-pressure booster pump is manually operated to control the fluid pressure. A built-in core holder heating device and temperature control system are used to simulate the strata temperature. Precision displacement sensors at both ends of the core holder (with an accuracy of 0.001 mm) record compaction displacements. The intermediate container group is used to contain the fluid required for the experiment. The CCCV pump provides the displacement pressure to displace fluid and pore water toward the fluid-receiving scale at a constant flow rate. The automatic control system is used for measuring and recording sandstone permeability in real time.

3 Experimental design

3.1 Experimental samples

3.1.1 Sample collection

The samples selected for experiments were modern unconsolidated sands ranging from 5 to 20 cm below from the surface at Golden Beach and Silver Beach, Qingdao; Yellow River Estuary, Dongying; point bar, Mazhan River, Weifang; and mouth bar, Feng River, Jiaonan eastern China. During the sampling process, in order to keep the original packing state (grain combination sequence, sorting, etc.), we used copper tubes with a length of 140 mm and an inner diameter of 25.7 mm to take samples (Fig. 2). Two samples were collected within a distance of about 15 mm in each group, of which one was used for sandstone

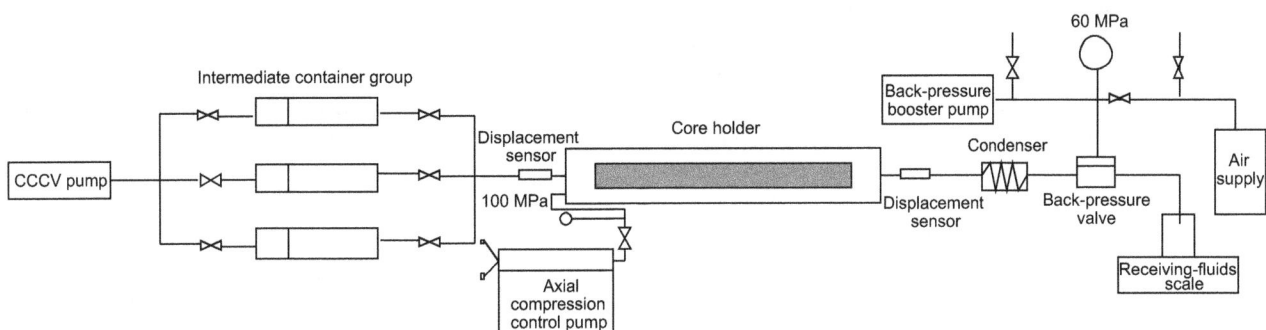

Fig. 1 Sandstone mechanical compaction simulation diagram

Fig. 2 Collection of experimental samples

Table 1 The parameters of experimental samples with different sources

Sample sources	Sample no.	Median particle size (Md), mm	Average particle size (M), mm	Sorting coefficient (So)	Initial length of samples (L_0), mm	Primary porosity (φ_0), %	Primary permeability (K_0), $\times 10^{-3} \mu m^2$
Silver Beach of Qingdao	I	0.215	0.219	1.556	88.5	42.96	696.7
Silver Beach of Qingdao	II	0.260	0.262	1.575	88.8	42.92	1779.9
Golden Beach of Qingdao	III	0.280	0.293	1.860	84.7	37.37	570.56
Yellow River Estuary of Dongying	IV	0.076	0.079	1.618	88.5	43.72	372.9
Mazhan River of Weifang	V	1.25	1.28	3.571	90.5	33.62	733.34
Silver Beach of Qingdao	VI	0.217	0.221	1.50	87.6	41.75	445.15
Yellow River Estuary of Dongying	VII	0.076	0.078	1.622	88.7	43.72	389.54
Feng River Estuary of Jiaonan	VIII	0.952	0.964	2.52	86.9	35.87	437.7

mechanical compaction simulation experiments and the other was for granularity parameter analysis. It was assumed that granularity parameters of the two samples were the same.

3.1.2 Laboratory analysis of samples

Firstly, the sand samples selected for simulation experiments were processed to 90 mm long with both ends flat, and they were covered by metal filters to prevent sands from becoming loose and sliding. Then the sand samples of each group were dried using a 101-1A-type electrothermal drying box. Finally, after being dried completely, one sand sample of each group for the mechanical compaction simulation experiment was put into the core holder, and then the primary porosity and permeability were measured using the porosity and permeability testing module (Table 1). Meanwhile, the corresponding other sand sample was used for granularity parameter analysis using sieve analysis method to obtain parameters such as median particle size (Md), average particle size (M), and sorting coefficient (So) (Table 1).

The sediments from Golden Beach and Silver Beach, Qingdao are mainly feldspar and quartz, with a relatively low content of volcanic rock debris and little biotite and

magnetite, and the content of feldspar is higher than that of quartz. The content of rigid particles, e.g., feldspar and quartz, ranges from 70 % to 80 %, while the content of ductile particles like eruptive rock and mica is generally less than 20 %. The content of quartz is relatively higher than that of feldspar in the sediments from the Mazhan and Feng Rivers, with a low content of volcanic rock debris and visible chert. The content of rigid particles is over 85 % and that of ductile particles is less than 15 %. Sediments from Yellow River Estuary, Dongying are fine grained, with a low content of rigid particles and a high clay mineral content, which results in strong ductility.

3.2 Experimental conditions

To simulate the geological conditions of the Dongying Sag of the Jiyang Depression, Bohai Bay Basin in eastern China, the experimental conditions were set as follows: the geothermal gradient is 3.5 °C/100 m (average paleo-geotherm gradient), the average formation density is about 2.4 g/cm³, the average surface temperature is 18 °C, and the pressure coefficient under normal compaction is 1.0 (Liu et al. 2006). In order to simulate a pure mechanical compaction, distilled water was used as the fluid medium. Previous studies have shown that overpressure can develop

Table 2 Reference list of temperature and pressure experimental conditions

Simulated burial depth, m	Overburden pressure, MPa			Framework pressure, MPa	Fluid pressure, MPa			Temperature, °C
	Pressure coefficient 1.0	Pressure coefficient 1.2	Pressure coefficient 1.4		Pressure coefficient 1.0	Pressure coefficient 1.2	Pressure coefficient 1.4	
0	0	0	0	0	0	0	0	18
100	2.18	2.18	2.18	1.2	0.98	0.98	0.98	21.5
200	4.36	4.36	4.36	2.4	1.96	1.96	1.96	25
400	8.72	8.72	8.72	4.8	3.92	3.92	3.92	32
600	13.08	13.08	13.08	7.2	5.88	5.88	5.88	39
800	17.44	17.44	17.44	9.6	7.84	7.84	7.84	46
1000	21.8	21.8	21.8	12	9.8	9.8	9.8	53
1200	26.16	26.16	26.16	14.4	11.76	11.76	11.76	60
1400	30.52	30.52	30.52	16.8	13.72	13.72	13.72	67
1600	34.88	38.016	41.152	19.2	15.68	18.816	21.952	74
1800	39.24	42.768	46.296	21.6	17.64	21.168	24.696	81
2000	43.6	47.52	51.44	24	19.6	23.52	27.44	88
2200	47.96	52.272	56.584	26.4	21.56	25.872	30.184	95
2400	52.32	57.024	61.728	28.8	23.52	28.224	32.928	102
2600	56.68	61.776	66.872	31.2	25.48	30.576	35.672	109
2800	61.04	66.528	72.016	33.6	27.44	32.928	38.416	116
3000	65.4	71.28	77.16	36	29.4	35.28	41.16	123

below 1600 m in the Dongying Sag (Liu et al. 2009). According to the stress–burial depth conversion formula 0.02262 MPa = 1 m (Gluyas and Cade 1999), the strata pressure at 1600 m is approximately 36.16 MPa. Therefore, taking the above geological factors into account, as well as the applicable temperature and pressure conditions of the facility, we designed a temperature and pressure reference list for sandstone mechanical compaction under normal compaction conditions with a pressure coefficient of 1.0 and under overpressure conditions with pressure coefficients of 1.2 and 1.4 for contrast experiments. In this way, the evolution of physical properties and the influencing factors during the simulation process of sandstone mechanical compaction can be analyzed (Table 2).

3.3 The calculation of porosity and permeability

3.3.1 Porosity calculation

The porosity calculation method during the simulation process of sandstone mechanical compaction is as follows:

$$S_0 = \pi r^2; \tag{1}$$

$$V_0 = L_0 S_0; \tag{2}$$

$$V_\Phi = V_0 - V_g; \tag{3}$$

$$\Phi_0 = V_\Phi/V_0 \times 100 \%. \tag{4}$$

Here, r is the cross-sectional radius of sample, cm, L_0 is the initial length of sample, cm, S_0 is the cross-sectional area of sample, cm^2, V_0 is the initial volume of sample, cm^3, V_Φ is the initial pore volume of sample, cm^3, V_g is the framework volume of sample, cm^3, and Φ_0 is the primary porosity, %.

Primary porosity Φ_0 could be measured by the porosity and permeability testing module of the diagenetic simulation system (Table 1).

Therefore, sample volume at each pressure point during compaction could be calculated with the recorded compaction displacement:

$$V = S_0(L_0 - L_1). \tag{5}$$

Here, L_1 is the recorded compaction displacement, cm, V is the sample volume at each pressure point during compaction, cm^3.

The loss of sample volume during compaction mainly consists of intergranular pore volume during the experiment by assuming the framework volume as a constant, thus:

$$V_0 - V_\Phi = V - V \times \Phi, \tag{6}$$

$$\Phi = (V - V_0 + V_\Phi)/V \times 100\%. \tag{7}$$

3.3.2 Permeability calculation

Permeability during the simulation of sandstone mechanical compaction can be calculated using Darcy's law.

$$K = Q\mu L/(\Delta PS_0). \tag{8}$$

That is

$$K = Q\mu(L_0 - L_1)/(\Delta PS_0). \tag{9}$$

Here, Q is the quantity of flow through the sample per unit time, cm^3/s, S_0 is the cross-sectional area of the sample, cm^2, μ is the fluid viscosity, $\times 10^{-3}$Pa s, L_0 is the original length of the sample, cm, L_1 is the recorded compaction displacement, cm, and ΔP is the pressure differential before and after fluid flowing through the sample, MPa.

The above K in Eq. (9) is the sample's permeability. It shows fluid flow capacity through the sample within a certain pressure differential.

Primary permeability K_0, permeability without compaction, could be measured by the porosity and permeability testing module after constant fluid passing through (Table 1). During the experiment, the fluid viscosity was set as 1×10^{-3} Pa s and other parameters were recorded by the automatic control system in real time, and then permeability variations could be calculated.

There is a close relationship between fluid viscosity (μ) and temperature (T). Therefore, on the basis of reviewing water viscosities at different temperatures (Yuan 1985), an empirical formula can be fitted as follows:

$$T = 1.056233e^{-0.018118\mu} \quad R^2 = 0.976362. \tag{10}$$

We calculated fluid viscosity at different temperatures using Eq. (10) and then corrected the recorded permeability values.

3.4 Data acquisition and processing

Detailed procedures of data acquisition and processing during sandstone mechanical compaction simulation experiment are as follows:

First, according to the reference list of temperature and pressure conditions, we set the experimental temperature and pressure and then conducted mechanical compaction simulation experiments. After compaction at each pressure point was stable (the compaction displacement was a constant), we recorded the data with a fixed time interval of 2 min, and the record time of each pressure point was about 120 min, that is, there were 60 sets of record data. The recorded experimental parameters included experimental time, overburden pressure, fluid pressure, fluid flow, pressure differential between the ends of the core, experimental

temperature, compaction displacement, and permeability. In the normal compaction simulation experiments, the upstream pressure on the sample is higher than the downstream pressure, so the fluid can be discharged onto the fluid-receiving scale in time. In the undercompaction simulation experiments, the differential between upstream pressure and downstream pressure was respectively set according to the pressure coefficients of 1.2 and 1.4, making the downstream pressure higher than the upstream pressure. In this way, fluid discharge was blocked and abnormally high pressure was formed.

Second, according to the compaction displacement, we calculated the corresponding porosity value of each data point.

Third, according to the relationship between viscosity and temperature, we corrected the corresponding permeability value of each data point.

Fourth, we precisely analyzed the data of overburden pressure, fluid pressure, temperature, porosity and permeability, and excluded abnormal data points caused by system errors.

Fifth, we calculated the average value of each parameter including overburden pressure, fluid pressure, temperature, porosity, and permeability after removing the abnormal data points, and regarded the average value as the experimental result of each pressure point to conduct experimental analysis and discussion.

4 Experimental results

According to experimental purposes, normal compaction simulation experiments were conducted on samples I, II, III, IV, and V, while with the simulated burial depth below 1600 m (overburden pressure was approaching 36.16 MPa), undercompaction simulation experiments with pressure coefficients of 1.2 and 1.4 were conducted on samples VI and VII, respectively. After data acquisition and processing of each sample, the experimental results can be obtained as follows (Tables 3, 4).

5 Discussion

5.1 Sandstone primary porosity analysis

Primary porosity is defined as the porosity of newly formed sediment. It is the starting point of porosity evolution during the reservoir burial process, and directly influences the accuracy of study of porosity evolution and is of great significance to reservoir porosity prediction.

Table 3 Experimental results of several samples under normal compaction

Sample I

Overburden pressure, MPa	Fluid pressure, MPa	Porosity, %	Permeability, $\times 10^{-3}$ μm^2
0.00	0.00	42.96	696.7
2.05	0.95	42.22	386.63
4.4	2.05	41.81	198.17
8.62	4.12	41.41	126.17
12.97	6.07	40.9	103.69
17.47	8.14	40.45	65.67
21.38	10	40.07	43.92
26.19	12.3	39.78	31.78
30.1	14.2	39.44	23.63
34.88	16.3	39.09	21.6
39.15	17.9	38.72	18.53
43.44	20	38.46	14.08
48.53	22.2	38.16	13.24
52.09	23.9	37.93	12.23
55.45	26.4	37.57	8.45
63.25	29.4	37.12	7.24
65.16	30	36.67	5.44

Sample II

Overburden pressure, MPa	Fluid pressure, MPa	Porosity, %	Permeability, $\times 10^{-3}$ μm^2
0.00	0.00	42.92	1779.9
2.39	1.01	42.4	888.7
4.71	2.01	42.16	744.4
9.24	4.47	41.79	218.2
12.85	5.77	41.45	152.4
16.68	7.35	41.43	143.1
21.96	9.9	40.94	97.9
26.16	11.95	40.53	76.02
30.47	13.87	40.17	50.6
34.62	16.03	39.97	44.9
39.09	18.31	39.69	32.9
43.66	20.09	39.46	23.8
48.47	22.43	39.2	19.6
52.38	23.99	38.99	14.6
57	26.03	38.8	12.9
61.09	27.73	38.59	10.9
65.45	29.47	38.49	9.4

Sample III

Overburden pressure, MPa	Fluid pressure, MPa	Porosity, %	Permeability, $\times 10^{-3}$ μm^2
0.00	0.00	37.37	570.56
2.3	1.11	37.08	391.05
5.29	2.6	36.48	202.53
8.61	4.11	34.62	134.69
12.84	6.1	31.98	103.8
17.63	8.43	31.26	56.69
21.18	10.09	31.1	46.39
25.82	12.1	30.84	37.47
30.15	14.06	30.56	30.06
34.35	16.21	30.32	25.58
39.09	18.48	30	22.1
42.46	20.12	29.7	19.3
46.72	22.1	29.45	14.41
51.85	24.25	29.28	11.47
57.19	26.21	28.94	11.28
–	–	–	–
–	–	–	–

Sample IV

Overburden pressure, MPa	Fluid pressure, MPa	Porosity, %	Permeability, $\times 10^{-3}$ μm^2
0.00	0.00	43.72	372.9
1.99	0.95	43.09	286.02
4.25	2.13	42.82	240.53
8.56	4.05	42.37	151.02
12.63	6.04	42.03	111.01
17.83	8.64	41.66	74.2
21.38	9.91	41.45	64.15
25.76	12.09	41.19	53.7
30.39	13.93	40.82	40.33
34.88	16.02	40.54	36.41
39.29	18	40.23	33.2
43.4	19.95	39.99	20.63

Sample V

Overburden pressure, MPa	Fluid pressure, MPa	Porosity, %	Permeability, $\times 10^{-3}$ μm^2
0.00	0.00	33.62	733.72
2.31	0.95	33.51	353.34
4.5	2.13	32.66	230.81
8.88	4.15	31.24	125.35
13.24	5.97	30.3	112.54
17.54	7.91	29.58	24.4
21.84	9.8	28.89	17.82
26.1	11.91	28.08	15.79
30.78	13.91	27.36	13.06
35.2	15.91	26.74	11.34
39.21	17.97	26.35	9.82
43.23	19.9	25.82	8.33

Table 3 continued

Sample IV				Sample V			
Overburden pressure, MPa	Fluid pressure, MPa	Porosity, %	Permeability, $\times 10^{-3}\,\mu m^2$	Overburden pressure, MPa	Fluid pressure, MPa	Porosity, %	Permeability, $\times 10^{-3}\,\mu m^2$
46.74	22.19	39.73	17.44	48.4	21.78	25.38	6.95
52.59	23.99	39.45	15.85	52.88	24.02	24.95	6.43
56.29	26.14	39.26	13.01	56.92	25.95	24.5	4.92
63.41	29.69	39.02	10.56	61.6	28.32	24.24	4.28
70.32	31.85	38.61	6.74	66	29.79	23.93	3.49

Previous studies have shown that primary porosity is influenced by a combination of parameters such as particle size, sorting, and sphericity, of which particle size and sorting are the most important factors influencing primary porosity and particularly the influence of sorting is more pronounced (Beard and Weyl 1973; Folk and Ward 1957; Rogers and Head 1961). Therefore, the average particle size and sorting coefficient were analyzed for their influences on primary porosity. Results show that there is a logarithmic relationship between primary porosity and particle size as well as sorting (Fig. 3).

$$\Phi_0 = -3.652\ln(M) + 35.744 \quad R^2 = 0.7755, \tag{11}$$

$$\Phi_0 = -12.61\ln(So) + 48.508 \quad R^2 = 0.8621, \tag{12}$$

where M is average particle size, So is sorting coefficient, and Φ_0 is primary porosity.

The primary sandstone porosity is influenced by a combination of sorting coefficient and average particle size. Moreover, predicting or estimating the dependent variable using the optimal combination of multiple independent variables is more effective and realistic than using only one independent variable (Hu et al. 2013). Therefore, the binary function relationship among primary porosity and average particle size and sorting coefficient is established with a stepwise regression method.

$$\Phi_0 = -3.0413So - 3.4907M + 47.828123. \tag{13}$$

Stepwise regression results show that the multiple correlation coefficient is 0.9031177, the average deviation is small and the precision is high (Table 5).

5.2 Physical property evolution during sandstone mechanical compaction

According to the above experimental conditions and the relation of overburden pressure and depth (Gluyas and Cade 1999), the overburden pressure can be converted into approximate depth (Fig. 4). The analysis shows that the evolution of porosity and permeability has a segmentation characteristic with the increasing overburden pressure and depth during mechanical compaction, and the evolution trend line can be divided into two stages. During the earlier stage of mechanical compaction, when the overburden pressure is less than 12 MPa and the equivalent burial depth is shallower than 600 m, with pressure increasing, detrital particles slide, displace, and rotate to rearrange and adjust their positions so that they can achieve a close-packing state with minimum potential energy. At this stage, i.e., the rapid compaction stage, porosity and permeability decrease rapidly (Fig. 4). Afterwards, detrital particles reach a stable packing state. With increasing pressure, the degree of close packing increases, and

Table 4 Experimental results of several samples under overpressure conditions

Sample VII				Sample VI			
Overburden pressure, MPa	Fluid pressure, MPa	Porosity, %	Permeability, $\times 10^{-3}\ \mu m^2$	Overburden pressure, MPa	Fluid pressure, MPa	Porosity, %	Permeability, $\times 10^{-3}\ \mu m^2$
0	0	43.72	389.54	0	0	41.75	445.15
1.93	0.97	42.53	147.852	2.36	1.22	41.25	399.29
4.74	2.36	42.09	103.387	4.28	2.01	41.07	307.98
8.64	4.21	41.56	76.359	8.67	4.12	40.75	221.32
17.33	8.02	40.12	43.347	13.56	6.36	40.41	130.06
21.86	10.01	39.85	36.832	17.45	8.02	40.19	88
26.03	11.95	39.58	26.841	19.39	9.01	40.04	–
29.97	13.97	39.30	23.693	28.94	13.68	38.99	56.86
41.49	22.38	39.01	16.971	35.38	16.09	38.83	30.57
46.55	25.39	38.87	14.515	39.89	21.56	38.77	33.12
51.51	27.96	38.72	12.882	43.62	23.65	38.60	24.24
57.12	30.82	38.58	10.075	48.21	26.37	38.44	21.86
62.01	33.73	38.42	8.33	51.32	28.52	38.23	16.42
68.19	36.47	38.23	5.557	56.98	31.34	38.09	–
71.89	38.63	38.08	4.078	57.41	34.23	37.64	10.88
–	–	–	–	68.64	35.72	37.63	8.8

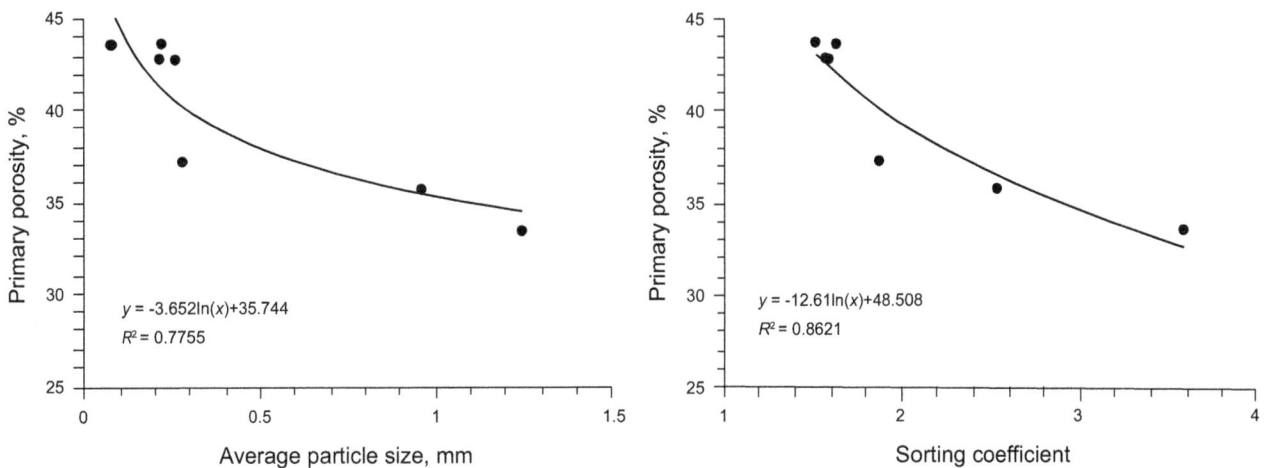

Fig. 3 Relationship between primary porosity and particle size as well as sorting coefficient

porosity and permeability decrease slowly. This is, the slow compaction stage (Fig. 4).

The experimental data from the slow compaction stage can be used to analyze the evolution of porosity and permeability with increasing overburden pressure during normal compaction (Liu et al. 2006). Regression analysis results show that there is an exponential relationship between porosity and overburden pressure, and the relationship can be expressed as $y = Ae^{Bx}$; while a power function relationship exists between permeability and

overburden pressure, that is $y = Cx^D$. Meanwhile, the relationship between porosity and permeability is exponential (Table 6).

According to the comparison of experimental results of different samples, the coefficient A of the functional relationship $y = Ae^{Bx}$ between overburden pressure and porosity mainly depends on the primary porosity value, which is mainly controlled by the sorting coefficient. While the coefficient B is principally influenced by the average particle size. The coefficient C of the functional

Table 5 Stepwise regression results of sandstone primary porosity and average particle size and sorting coefficient

Sample no.	Sorting coefficient So	Average particle size M, mm	Regression value	Original value	Deviation
I	1.556	0.219	42.345378	42.959999	0.614621
II	1.575	0.262	41.193932	37.369999	−3.823933
III	1.860	0.293	42.130514	42.919998	0.789484
IV	1.618	0.079	42.642020	43.720001	1.077982
V	3.571	1.28	32.604338	33.619999	1.015661
VI	1.50	0.221	42.494747	41.750000	0.744747
VII	2.52	0.964	36.799067	35.869999	−0.929068

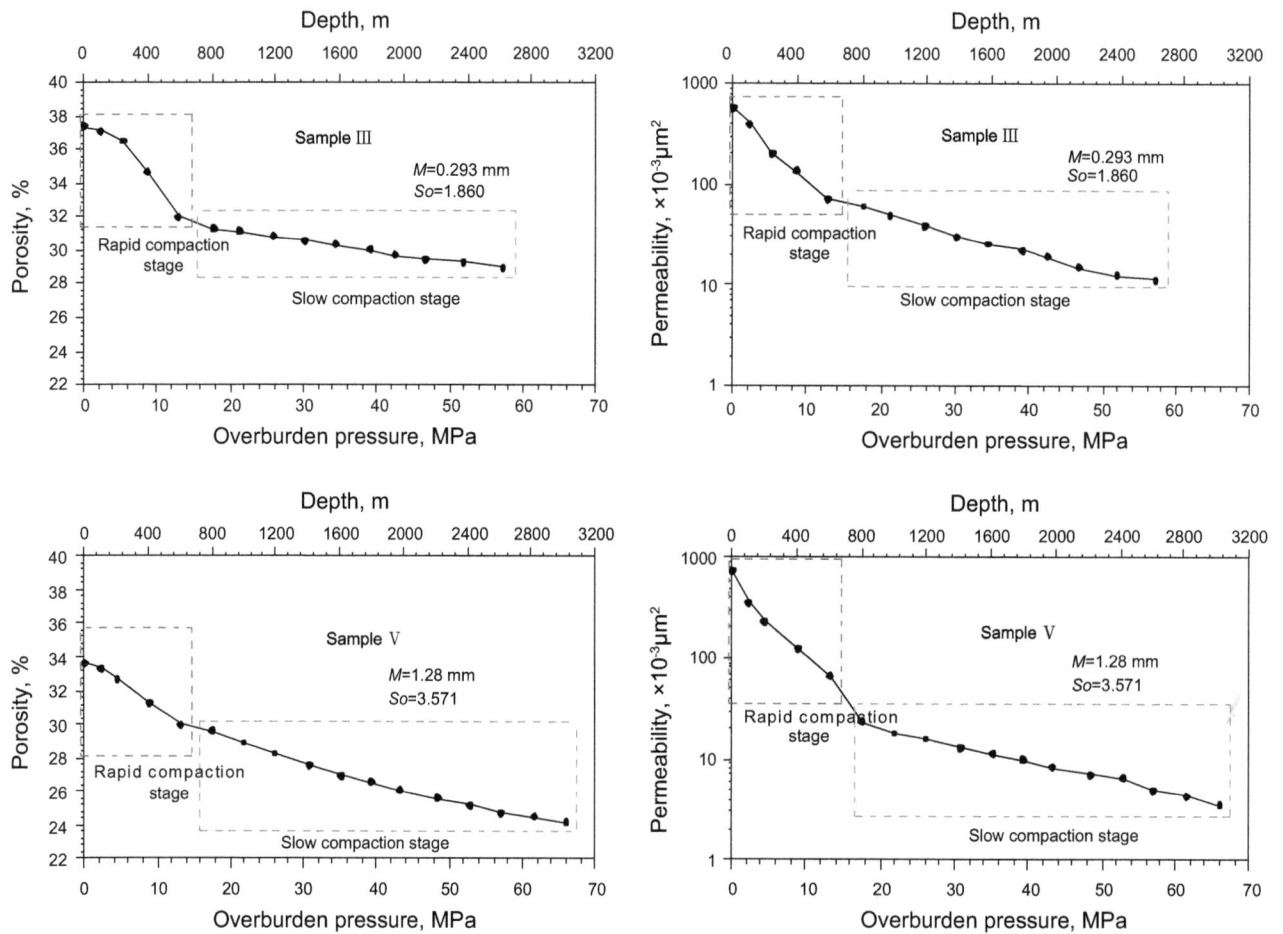

Fig. 4 Variation of porosity and permeability with increasing overburden pressure and depth during normal compaction

relationship $y = Cx^D$ is mainly controlled by the average particle size, while the coefficient D is greatly influenced by the sorting coefficient.

5.3 The influence of particle size on sandstone mechanical compaction

Sample I and sample II are characterized by the same composition, similar sorting, but different average particle sizes. Sample I is fine sand, but sample II is medium sand. Simulation experiment results show that two samples have approximately equal primary porosity (Table 1). During the rapid compaction stage, the evolution processes of two samples are almost the same. After entered into the slow compaction stage, the decrease rate of porosity of sample II with coarser average particle size is obviously smaller than that of sample I with finer average particle size during the increasing overburden pressure process. Meanwhile, the

Table 6 Evolution of porosity and permeability with overburden pressure under normal compaction

Sample no.	The relation between overburden pressure (x) and porosity (y)	The relation between overburden pressure (x) and permeability (y)	The relation between porosity (x) and permeability (y)
I	$y = 41.82372e^{-0.00193x}$	$y = 8{,}761.51262x^{-1.70980}$	$y = 8E - 10e^{0.6146x}$
	$R^2 = 0.99456$	$R^2 = 0.97494$	$R^2 = 0.9796$
II	$y = 42.18101e^{-0.00148x}$	$y = 61{,}210.68436x^{-2.08359}$	$y = 1E - 15e^{0.9534x}$
	$R^2 = 0.98048$	$R^2 = 0.98755$	$R^2 = 0.9939$
III	$y = 32.43762e^{-0.00201x}$	$y = 3777.39155x^{-1.43168}$	$y = 2E-08e^{0.6991x}$
	$R^2 = 0.99545$	$R^2 = 0.97816$	$R^2 = 0.9816$
IV	$y = 42.72003e^{-0.00148x}$	$y = 22{,}104.72226x^{-1.86625}$	$y = 3E-13e^{0.7989x}$
	$R^2 = 0.99416$	$R^2 = 0.93290$	$R^2 = 0.9827$
V	$y = 31.48912e^{-0.00436x}$	$y = 2{,}491.68733x^{-1.53245}$	$y = 0.00094e^{0.34938x}$
	$R^2 = 0.98478$	$R^2 = 0.97417$	$R^2 = 0.97009$

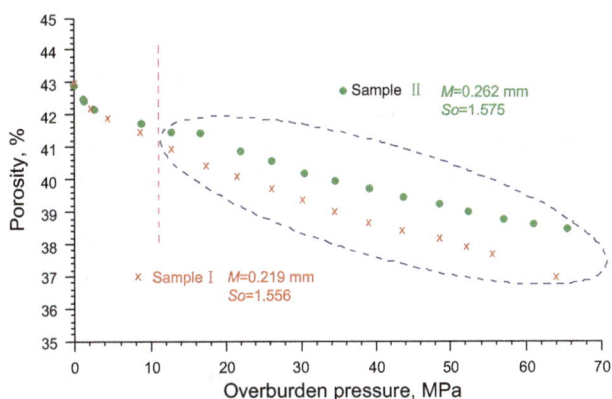

Fig. 5 Influence of particle size on sandstone mechanical compaction

difference of the remaining porosity of the two samples is more significant (Fig. 5).

By quantitative analysis of experimental data, the porosity reduction of sample I is 6.29 %, which is obviously larger than that of sample II 4.43 % when the overburden pressure increases to 65 MPa. Studies of different compaction stages show that during the rapid compaction stage, the average porosity reduction of sample I and sample II is 0.339 % and 0.334 %, respectively, with

2.3 MPa increase of the overburden pressure (about 100 m of the burial depth, the same below), which is almost the same. However, during the slow compaction stage, the porosity reduction of sample I is 0.197 %, which is larger than that of sample II 0.156 % with 2.3 MPa increase of the overburden pressure (Table 7).

Therefore, the influence of particle size on mechanical compaction is mainly in the slow compaction stage during the burial process of sandstone. The coarser the particle size, the slower the compaction rate and the larger the final porosity. The influence of the particle size on mechanical compaction is more pronounced with increasing burial depth and overburden pressure. After experienced the rapid compaction, detrital particles are generally in contact with each other. The sample with finer particle size has a larger specific surface area and thus a smaller force per unit area. Sliding deformation does not occur easily with increasing overburden pressure which is mainly used to squeeze the pore space, resulting in the rapid loss of porosity. While the sample with coarser particle size has a smaller specific surface area and thus a larger force per unit area, so sliding deformation occurs with increasing overburden pressure which offsets a portion of force squeezing the pore space, and the rate of porosity loss decreases.

Table 7 Data about the influence of particle size on sandstone mechanical compaction

Sample no.	I	II
Average particle size M, mm	0.219	0.262
Sorting coefficient So	1.556	1.575
Primary porosity Φ_0, %	42.96	42.92
Porosity loss with 65 MPa overburden pressure, %	6.29	4.43
Porosity loss during rapid compaction stage, %/2.3 MPa	0.339	0.334
Porosity loss during slow compaction stage, %/2.3 MPa	0.197	0.156

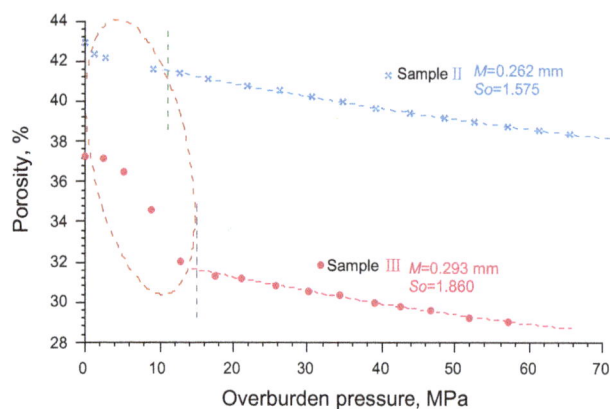

Fig. 6 Influence of sorting on sandstone mechanical compaction

5.4 The influence of sorting on sandstone mechanical compaction

Sample II and sample III selected from the Golden Beach and Silver Beach, Qingdao, eastern China are characterized by the same composition, similar particle size of medium sand, but different sorting, of which sample III has poorer sorting. Simulation experiment results show that sample III with poorer sorting has a smaller primary porosity (Table 1). During the rapid compaction stage, a great difference exists between the porosity evolution processes of the two samples, and the porosity reduction rate of sample III is significantly greater than that of sample II. After entered into the slow compaction stage, the evolution processes of the two samples are almost the same with increasing overburden pressure (Fig. 6).

The experimental results of different compaction stages show that during the rapid compaction stage, the average porosity loss of sample II is 0.334 % per 2.3 MPa increase of overburden pressure and the average porosity loss of sample III is 0.886 %, which is 2.65 times greater than that of sample II. However, the average porosity reduction of sample II and sample III is 0.156 % and 0.127 %, respectively, per 2.3 MPa increase of overburden pressure during the slow compaction stage, which is similar, and the porosity loss is in a negative relation to the average particle size (Table 8).

Therefore, the influence of sorting on sandstone mechanical compaction is mainly in the rapid compaction stage. The poorer the sorting, the higher the compaction rate and the more distinct the difference between the rapid compaction stage and the slow compaction stage. Also the dividing overburden pressure of the two stages will be larger (Fig. 6), that is, the rapid compaction stage lasts longer, and the equivalent burial depth is deeper. During the rapid compaction stage, for the sandstone sample with poorer sorting, when the position adjustment and rearrangement of detrital particles occur, finer particles will easily fill in the pore space formed by the arrangement of coarse particles, which results in a rapid loss of porosity. However, after entering the slow compaction stage, particles have a stable packing state, compaction further increases the tightness of particles, and the compaction rate is mainly influenced by the particle size.

5.5 The influence of abnormally high pressure on sandstone mechanical compaction

During the burial process of clastic sediments from early deposition to mid-deep strata, abnormally high pressure is mainly formed by tectonic evolution, disequilibrium compaction, hydrothermal pressurization, clay mineral transformation, and hydrocarbon generation (Akrout et al. 2012). It inhibits compaction and protects primary pores and is of great significance to the development of mid-deep high quality reservoirs (Hunt 1990; Ma et al. 2011; Bloch et al. 2002; Cao et al. 2014).

In this paper, undercompaction simulation experiments with pressure coefficients of 1.2 and 1.4 were conducted on sample VI from the Silver Beach of Qingdao, eastern China (the same parameters as sample I) and sample VII from the Yellow River Estuary of Dongying, eastern China (the same parameters as sample IV), respectively, under an overburden pressure larger than 34 MPa. Then the experimental results were compared with those of sample I and sample IV under normal compaction, and the influence of abnormally high pressure on sandstone mechanical compaction was analyzed.

Experimental results show that the mechanical compaction rate under abnormally high pressure is obviously smaller than that under normal compaction; moreover, the higher the pressure coefficient, the greater the difference between the evolution trend line of actual porosity under

	Sample no.	II	III
Table 8 Data about the influence of sorting on sandstone mechanical compaction	Average particle size M, mm	0.262	0.293
	Sorting coefficient So	1.575	1.86
	Primary porosity Φ_0, %	42.92	37.37
	Porosity loss during rapid compaction stage, %/2.3 MPa	0.334	0.866
	Porosity loss during slow compaction stage, %/2.3 MPa	0.156	0.127

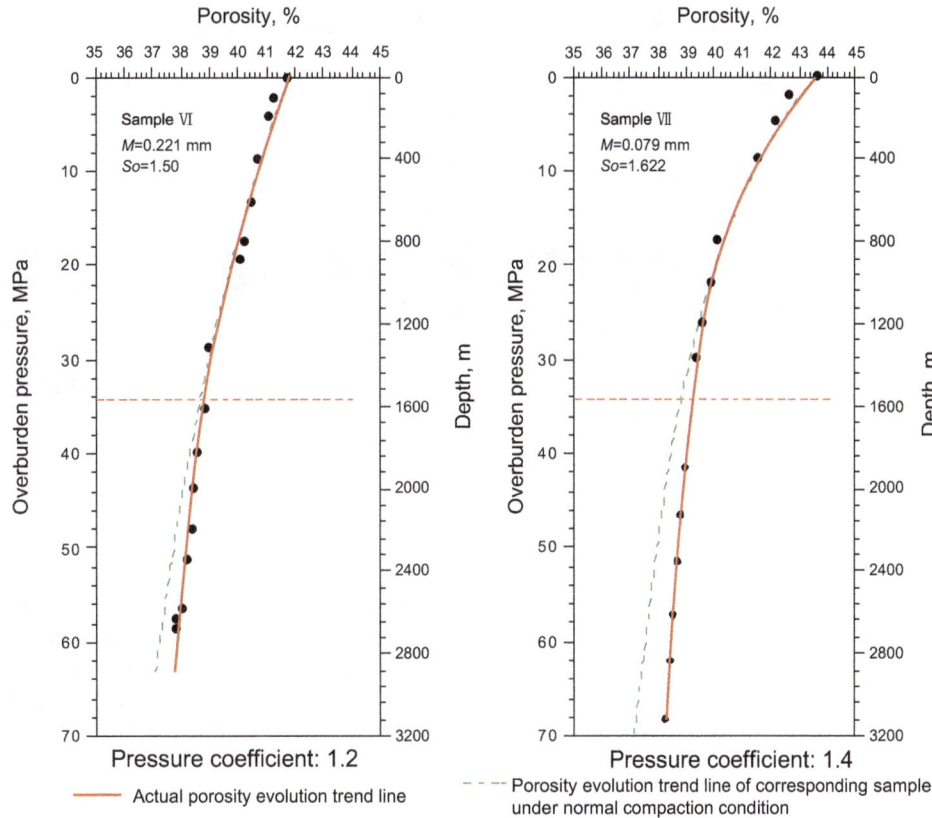

Fig. 7 Influence of formation overpressure on sandstone mechanical compaction

Table 9 Data about the influence of abnormally high pressure on sandstone mechanical compaction

Sample no.	VII	VI
Average particle size M, mm	0.079	0.221
Sorting coefficient So	1.622	1.5
Pressure coefficient	1.4	1.2
Average porosity reduction of normal compacted sample, %/2.3 MPa	0.182	0.124
Average porosity reduction of undercompacted sample, %/2.3 MPa	0.0708	0.094

abnormally high pressure and that under normal compaction (Fig. 7).

The development of abnormally high pressure is in the slow compaction stage. Experimental results show that when the pressure coefficient is 1.4, the average porosity loss of the normally compacted sample and the undercompacted sample is 0.182 % and 0.0708 %, respectively, per 2.3 MPa increase of overburden pressure. The porosity loss of the normally compacted sample is 2.571 times that of the undercompacted sample. When the pressure coefficient is 1.2, the average porosity loss of the normally compacted sample and the undercompacted sample is 0.124 % and 0.094 %, respectively, per 2.3 MPa increase of overburden pressure. The porosity loss of the normal compacted sample is 1.305 times that of the undercompacted sample (Table 9).

Therefore, abnormally high pressure inhibits sandstone mechanical compaction. The higher the pressure coefficient, the slower the compaction rate. The compaction rate increases twofold when the pressure coefficient decreases by 0.2.

Further studies show that the compaction rate of sample VII with finer particle size and poorer sorting is larger than that of sample VI with coarser particle size and better sorting under normal compaction condition, which accords with particle size and sorting influencing sandstone mechanical compaction. However, when abnormally high pressure develops, the compaction rate of sample VII with a pressure coefficient of 1.4 is smaller than that of sample VI with a pressure coefficient of 1.2 in the undercompaction condition. Sandstone mechanical compaction is

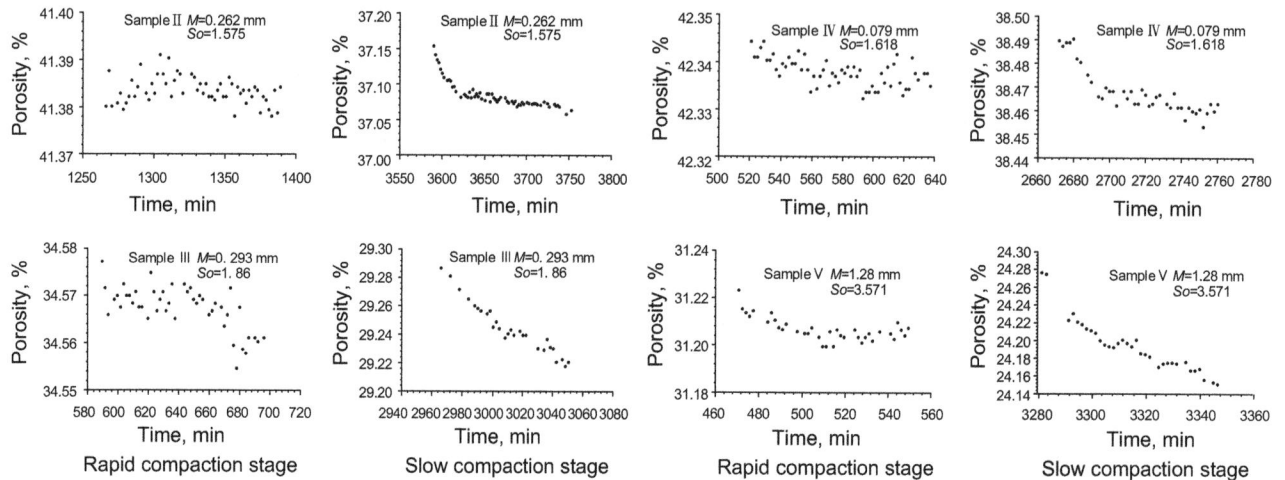

Fig. 8 Influence of burial time on sandstone mechanical compaction

Table 10 Data about the influence of burial time on sandstone mechanical compaction

Sample no.	II	III	IV	V
Average particle size M, mm	0.262	0.293	0.079	1.28
Sorting coefficient So	1.575	1.86	1.618	3.571
Porosity loss, %/h	0.389	0.349	0.229	0.119

mainly controlled by abnormally high pressure, which does not accord with particle size and sorting influencing mechanical compaction. Therefore, the control of abnormally high pressure on mechanical compaction is stronger than that of particle size and sorting.

5.6 The influence of burial time on sandstone mechanical compaction

Clastic sediments experiencing different burial time suffer different compaction effects under the same overburden pressure (Liu et al. 2007). By analyzing the experimental data at the final pressure point of rapid compaction stage and slow compaction stage during the simulation experiment, it is concluded that there is no evident correlation between porosity and compaction time during the rapid compaction stage, while a good negative relation exists between them during the slow compaction stage. Moreover, the longer the compaction time, the slower the rate of porosity reduction under the constant overburden pressure (Fig. 8). Quantitative calculations show that when the overburden pressure of the final pressure point during the slow compaction stage is constant, the porosity reduction caused by compaction per hour can be as high as 0.119 %–0.389 %, and porosity reduction caused by compaction is mainly controlled by the average particle size (Table 10). Therefore, during geological time, burial time and burial

depth are the two equivalently important factors influencing sandstone mechanical compaction, and the influence of time is mainly reflected in the slow compaction stage. Only after the particles are in close contact with each other, can the creep characteristics of formation be shown with increasing compaction time.

6 Conclusions

1. There is a logarithmic relationship between primary porosity and particle size as well as sorting, and the binary function relation between primary porosity and the two factors is $\Phi_0 = -3.0413 So - 3.4907 M + 47.828123$, which can provide reference for the calculation of primary porosity of sandstone reservoir.

2. During sandstone mechanical compaction, the evolution of porosity and permeability has a segmentation characteristic with the increasing overburden pressure and depth, and the evolution curves can be divided into two sections, i.e., a rapid compaction stage with a steep slope at the earlier stage and a slow compaction stage at the later stage. The dividing pressure of two sections is about 12 MPa. During the slow compaction stage, there is an exponential relationship between porosity and overburden pressure, while a power function relationship exists between permeability and overburden pressure. The relationship between porosity and permeability is exponential.

3. The influence of particle size on mechanical compaction is mainly reflected in the slow compaction stage. The coarser the particle size, the slower the compaction rate and the larger the final porosity. The influence of particle size is more pronounced with increasing depth and overburden pressure.

4. The influence of sorting on sandstone mechanical compaction is mainly in the rapid compaction stage. The poorer the sorting, the higher the compaction rate and the more distinct the difference between the rapid compaction stage and the slow compaction stage. The dividing overburden pressure value of the two stages will be larger.

5. Abnormally high pressure inhibits sandstone mechanical compaction. The higher the pressure coefficient, the slower the compaction rate. The control of abnormally high pressure on sandstone mechanical compaction is stronger than that of particle size and sorting.

6. During geological time, burial time and burial depth are the two equivalently important factors influencing sandstone mechanical compaction. The influence of burial time is mainly reflected in the slow compaction stage, and porosity reduction caused by compaction is mainly controlled by the average particle size.

7. Although the experiments cannot be compared completely with real geological processes, they can provide some useful guidance for understanding the real geological processes. Further study should focus on simulating longer geological time by changing the pressure and temperature conditions.

Acknowledgments This study is co-funded by the National Natural Science Foundation of China (Grant No. U1262203), the National Science and Technology Special Grant (Grant No. 2011ZX05009-003), the Fundamental Research Funds for the Central Universities (Grant No. 14CX06013A), and the Chinese Scholarship Council (No. 201406450019).

References

Ajdukiewicz JM, Nicholson PH, Esch WL. Prediction of deep reservoir quality using early diagenetic process models in the Jurassic Norphlet Formation, Gulf of Mexico. AAPG Bull. 2010;94(8):1189–227.

Akrout D, Ahmadi R, Mercier E, et al. Natural hydrocarbon accumulation related to Formation overpressured interval; study case is the Saharan platform (Southern Tunisia). Arab J Geosci. 2012;5(4):849–57.

Aplin AC, Matenaar IF, McCarty DK, et al. Influence of mechanical compaction and clay mineral diagenesis on the microfabric and pore-scale properties of deep-water Gulf of Mexico mudstones. Clays Clay Miner. 2006;54(4):500–14.

Beard DC, Weyl PK. Influence of texture on porosity and permeability of unconsolidated sand. AAPG Bull. 1973;57(2):349–69.

Bloch S, Lander RH, Bonnell L. Anomalously high porosity and permeability in deeply buried sandstone reservoirs: origin and predictability. AAPG Bull. 2002;86(2):301–28.

Cao YC, Yuan GH, Li XY, et al. Characteristics and origin of abnormally high porosity zones in buried Paleogene clastic reservoirs in the Shengtuo area, Dongying Sag, East China. Petrol Sci. 2014;11(3):346–62.

Chester JS, Lenz SC, Chester FM, et al. Mechanisms of compaction of quartz sand at diagenetic conditions. Earth Planet Sci Lett. 2004;220(3–4):435–51.

Dai JX, Ni YY, Wu XQ. Tight gas in China and its significance in exploration and exploitation. Petrol Explor Dev. 2012;39(3):277–84.

Folk RL, Ward WC. Brazos River bar: a study in the significance of grain size parameters. J Sediment Petrol. 1957;27(1):3–26.

Gluyas J, Cade CA. Prediction of porosity in compacted sands. AAPG Memoir. 1999;69:19–27.

Hart BS. Seismic expression of fracture-swarm sweet sports, Upper Cretaceous tight-gas reservoirs, San Juan Basin. AAPG Bull. 2006;90(10):1519–34.

Hu ZW, Huang SJ, Wang DH, et al. Application of multiple stepwise regression to influential evaluation of pore-throat size on low-permeability sandstone reservoirs. J Guilin Univ Technol. 2013;33(1):21–5 (**in Chinese**).

Hunt JM. Generation and migration of petroleum from abnormally pressured fluid compartments. AAPG Bull. 1990;74(1):1–12.

Jia CZ, Zheng M, Zhang YF. Unconventional hydrocarbon resources in China and the prospect of exploration and development. Petrol Explor Dev. 2012;39(2):139–46.

Liu GY, Jin ZJ, Zhang LP. Simulation study on clastic rock diagenetic compaction. Acta Sedimentol Sin. 2006;24(3):408–13 (**in Chinese**).

Liu H, Cao YC, Jiang ZX, et al. Distribution characteristics of evaporates and Formation pressure of the fourth member of the Shahejie Formation in the Dongying Sag, the Bohai Bay Basin. Oil Gas Geol. 2009;30(3):287–93 (**in Chinese**).

Liu MJ, Liu Z, Liu JJ, et al. Coupling relationship between sandstone reservoir densification and hydrocarbon accumulation: a case from the Yanchang Formation of the Xifeng and Ansai areas, Ordos Basin. Petrol Explor Dev. 2014;41(2):185–92.

Liu Z, Shao XJ, Jin B, et al. Co-effect of depth and burial time on the evolution of porosity for clastic rocks during the stage of compaction. Geoscience. 2007;21(1):125–32 (**in Chinese**).

Lv ZX, Liu SB. Ultra-tight sandstone diagenesis and mechanism for the Formation of relatively high-quality reservoir of Xujiahe Group in western Sichuan. Acta Petrol Sin. 2009;25(10):2373–83 (**in Chinese**).

Ma XM, Zhao ZY, Liu HW. Influences of abnormal overpressure on super-low permeability reservoirs in Chexi depression in Shandong. J Central South Univ (Sci Technol). 2011;42(8):2507–13 (**in Chinese**).

Maast TM, Jahren J, Bjorlykke K. Diagenetic controls on reservoir quality in Middle to Upper Jurassic sandstones in the South Viking Graben, North Sea. AAPG Bull. 2011;95(11):1937–58.

Rogers JJ, Head WB. Relationships between porosity, median size, and sorting coefficients of synthetic sands. J Sediment Petrol. 1961;31(3):467–70.

Taylor TR, Giles MR, Hathon LA, et al. Sandstone diagenesis and reservoir quality prediction: models, myths, and reality. AAPG Bull. 2010;94(8):1093–132.

Tobin RC, McClain T, Lieber RB, et al. Reservoir quality modeling of tight-gas sands in Wamsutter field: integration of diagenesis, petroleum systems, and production data. AAPG Bull. 2010;94(8):1229–66.

Wang B, Feng Y, Zhao YQ, et al. Determination of hydrocarbon charging history by diagenetic sequence and fluid inclusions: a case study of the Kongquehe area in the Tarim Basin. Acta Geol Sin. 2015;89(3):876–86.

Wang YJ, Tian ZY. Oil and gas exploration potential and prospect of basins in eastern area of North China. Acta Petrol Sin. 2003;24(4):7–12 (in Chinese).

Wang ZM, Liu LF, Yang HJ, et al. Characteristics of Paleozoic clastic reservoirs and the relationship with hydrocarbon accumulation in the Tazhong area of the Tarim Basin, west China. Petrol Sci. 2010;7(2):192–200.

Worden RH, Mayall M, Evans IJ. The effect of ductile-lithic sand grains and quartz cement on porosity and permeability in Oligocene and lower Miocene clastics, South China Sea: prediction of reservoir quality. AAPG Bull. 2000;84(3):345–59.

Yuan EX. Engineering fluid mechanics. Beijing: Petroleum Industry Press; 1985. p. 8 (in Chinese).

Zhang N, Tian ZJ, Wu SH, et al. Study Xujiahe reservoir diagenetic process, Sichuan Basin. Acta Petrolog Sin. 2008;24(9):2179–84 (in Chinese).

Zhang Q, Zhu XM, Ronald JS, et al. Variation and mechanisms of clastic reservoir quality in the paleogene shahejie Formation of the Dongying Sag, Bohai Bay Basin, China. Petrol Sci. 2014;11(2):200–210.

Zhang SC, Zhang BM, Li BL, et al. History of hydrocarbon accumulations spanning important tectonic phases in marine sedimentary basins of China: taking the Tarim Basin as an example. Petrol Explor Dev. 2011;38(1):1–15.

Zhu HH, Zhong DK, Li QR, et al. Characteristics and controlling factors of upper Triassic Xujiahe tight sandstone reservoir in southern Sichuan Basin. Acta Sedimentol Sin. 2013;31(1):167–75 (in Chinese).

Zhu RK, Zou CN, Zhang N, et al. Diagenetic fluids evolution and genetic mechanism of tight sandstone gas reservoirs in Upper Triassic Xujiahe Formation in Sichuan Basin, China. Sci China Ser D. 2008;51(9):1340–53.

Zou CN, Zhang GS, Yang Z, et al. Geological concepts, characteristics, resource potential and key techniques of unconventional hydrocarbon: on unconventional petroleum geology. Petrol Explor Dev. 2013;40(4):385–99 (in Chinese).

17

Genetic mechanism and development of the unsteady Sarvak play of the Azadegan oil field, southwest of Iran

Yang Du[1,2] · Jie Chen[2] · Yi Cui[3] · Jun Xin[2] · Juan Wang[2] · Yi-Zhen Li[2] · Xiao Fu[2]

Abstract The upper Cretaceous Sarvak reservoir in the Azadegan oil field of southwest Iran has its oil–water contact nearly horizontal from the north to the center and dips steeply from the center to the south. The purpose of this paper is to interpret this abnormal reservoir feature by examining the accumulation elements, characteristics, and evolution based on the 3D seismic, coring, and well logging data. Generally, in the field, the Sarvak reservoir is massive and vertically heterogeneous, and impermeable interlayers are rare. The distribution of petrophysical properties is mainly dominated by the depositional paleo-geomorphology and degrades from north to south laterally. The source is the lower Cretaceous Kazhdumi Formation of the eastern Dezful sag, and the seal is the muddy dense limestone of the Cenozoic Gurpi and Pebdeh Formations. Combined with the trap evolution, the accumulation evolution can be summarized as follows: the Sarvak play became a paleo-anticlinal trap in the Alpine tectonic activity after the late Cretaceous (96 Ma) and then was relatively peaceful in the later long geologic period. The Kazhdumi Formation entered in the oil window at the early Miocene (12–10 Ma) and charged the Sarvak bed, thus forming the paleo-reservoir. Impacted by the Zagros Orogeny, the paleo-reservoir trap experienced a strong secondary deformation in the late Pliocene (4 Ma), which shows as the paleo-trap shrank dramatically and the pre-low southern area uplifted and formed a new secondary anticline trap, hence evolving to the current two structural highs with the south point (secondary trap) higher than the north (paleo-trap). The trap deformation broke the paleo-reservoir kinetic equilibrium and caused the secondary reservoir adjustment. The upper seal prevented vertical oil dissipation, and thus, the migration is mainly in interior Sarvak bed from northern paleo-reservoir to the southern secondary trap. The strong reservoir heterogeneity and the degradation trend of reservoir properties along migration path (north to south) made the reservoir readjustment extremely slow, plus the short and insufficient re-balance time, making the Sarvak form an "unsteady reservoir" which is still in the readjustment process and has not reached a new balance state. The current abnormal oil–water contact versus the trap evolutionary trend indicates the secondary readjustment is still in its early stage and has only impacted part of paleo-reservoir. Consequently, not all of the reservoir is dominated by the current structure, and some parts still stay at the paleo-reservoir form. From the overview above, we suggest the following for the future development: In the northern structural high, the field development should be focused on the original paleo-reservoir zone. In the southern structural high, compared with the secondary reservoir of the Sarvak with the tilted oil–water contact and huge geologic uncertainty, the lower sandstone reservoirs are more reliable and could be developed first, and then the deployment optimized of the upper Sarvak after obtaining sufficient geological data. By the hints of the similar reservoir characteristics and tectonic inheritance with Sarvak, the lower Cretaceous Fahliyan

✉ Yang Du
157762166@qq.com

1   School of Geoscience and Technology, Southwest Petroleum University, Chengdu 610500, Sichuan, China

2   Geology and Exploration Research Institute, CNPC Chuanqing Drilling Engineering Company Limited, Chengdu 610051, Sichuan, China

3   Iraq Branch Company of CNODC, CNPC, Dubai 500486, United Arab Emirates

Edited by Jie Hao

carbonate reservoir is also proved to be an unsteady reservoir with a tilted oil–water contact.

Keywords Iran · Azadegan oil field · Sarvak · Oil–water contact · Accumulation elements · Accumulation evolution · Unsteady reservoir · Development suggestion

1 Introduction

Azadegan oil field, which lies adjacent to the Iran–Iraq border area of Khuzestan Province in the southwest of Iran, is currently the largest untapped oil field in the world (Liu et al. 2013a; Du et al. 2015a, b) (Fig. 1). Four Cretaceous reservoirs have been found: Sarvak, Kazhdumi (Burgan sandstone), Gadwan (Zubair sandstone), and Fahliyan (Fig. 2). The Sarvak is the main development zone, which accounts for 91.8 % of the total reserves. Preliminary exploration proves that the oil–water contact (OWC) of Sarvak is nearly horizontal in the north-central zone of the

field but tilts steeply up from the center to the south along the major axis of the structure, and the height difference can reach over 300 m according to the drilling (Fig. 3). Additionally, this is not the only case, and neighboring oil fields show similar phenomenon as well. For example (oil field locations below are shown in Fig. 1), the tilted OWC and the 150 m height difference were discovered in the Sarvak of the eastern Yadavaran oil field (Xu et al. 2010), while the OWC of the upper Cretaceous Sarvak, Ilam reservoir in the eastern Ab-e Teymur, Mansuri, Ahwaz fields tilts from SW to NE. The different OWCs of the Mishrif (upper Sarvak) reservoir which are 2710, 2750, and 2680 m deep, respectively, were proven in three wells of the western Iraqi Majnoon field. The lower part of the upper Cretaceous Yamama reservoir also has an OWC with a height difference of nearly 150 m and a tilt angle of 3°. The depth of the oil column in the upper Cretaceous Yamama reservoir of the Umr Nahr field in the west is close to 200 m. The Mishrif of Western Missan field is also a tilted one, and the height difference is 100 m. A good

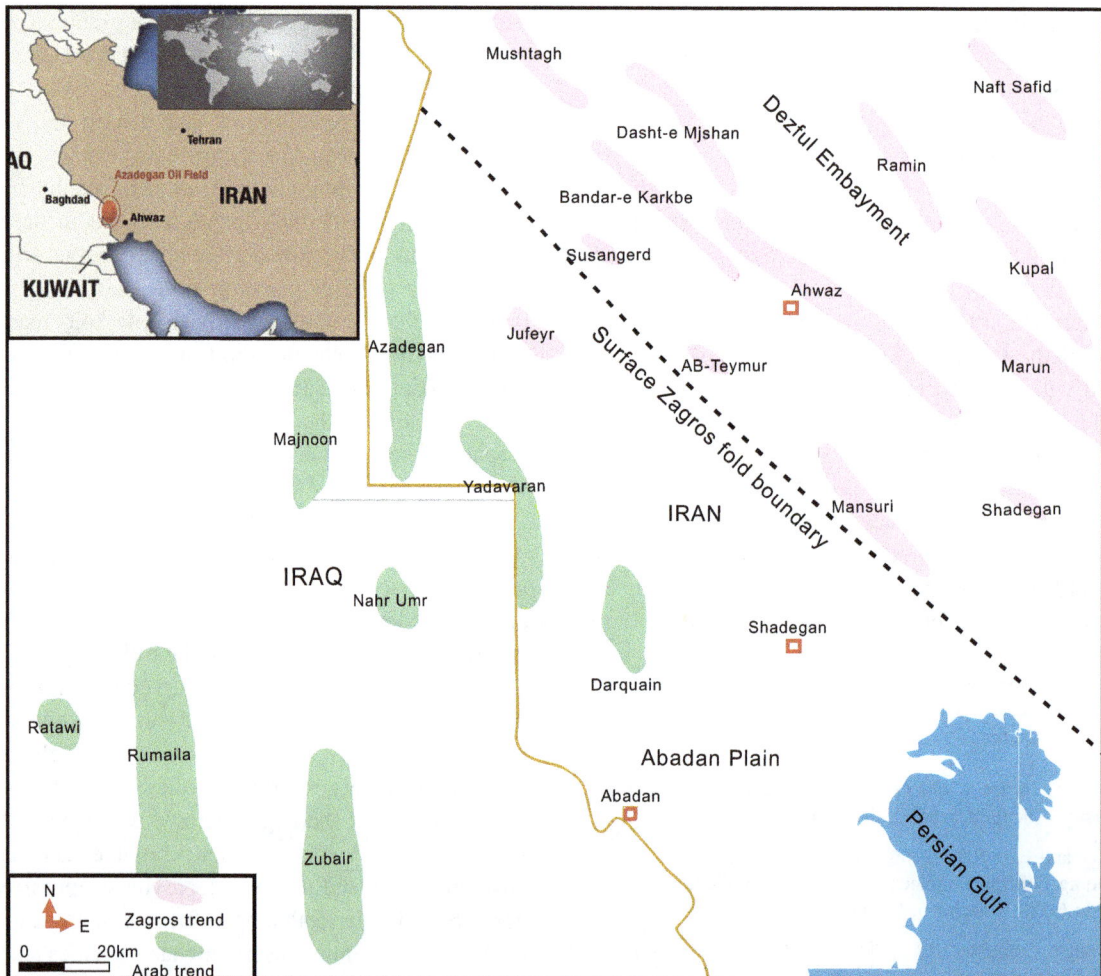

Fig. 1 Regional location map of the Azadegan oil field in southwest Iran

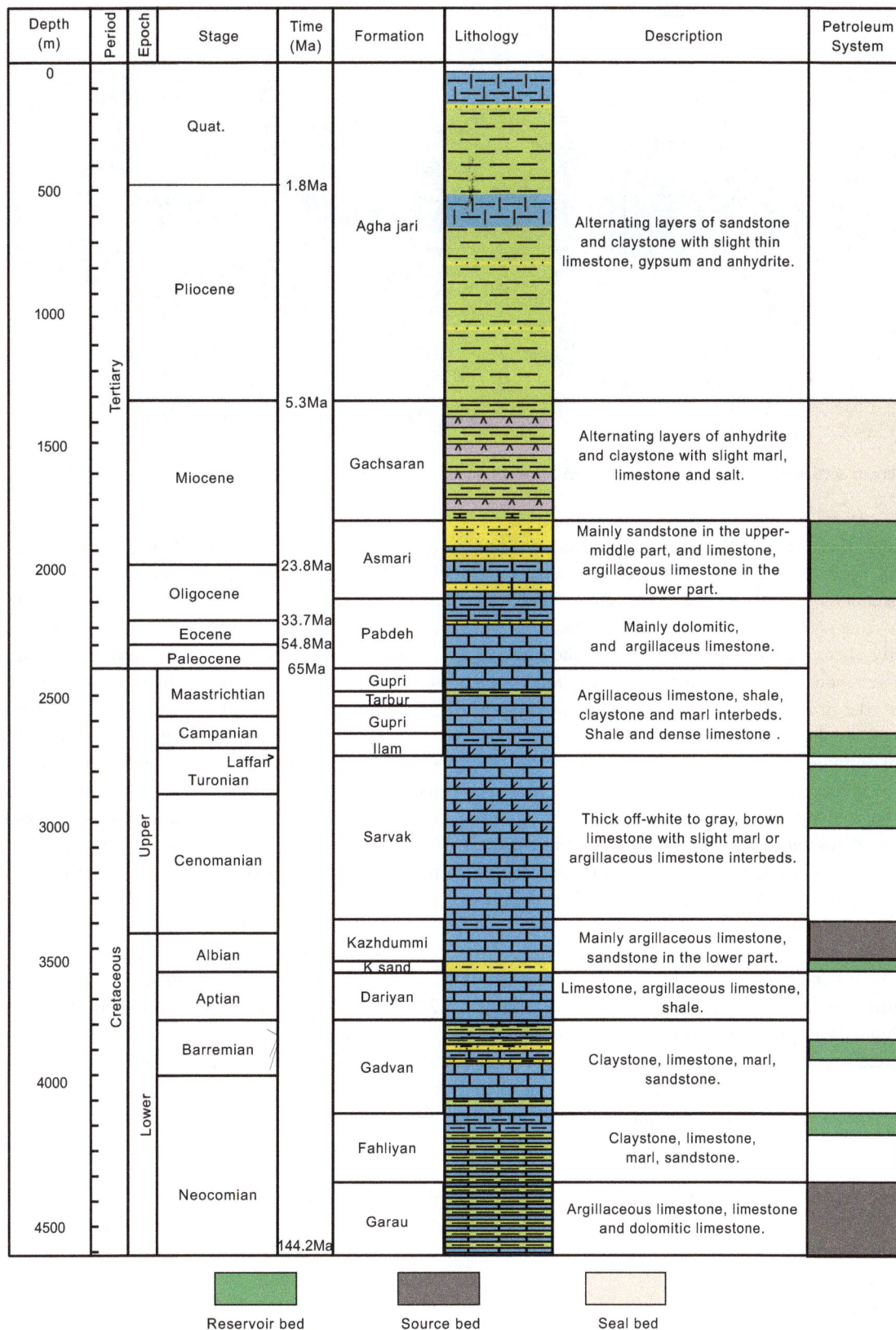

Fig. 2 The stratigraphic column section of the Azadegan oil field in southwest Iran

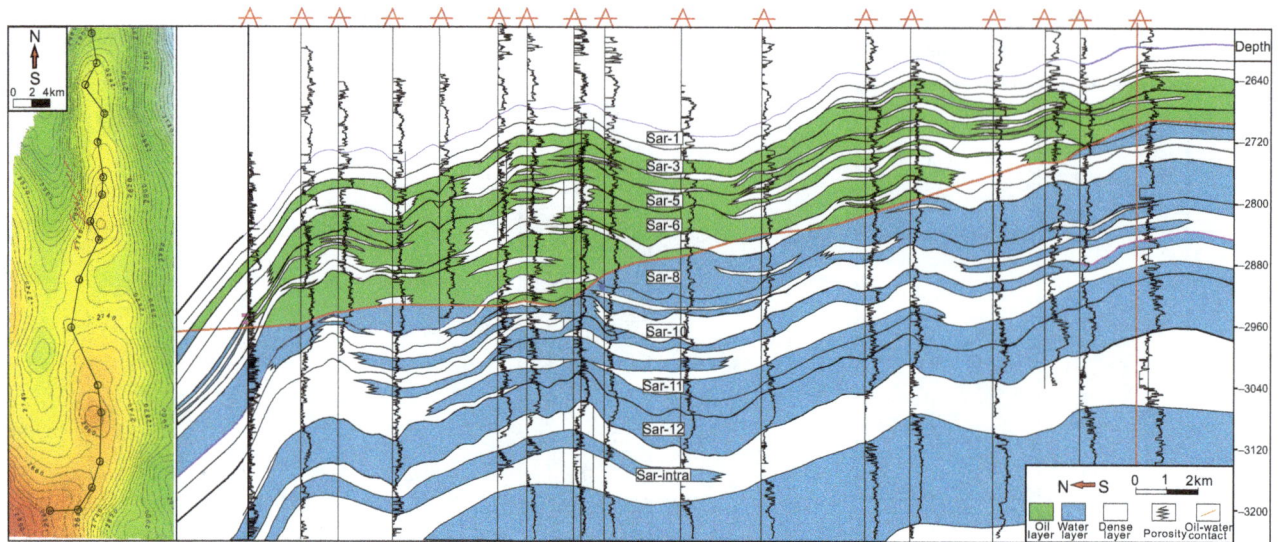

Fig. 3 Reservoir section of the upper Cretaceous Sarvak Formation of the Azadegan oil field (N-S), Iran

knowledge of these abnormal reservoir characteristics and the genesis of the tilted OWC is closely related to the reserve calculations, well spacing, well pattern, and well type, with great significance for field development.

Currently, there is no certain conclusion on the cause of such phenomenon. Some scholars once explained the genesis of the irregular OWC and built accumulation models such as the "differential entrapment" and "leak differential entrapment" (Gussow 1954; Schowalter 1979). However, it is hard to use them to interpret the Azadegan case. We have analyzed and confirmed that faults, hydrodynamics, and reservoir heterogeneity are not the causes, and propose the very late trap deformation by the Zagros orogeny as the main reason for the tilted OWC in Azadegan (Du et al. 2015a). Based on the previous findings, this paper uses the new geological data to describe the accumulation factors including the reservoir, source, and seal, and reconstructs the field paleo-structure and trap evolution history. By analyzing the relationship between the tectonic evolution and reservoir accumulation, the aim is to clarify the genetic mechanism of the unsteady reservoir and then make suggestions for the field development.

2 Tectonic features and evolution

2.1 Tectonic features

Tectonically, Azadegan oil field is situated in the southwest of the Zagros overthrust fault zone and in the transition zone between the Zagros foreland basin and Arabian platform (Soleimani 2013). Two different types of traps with different forming mechanisms are confirmed

here: One type is called the Zagros Trend, located in the foothill zone (Zagros folded zone) with an NW–SE strike, and with similar trends as the anticlinal structure elongated in the Zagros Mountain range. The Miocene Asmari Formation, which is dominantly limestone and partially sandstone, is the main reservoir of the Zagros Trend oil fields (Wang et al. 2011; Zhang et al. 2012; McQuillan 1973, 1974; Bordenave and Hegre 2005). The other type, known as the Arabian Trend—uplifting caused by basement fault "resurrection" and salt flow, widely distributed in a number of oil fields which are located in southeast of Iraq, Kuwait, and northeast of Saudi Arabia, with N-S trending anticlinal structures. In this type of oil field, the reservoirs are mainly Cretaceous Formations, i.e., Ilam, Sarvak in Iran and Mishrif, Rumalia in Iraq (Alsharhan 1995; Alsharhan and Nairn 1997; Beydoun 1991; Sadooni and Aqrawi 2000; Sadooni 2005; Bordenave and Hegre 2005). As shown in Fig. 1, Majnoon, NahrUmr, West Qurna, Rumaila, and Zubair oil fields in Iraq are classified as the Arabian Trend, whereas the oil fields of the Ahwaz area in the east of the Azadegan are identified as the Zagros Trend.

It is understood that the Azadegan field is a huge long-axis N-S trend anticline of the Arabian Trend. The formation is continuous without any huge strike fault (Fig. 4a). Two domes are situated in the north and south, respectively, and are connected by the middle saddle area. Related to the dome elevation, the uplift degree of the southern dome is higher than that of the northern one (Fig. 4b). The northern structure is an N-S trend small-size anticline with steep limbs. The southern structure is in the N-S trend with larger oil field area and it extends southwestward into Iraq.

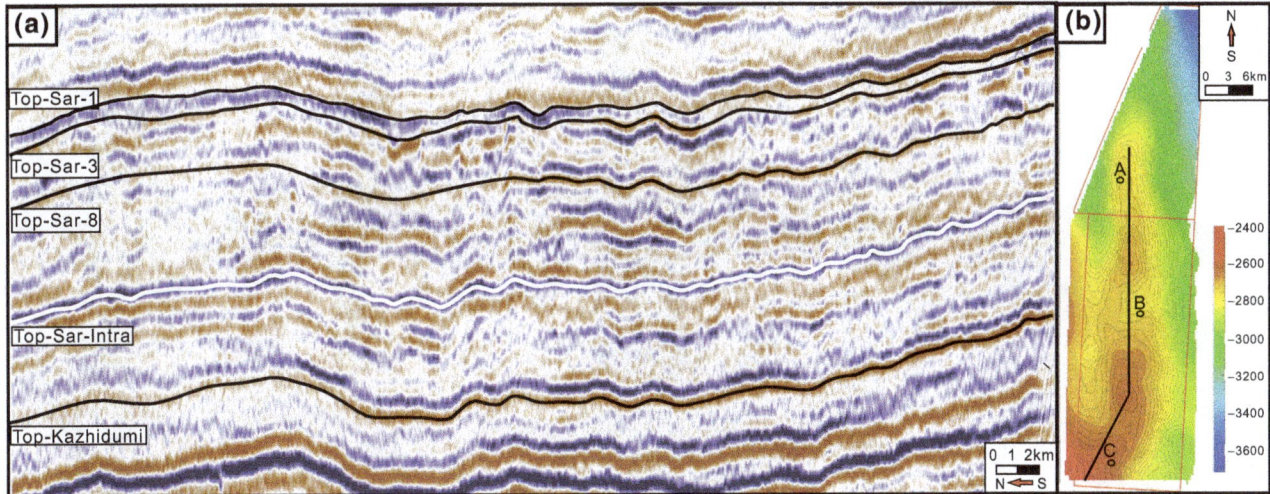

Fig. 4 **a** The seismic cross section of Sarvak along the structural crest (*black line* in (**b**)), **b** the structural map of the Sarvak top surface

2.2 Trap evolution

The Azadegan oil field is located in a part of the Arabian platform. During most of its geological history, it was in a stable subsidence process except for a short period of traceable regional instability in the upper Cretaceous Turonian Stage (Murris 1980; Berberian and King 1981; Koop and Kholief 1982; Alsharhan and Nairn 1997; Alavi 2004). Consequently, in this study, the seismic flattening technique is used to reconstruct the structural trap evolution of Azadegan. According to the previous findings of regional tectonic activities and deposition time of the flattened formations, the genesis and stages of the trap evolution are analyzed and determined.

The available data show that the Arabian Platform was a part of the Gondwana super-continent during the Precambrian and early Paleozoic (Stöcklin 1968; Berberian and King 1981). By the end of the Precambrian, the Arabian Plate was characterized by the extension of a discontinuous subsiding basin with very thick evaporite deposition. The Hercynian Orogeny in the late Paleozoic had caused the formation of the N-S trending horsts and grabens (Stöcklin 1968; Murris 1980). From the late Carboniferous to the early Permian, the northeastern margin of intra-continental rifting and seafloor spreading along the Zagros belt formed the new Tethys Ocean and led to the separation of the Iranian and Arabian plates (Alavi 1994; Glennie 2000; Sherkati and Letouzeh 2004; Sepehr and Cosgrove 2004). After entering the middle of the Cretaceous, with the Arabian plate drifting to the north, the plate moved from the passive continental margin into the active continental margin development stage (Alavi 2004) and experienced two periods of tectonic activity.

The first period is the Alpine tectonic activity of the upper Cretaceous. From early to middle Cretaceous, the Arabian plate subduction caused tectonic movement, resulting in the regional sedimentary evolution with active tectonic extrusion and Neo-Tethys contraction. This tectonic change turned the area into an active and compressional tectonic setting from the Cenomanian of the Cretaceous, which caused a readjustment of the fault system to the N-S strike during the early Hercynian and brought an increment in salt movement, such as salt plugs, swells, ridges, and relevant domes of the N-S Arabian trend. Regionally, these structures are abundant in the Persian Gulf, Saudi Arabia, Kuwait, southeast Iraq, and southwest Iran (Murris 1980; Koop and Kholief 1982; Glennie 2000; Sherkati and Letouzeh 2004). In the Azadegan field, the Sarvak occurs as a result of carbonate shelf deposition in the mid-Cretaceous (Murris 1980; Alsharhan and Nairn 1997; Bordenave and Hegre 2005). In the early Paleocene (65 Ma, Fig. 5a), a wide and gentle paleo-fold has emerged in the current northern field area. Towards the early Miocene (20 Ma, Fig. 5b), it had evolved into the N-S strike anticline by the impact of plate extrusion. The paleo-anticline was higher in the north and lower in the south, which is in contrary to the current structural feature of south higher than north.

The second period is the Zagros orogeny of the early Miocene. Due to the second collision of the Arabian Plate and central Iranian Plate in the Miocene, the Neo-Tethys Ocean closed and the Zagros Fold Belt formed. This tectonic activity began approximately 20–16 Ma ago, from the Zagros Mountains piedmont region to the Dezful Depression where the oil field is located, which contributed to the typical Zagros anticlines manifested in the form of large-amplitude asymmetric whaleback-shaped mountains (Colman 1978; Berberian and King 1981; Sherkati and Letouzeh 2004; Alavi 1982, 1994, 2004). The extension of this NW–SE strike fold, named the Zagros Trend, has been

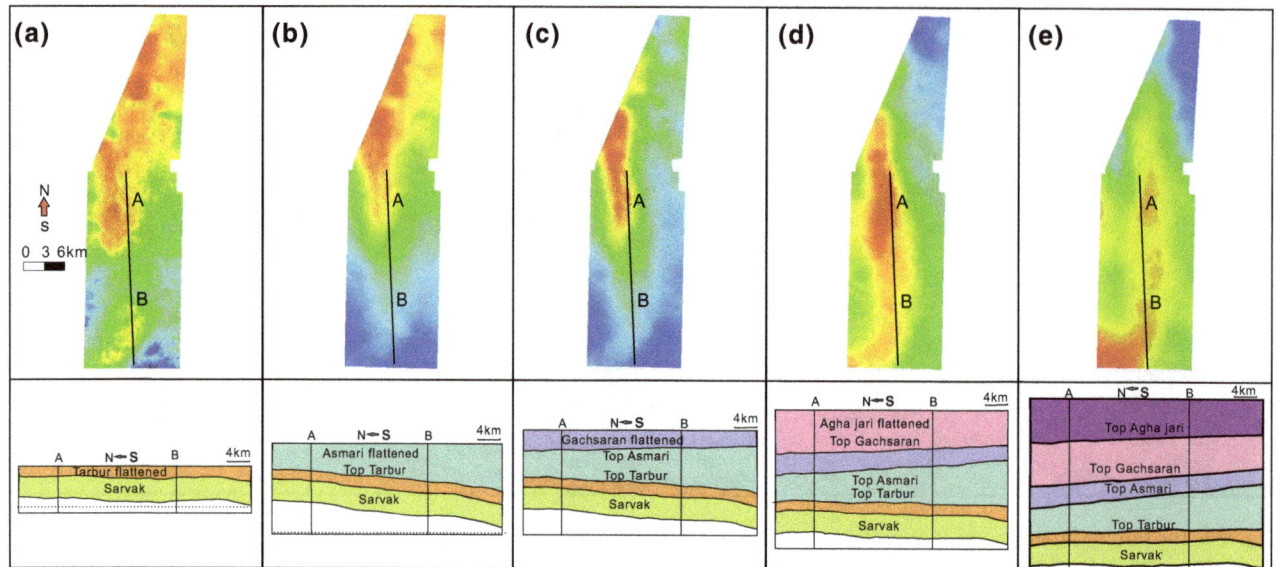

Fig. 5 Tectonic evolution section of Sarvak Formation (*light color* represents the high structure, and the *dark color* represents the low structure) and the depositional formations in different geological stages which are depicted by the flattened seismic balanced section crossing the line of the A and B wells. **a** Paleo-structure of the Sarvak in the Paleocene (flattened by the top of the Tarbur Formation in the Paleocene, 65 Ma); **b** Paleo-structure of the Sarvak in the Early Miocene (flattened by the top of the Asmari Formation in the middle Miocene, 20 Ma); **c** Paleo-structure of the Sarvak in the Late Miocene (flattened by the top of the Gachsaran Formation in the Late Miocene, 6 Ma); **d** Paleo-structure of the Sarvak in the Pliocene (flattened by the top of the Aghajari Formation in the Pliocene, 2 Ma); **e** Current structure of the top of the Sarvak Formation

formed continuously throughout time from NE to SW and is limited by the Iranian shoreline and Iran-Iraq border in the northwest. Besides this, due to the extrusion stress decreasing, the folding amplitude as well as the deformation intensity decreased progressively following the same trend (Hessami et al. 2001; Alavi 2004). The borderline along the low-amplitude Zagros trend traps, such as Ab-e Teymur and Mansuri (formed during 5–4 Ma) in the east of the Azadegan oil field, is deemed as the boundary of the surface Zagros folded zone (Hessami et al. 2001; Bordenave and Hegre 2005). This means that the effect of the Zagros Trend is hardly observed on the surface outside the line, where the buried structure zone exists (Fig. 1). In the buried zone, it is hard to observe the fold on the surface, but the subsurface formation is still tectonically active, and the structural activity is trending towards the southwest continuously. Under the control of early basement faults like the Najad fault system, a new subsurface Zagros fold formed and some early folds like the Arabian fold were subdued to secondary deformation (Sepehr and Cosgrove 2004). Up to now, it is an ongoing tectonic evolution trending to the southwest.

For the Azadegan field, since the Zagros orogeny took place, the trap deformation accelerated significantly shown as the trap scale shrinking dramatically and the limbs becoming steeper. Until the late Miocene (6 Ma, Fig. 5c), the fold have evolved from a wide and gentle anticline into a long-narrow one. Before about 3 Ma (the latest surface Zagros fold formed at nearly 4 Ma), the plate nappe stress reached the Azadegan oil field and led to the deep basement fault being reactivated (Sepehr and Cosgrove 2004), causing the trap to experience strong secondary deformation. The northern paleo-trap was squeezed continuously and evolved into the northern high of the field nowadays. The southern paleo-low part behaves as a "teeterboard," sharply uplifted and formed a new secondary trap. The top surface of the Pliocene Aghajari Formation which was deposited before 2 Ma is flattened (Fig. 5d), and it can be seen that the northern paleo-trap narrowed sharply, and formation uplifting occurred in the south, but the elevation was still lower than north. Now the structural amplitude in the south has surpassed that in the north, and two structural highs were formed where the south is higher than the north (Fig. 5e).

3 Accumulation factors

3.1 Reservoir

3.1.1 Sequence stratigraphy and reservoir characteristics

The Sarvak is a carbonate formation which deposited in a gently sloping shallow marine environment mostly during the Cenomanian to early Turonian (96-92) Ma (Murris 1980; Alsharhan and Nairn 1997). According to the high-resolution sequence stratigraphy (Xu et al. 2007), six significant sedimentary cycles were distinguished. Moreover, 12 subzones were further distinguished based on well logging and paleontology. The oil intervals primarily include the Sar-8 in SEQ-4 and the Sar-3, 4, 5, 6 in SEQ-5. Sar-1 and Sar-2 zones are the regional dense interlayers of marl, mudstone, and shale, and the Sar-7 is mud/wackestone containing poor oil (Fig. 6).

The Sarvak is mainly a pore-type reservoir without large-scale high-angle tectonic fractures. Diagenetic micro-fissures and stylolites can be found in the cores but only densely distributed in a few parts. Vertically, the lithology and physical properties of various subzones are different and have strong heterogeneity. The lithology of the Sar-3 and Sar-8 zones is mainly characterized by rudist grain/packstones which were deposited in high-energy sites (Fig. 7a). Karstification and in situ solution brecciation developed extensively during exposure of the upper sequences (Fig. 7g), while the pore types are mainly inter-granular dissolved pores, moldic pores, and vugs (Fig. 7b, c). The porosity and permeability are, respectively, 15 %–35 % and 10–$150 \times 10^{-3} \mu m^2$. The Sar-3 subzone has an average porosity of 17.7 % and permeability of $45.5 \times 10^{-3} \mu m^2$, and the average porosity and permeability of the Sar-8 are 17.1 % and $11.7 \times 10^{-3} \mu m^2$, respectively. The Sar-4, 5, 6 subzones present a similar lithology of mainly two types: the foraminifera and mollusk packstone/wackestone of moderate-energy shallow-water deposition (Fig. 7d) with porosity and permeability of 10 %–20 % and 1 to $10 \times 10^{-3} \mu m^2$, respectively (Fig. 8), as well as the planktonic foraminifera, echinoderm, and algae wackestone of low-energy deep-water deposition (Fig. 7f), and the porosity and permeability are 5 %–15 % and 0.1–$5 \times 10^{-3} \mu m^2$, respectively (Fig. 8). Pore types are mainly isolated intra-particle, residual inter-granular pores, lime mud matrix, and moldic pores (Fig. 7e). The reservoir qualities of the Sar-4, 5, 6 are controlled by the depositional environment and are poorer than those of the Sar-3, 8, and the strong reservoir heterogeneity also causes relatively poor oiliness (Fig. 7h). The average porosity is 10.8 %, and the permeability is $3.8 \times 10^{-3} \mu m^2$.

3.1.2 Reservoir distribution

The distribution of the upper Cretaceous rudist buildups in the Iraq-Iran border area is mainly controlled by the paleo-highs which were generated by the Alpine tectonic activity (Alsharhan 1995; Glennie 2000; Sadooni and Aqrawi 2000; Sadooni 2005; Du et al. 2015b). During the growth of the rudist reef, when it reached the wave base when sea level fell, the rudist sand was formed from the reef by strong current erosion. After being transported and re-deposited, a gentle hilly, paleo-high centered and continuous rudist bio-stratum composed of rudist clastic grain/pack/wackestone was formed (Aqrawi et al. 1998; Sadooni 2005; Du et al. 2015b). So the Sar-3 and Sar-8 rudist-bearing layers are continuous with weak heterogeneity and thinning from north to south controlled by the paleo-high (Fig. 9a). The Sar-4, 5, 6 were mainly deposited in an open platform and lagoon environment where the sediments are predominantly from in situ deposition (Ghabeishavi et al. 2010), while the reservoir quality is mainly controlled by the depositional site. The shallower the water environment is, the better the reservoir quality is. According to the reservoir correlation, the Sar-4, 5, 6 are relatively better in the northern paleo-high; conversely, with the deposition setting turning deeper, in the southern paleo-low, the shale content is increasing, and the dense limestone is thickening, and the reservoir quality is showing a degradation trend from north to south. In general, the Sarvak reservoir is a massive and interconnected heterogenetic reservoir. Vertically, the subzones with diverse lithology and properties are connected to each other without clear boundaries and are interbedded. Laterally, the reservoir quality is mainly controlled by the paleo-geomorphology and shows the degrading trend from north to south, and this is also shown by the new 3D seismic interpretation (Fig. 9b).

3.2 Source

The source bed of the Sarvak reservoir in the Azadegan oil field has not been confirmed due to the lack of regional exploration. The regional resource studies mainly focus on the late Jurassic and early Cretaceous beds (Bordenave and Burwood 1990, 1995; Bordenave and Huc 1995; Bordenave and Hegre 2005). As for Sarvak, the potential sources are the Cretaceous Garau and Kazhdumi Formations (Fig. 2).

The Garau is composed of a series of thick carbonate muds, argillaceous limestones, and organic matter formed during the early Cretaceous Valanginian stage. It is believed that the Garau reached the oil window at the end of the Paleocene to early Miocene (Bordenave and Burwood 1990; Bordenave and Hegre 2005) and covered the whole Dezful Embayment in Iran. It is assumed, however, without faults or large-scale vertical fractures, the hydro-carbon from the Garau cannot cross the Kazhdumi dense bed with high pore pressure, and hence, it cannot migrate to the Sarvak (Alavi 1982; Bordenave and Hegre 2005). The large difference of the gravity of the crude oil between the Sarvak and lower reservoirs in Azadegan also indicates they have a different source (Fig. 10). The Kazhdumi Formation, deposited at the Cretaceous Albian stage,

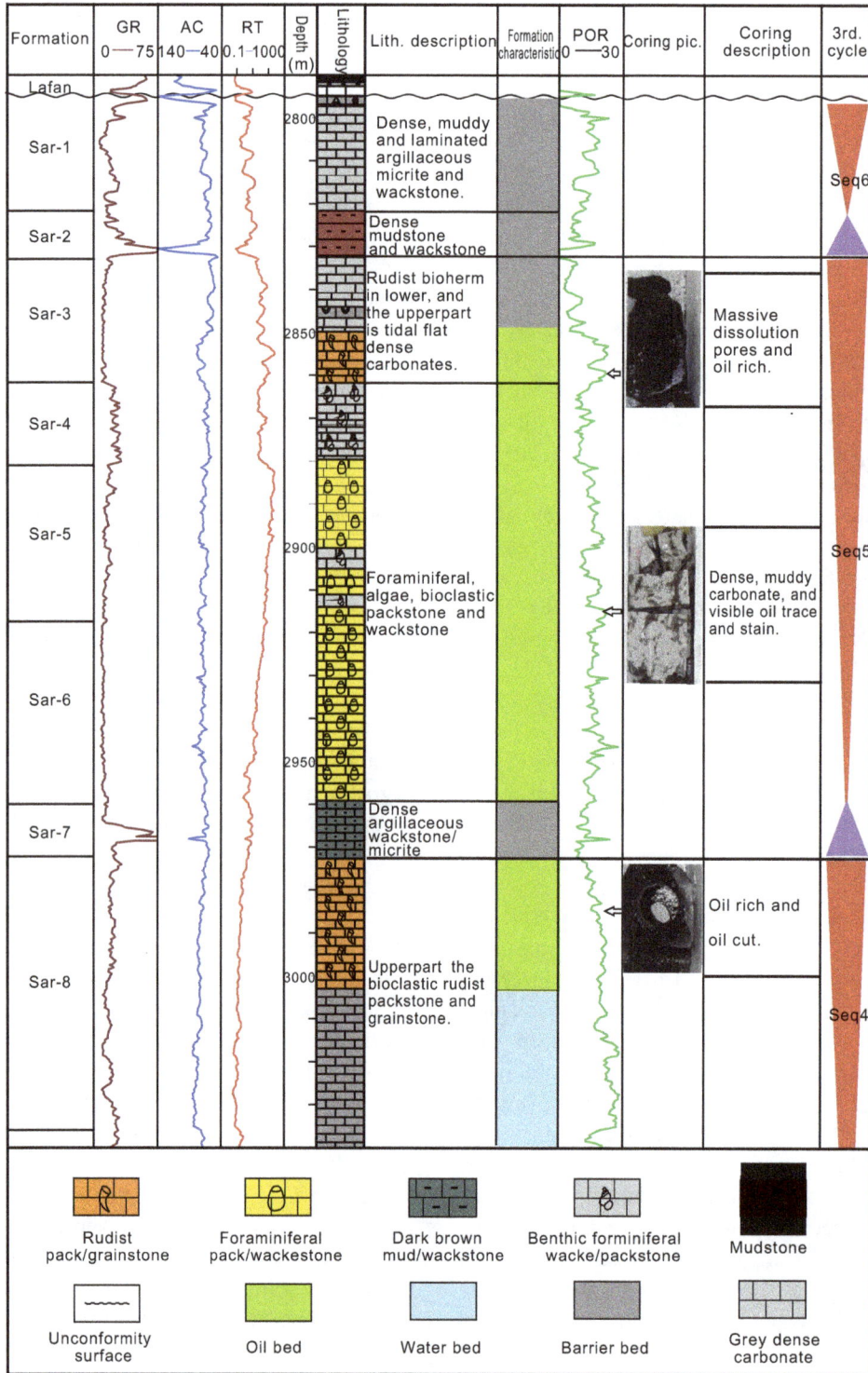

Fig. 6 Synthesized stratigraphic sequence histogram of the upper Sarvak in the Azadegan oil field

conformably contacts with the Sarvak (Murris 1980; Alsharhan and Nairn 1997; Bordenave and Burwood 1995; Bordenave and Huc 1995; Bordenave and Hegre 2005). Deep marine still water mudstone deposited in the central

part of Dezful Embayment has an excellent hydrocarbon generation potential and is the main source rock for the traps in the Zagros foreland basin (Murris 1980; Bordenave and Burwood 1990, 1995; Bordenave and Huc 1995;

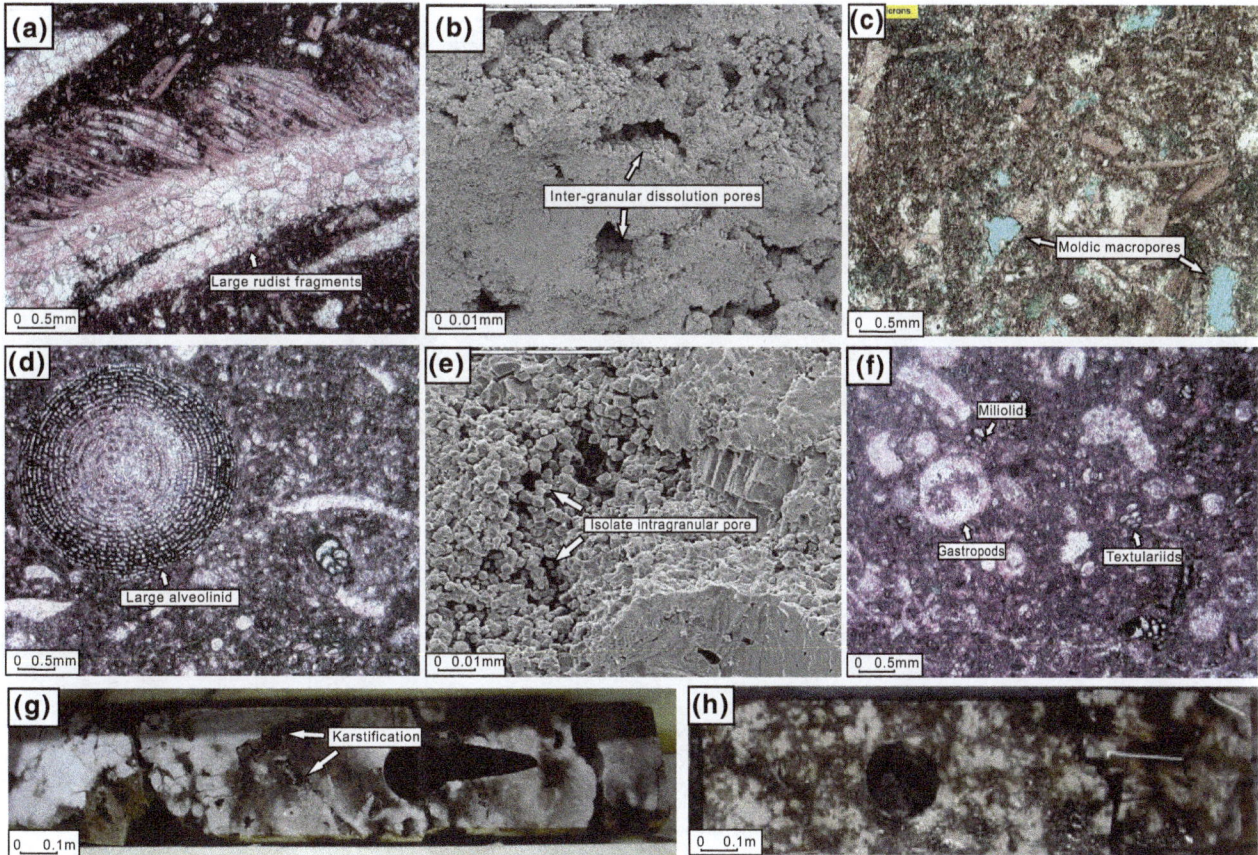

Fig. 7 The thin sections and SEM pictures of different subzones of Sarvak in the Azadegan oil field. **a** Sar-3, bioclastic packstone/wackestone containing rudist fragments. The large rudist fragment in the picture was originally bimineralic, consisting of calcite (*upper part*, microstructure well preserved) and aragonite (*lower part*, now blocky calcite spar); **b** Sar-3, SEM picture, macro-porosity comprises both primary inter- and intra-granular pores and secondary grain dissolution pores; **c** Sar-3, moldic macro-pores after leaching and dissolution of skeletal aragonite; **d** Sar-4, bioclastic/peloidal packstone containing large alveolinid foraminifera; **e** Sar-4, SEM picture, detailed view showing coarse blocky calcite cementing pores, and isolated micro-pores within the lime mud matrix; **f** Sar-5, bioclastic packstone containing gastropods, benthonic foraminifera (including textulariids and miliolids), and a variety of other skeletal grains; **g** Sar-3, core picture, the development of karstification, and in situ solution indicate the subaerial exposure of top of the Sar-3 subzone; **h** Sar-6, core, oil stain, and patch; the *shallow color* indicates the dense limestone and poor oiliness, and the *dark color* shows the zone with relatively good properties and oiliness

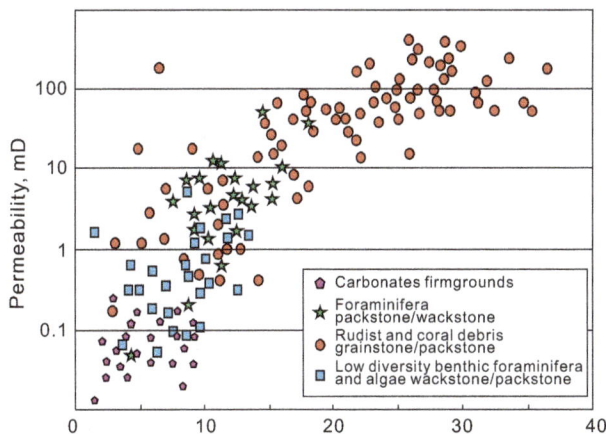

Fig. 8 The physical property cross-plot of coring in Sarvak with different lithologies

Bordenave and Hegre 2005; Zhang et al. 2012; Soleimani 2013). In the Azadegan oil field, dark brown and gray Kazhdumi marls were deposited without hydrocarbon potential. Regionally, the field is located in the western Dezful Embayment and far from the hydrocarbon generation center (Fig. 11).

The Japanese company Inpex collected crude oil API° data from fields in the Iranian Dezful Embayment and Iraqi Mesopotamian Basin (Fig. 10, and the fields' locations are shown in Figs. 1 and 11). After comparing and analyzing, it is considered that the Sarvak in Azadegan has the same source as the eastern Ab-e Teymur oil field due to their similar oil properties (Fig. 10). The accumulation process is interpreted as follows: when the Kazhdumi of Ab-e Teymur became mature at 5-1 Ma (Bordenave and Hegre 2005), the oil charged the Zagros trend traps near the

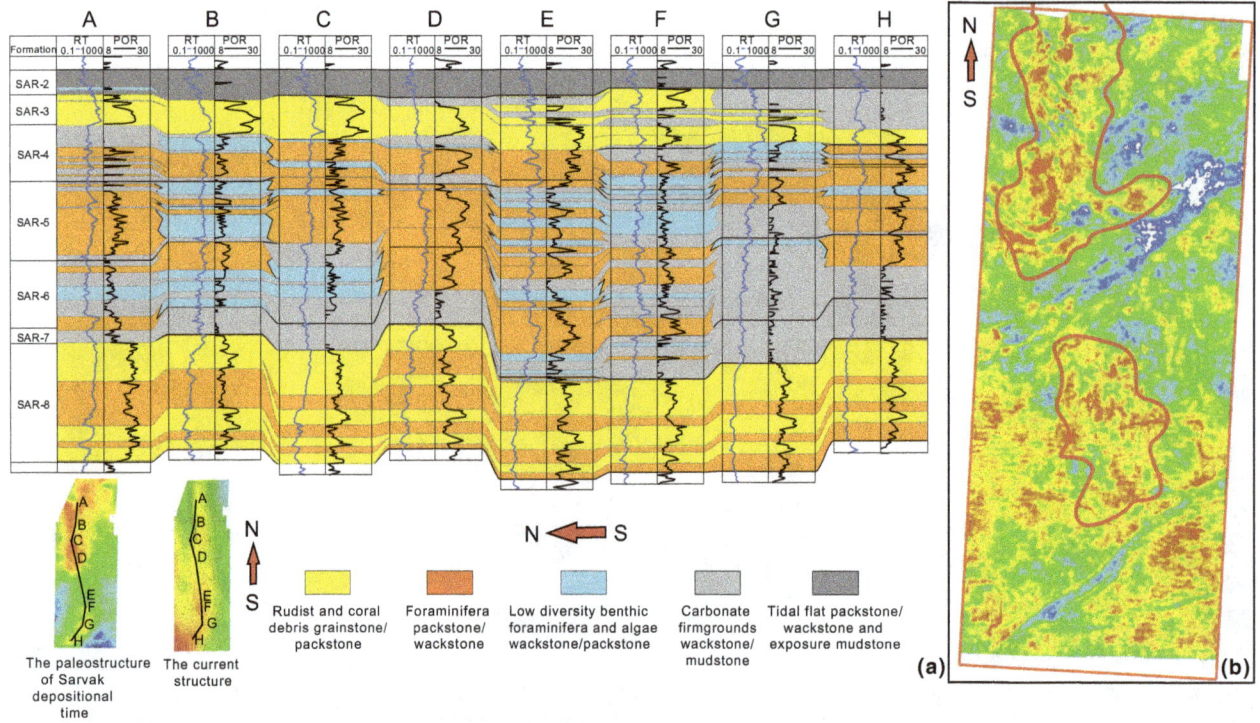

Fig. 9 The reservoir correlation figure of Sarvak in Azadegan field, Iran (**a**); the reservoir distribution prediction figure of South Azadegan oil field by newly 3-D seismic data; the *red circle* indicates the favorable reservoir zone (**b**)

		IRAN South Dezful Embayment						IRAQ-Mesopotamian Basin					
		Zargros Trend						Arabian Trend					
		PAYDER	AHWAZ	SUSANGERD	JUFEYR	AB-E TEYMUR	MANSUR	DARQUAIN	AZADEGAN	NAHRUM	MAJNOON	WEST QURNA	ZUBAIR
Tertiary	Miocene	Asmari 11-13	Asmari 32.5	Asmari 15-19			Asmari 28-30			Ghar small	Ghar 23-32		Ghar 10-20
	Oligocene												
	Eocene												
	Paleocene												
Cretaceous	Upper		Ilam 26		Ilam 22-23	Ilam 24	Ilam 24	Ilam 21			Khasib 27	Sadi 29-32	
	Middle	Sarvak 23	Sarvak 25		Sarvak 22-23	Sarvak 24	Sarvak 25-29		Sarvak 18-20	Mishrif 20-22	Mishrif 17-21	Mishrif 19-22	Mishrif 27-28
									Kazhdumi 29-30	Nahr Umr 41	Nahr Umr 30-32		
	Lower			Gadvan 25-39					Gadvan 34-38	Zubair 25-30	Zubair 31	Zubair 23-34	Zubair 36-42
												Ratawi 31-34	Ratawi 38-41
				Fahliyan show				Fahliyan 39	Fahliyan 33-38	Yamama 39-43	Yamama 35-38	Yamama 33-41	Yamama 40-43

Fig. 10 Oil gravity of the oil fields from the Cretaceous to Miocene in the Dezful Embayment (unit: API°)

generation center such as the Ab-e Teymur and Mansuri fields. After that, the oil charged the Arabian trend traps in the westwards basin margin like Azadegan by lateral migration. However, another two doubts need to be considered. Firstly, according to the findings of Bordenave and Burwood (1990, 1995) and Bordenave and Hegre (2005),

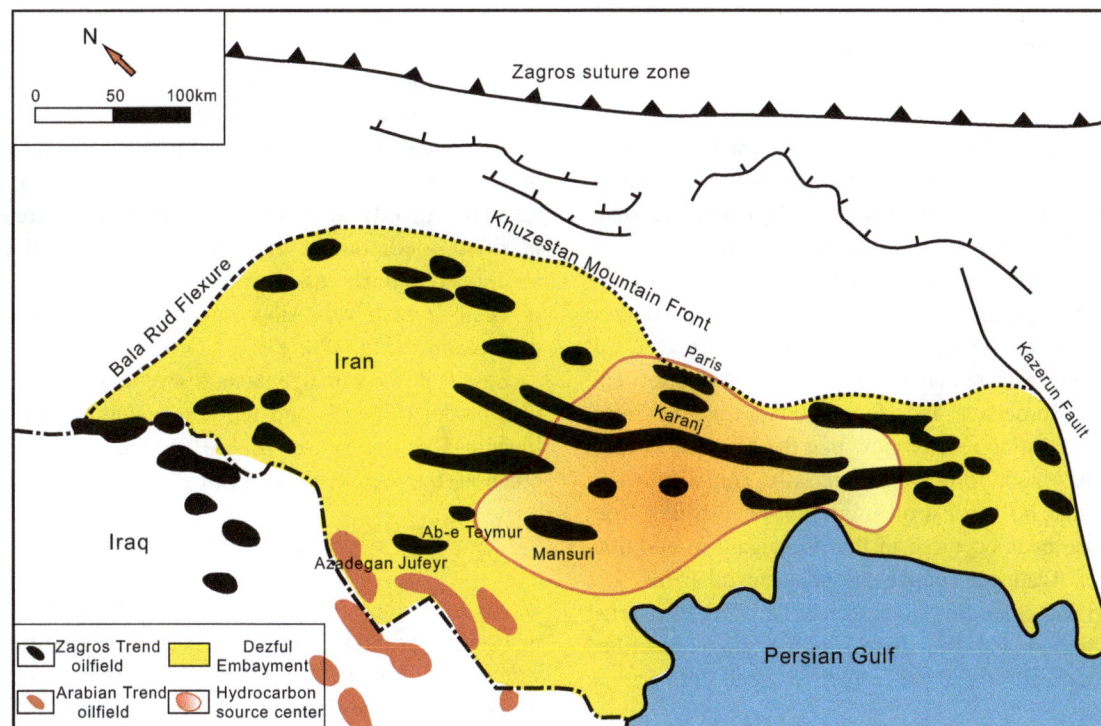

Fig. 11 Isopach map of source rock maturation in the Kazhdumi Formation (Bordenave and Hegre 2005)

most oil from Kazhdumi will first charge the nearby Zagros trend traps in Dezful Embayment. Assuming that the long-range lateral migration charging the Arabian trend traps will occur later, and according to the principal that the higher the source rock maturity is, the lower the crude oil gravity is, so the higher API° of crude oil should be present in Arabian trend oil fields. Nevertheless, the data show that the crude oil gravity in both the Azadegan and the other Arabian trend traps are lower than that in the Zagros trend traps (Fig. 10), contrary to that assumption. Moreover, the filling of the Sarvak reservoir in the Jufeyr field of the eastern of Azadegan is only partial. Thus, it is hard to explain how the oil crosses the Jufeyr during the migration and directly infills the Azadegan (Fig. 11).

Bordenave and Hegre (2005) has discussed the relative timing and chronology of oil expulsion from the Kazhdumi and the formation of the Zagros trap in the Dezful Embayment, and concluded that both hydrocarbon generation peak periods of Kazhdumi and Zagros trap formation stage happened during the period 3–8 Ma, showing a compatible matching relation. Therefore, close range migration is dominant in the Dezful Embayment. However, some studies (Bordenave and Hegre 2005) also proved that parts of the Kazhdumi in the Dezful Embayment also reached the oil window and the expulsion stage commenced before 10 Ma due to the rapid subsidence caused by the early folding, like the Karanj and Paris oil field areas (Fig. 11), which is earlier than the formation time of the

Zagros trend trap. Consequently, during that period, there were only Arabian trend traps in the west of Dezful Embayment, and the Zagros trend trap was not formed; nevertheless, parts of Kazhdumi source bed reached the oil window, and the low-maturity heavy crude oil was expelled and migrated long distances along the gently dipping ramps or the unconformity surface towards the western Arabian platform (Zagros Basin margin), charging the Arabian trend traps first. Afterwards, the source bed reached its peak time of hydrocarbon generation, while the Zagros traps were formed simultaneously. Thereafter, a short-range migration to the nearby Zagros trend traps took place. That is a valid explanation why the oil properties of Sarvak reservoir of Arabian trend traps show low API°, while Zagros trend traps have relatively low oil gravity with high source bed maturity.

3.3 Seal

In the Dezful Embayment of southwestern Iran, there are two favorable seal beds in the Cretaceous to Tertiary oil system, the Miocene Gachsaran which consists of salt and anhydrite rock as well as the Cenozoic Gurpi and Pebdeh consisting of marl, shale, and marly limestone (Fig. 2). Due to the fracturing caused by the strong tectonic compression, the seal efficiency of Gurpi and Pebdeh Formations in Zagros trend trap is limited and lowered, while the Cretaceous and Miocene reservoirs are connected by high-angle fractures and transformed into an unified reservoir whose main cap formation is the Gachsaran (McQuillan 1973, 1974; Alavi 1982; Wang et al. 2011; Zhang et al. 2012; Bordenave and Hegre 2005).

The Azadegan oil field is located in the transitional zone between the Arabian platform and Zagros foreland basin, which is relatively far away from the Zagros suture zone. Although the Zagros orogeny caused deformation of the subsurface layers, the shale and marl seals of Cenozoic formation (Gurpi and Pebdeh) did not develop intense fracturing due to the tectonic stress decreasing, yet still play a sealing role restraining vertical oil migration and dissipation (Bordenave and Hegre 2005). Logging interpretation and testing results show that the Miocene Asmari reservoir has good properties but only contains water, indicating that no vertical oil migration or charging from lower source beds occurred in the Azadegan. In addition, the Ilam of Azadegan, unlike the adjacent oil fields, is a dense marl and chalk limestone formation. With the Laffan shale formation and the Cenozoic formation described before, they all form an effective seal for Sarvak accumulation.

4 Genesis of Sarvak unsteady reservoir

Through combining the characteristics of the reservoir, source, seal, and the trap evolution, the Sarvak accumulation genesis and evolution are discussed.

The Sarvak Formation was deposited in the middle of the Cretaceous and evolved into the N-S Arabian trend anticline trap under the impact of Alpine tectonic activity and remained in a relatively stable state in the geological history from 65 to 20 Ma (Fig. 12a, b). The Zagros orogeny commenced about 20 Ma, and when the Zagros trend fold had not been completely formed (before about 20-8 Ma), parts of the Kazhdumi source bed of the Dezful Embayment reached the oil window and expelled low-maturity heavy crude oil. This charged the Arabian trend trap in the western Dezful Embayment by long-range migration and turned the Azadegan into a paleo-anticline accumulation (Fig. 12c). At approximate 3–4 Ma, the Zagros orogeny began impacting the Azadegan zone (margin area of the Zagros foreland basin) and led to the drastic secondary deformation of the paleo-trap. The paleo-trap shrank dramatically, while the previously low southern formation was uplifted and formed a new secondary anticline trap. This evolutionary trend lasted till the present and is now evolving into the current structures of two domes with the south higher than the north.

Tectonic activity altered the paleo-reservoir trap shape and broke the reservoir kinetic equilibrium. The Gurpi and Pebdeh Formation seal beds prevented vertical dissipation, causing intra-formational secondary re-migration and adjustment in the Sarvak. The impact factors are as follows: (1) trap deformation and secondary migration proceeding simultaneously; (2) a massive reservoir with strong vertical and lateral heterogeneity; (3) heavy crude oil of high viscosity (18–20 API°); and (4) reservoir quality that shows a degradation trend from north to south (along the migration path). All of these mentioned above make the accumulation adjustment proceeding at a very slow rate and lag behind the trap deformation (Fig. 12d, e).

Based on the homogeneous reservoir secondary migration model (Fig. 13) and the corresponding oil/gas equilibrium formula below (Li 2010), the time needed for reaching the secondary reservoir equilibrium is calculated:

$$t = \frac{\phi \mu_0 \Delta L}{2K \Delta \rho_{wo} g \sin \alpha} \ln \frac{\tan \theta_0}{\tan \theta} \approx \frac{\phi \mu_0 \Delta L}{2K \Delta \rho_{wo} g \sin \alpha} \ln \frac{\theta_0}{\theta},$$

where μ_0 is the crude oil viscosity, mPa s; K is the formation permeability, mD; $\Delta \rho_{wo}$ is the density difference of water and oil, g/cm^3; θ_0 is the initial dip angle of the oil–water contact; and θ is the equilibrium dip angle of the oil–water contact.

The computed result is 1.78 Ma for the Sar-3 and 10.86 Ma for the Sar 4–6. It should be noted that this is a rough static time length estimate based on the current structure condition and the reservoir properties are the average values from the core data. The calculation does not consider impacting factors such as the geologic synchronicity of the oil migration and trap deformation, and the strong reservoir heterogeneity. Otherwise, the time needed would be longer. The secondary trap deformation occurred after the formation of the surface Zagros trend fold (approx. 4 Ma), and the greater part of the re-migration occurred at an ultra-late stage. This is because the uplift of the southern secondary trap surpassed the northern paleo-trap at least after 2 Ma, and the higher elevation of the secondary trap would improve the upward buoyancy of the oil migration. The Sarvak reservoir could only experience a very short readjustment time span (perhaps only 1–2 Ma) less than the timeframe needed for the new reservoir equilibrium.

From the oil migration theory, during the migration process, the light component of crude oil will migrate first, and the heavy content will remain in the reservoir (Li 2004). The PVT analysis also proved that the fluid properties are diverse in different zones in the field (Liu et al. 2013b). From north to south (paleo-trap to secondary trap), the density is lighter and the viscosity becomes lower, which means the fluid mobility is getting better from north to south, and it also proves the reservoir is still in the re-migration process from the reservoir engineering standpoint (Fig. 14).

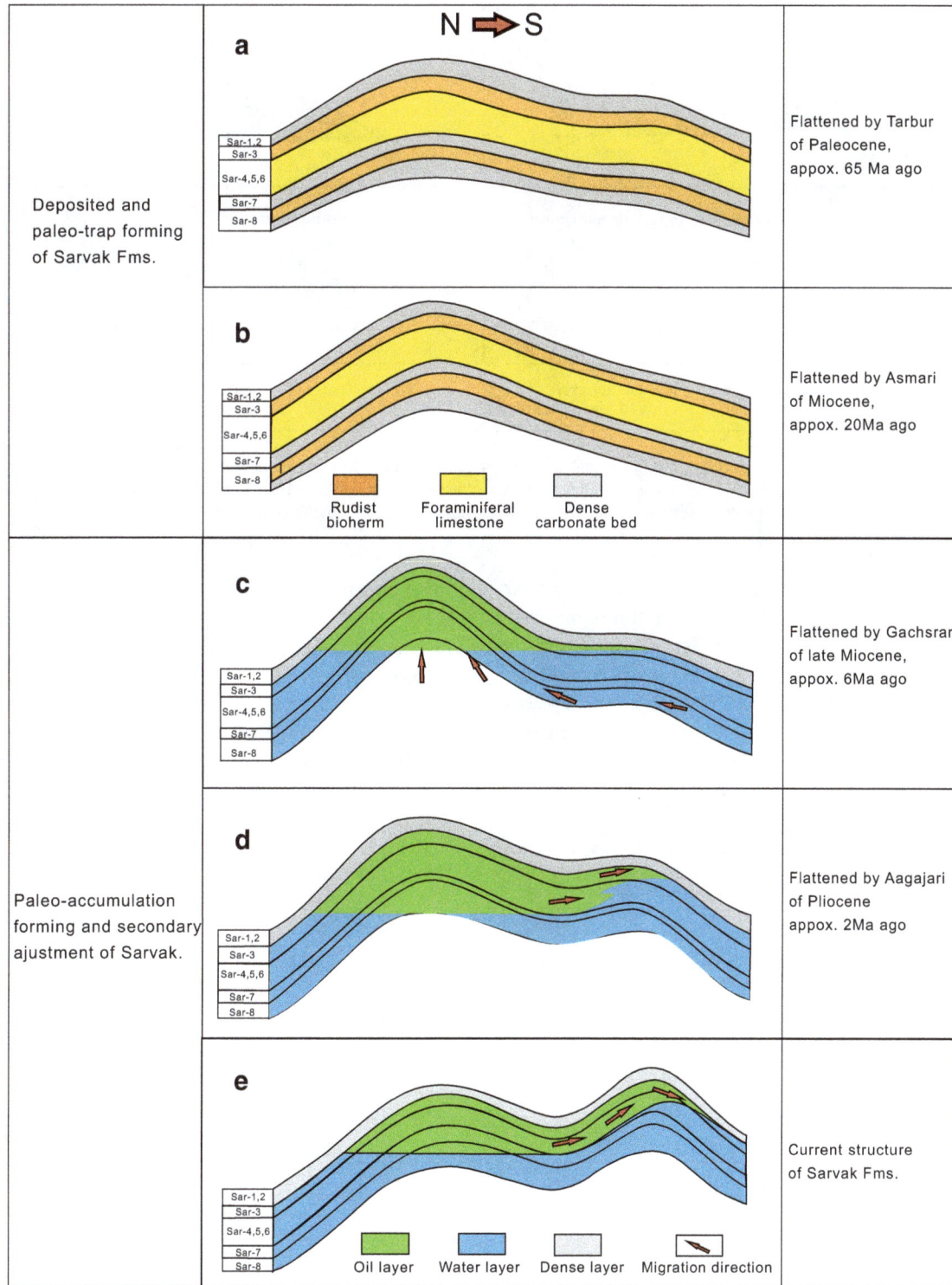

Fig. 12 Sketch map of Sarvak reservoir evolution of the Azadegan oil field in different geological stages (the accumulation evolution corresponds with the trap evolution of Fig. 5)

To sum up, the strong reservoir heterogeneity makes the secondary adjustment occur at a very low rate, while the very late trap deformation does not provide the time required, so these two main factors together caused the Sarvak reservoir to form an unsteady reservoir which has not reached a new equilibrium and is currently in an unstable process of re-migration, accumulation, and adjustment. The nearly stable horizontal OWC of the

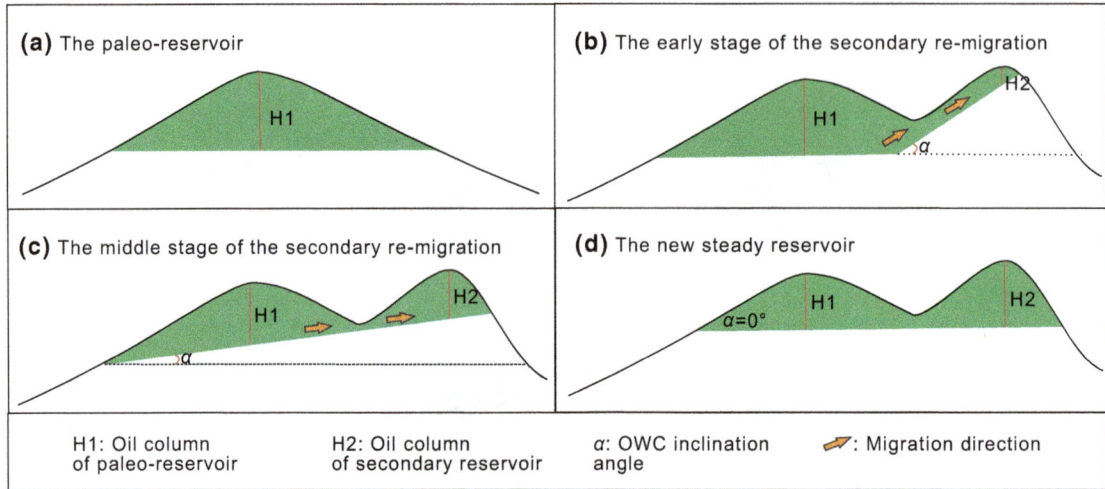

Fig. 13 A schematic figure of the secondary migration and adjustment process in the homogeneous reservoir model. This model ignores the formation time interval of the secondary trap. Generally, the reservoir adjustment can be described as the oil column decreasing in the paleo-reservoir and increasing in the secondary reservoir, along with a decreasing of the OWC inclination angle

northern crest wells (these wells are in the paleo-reservoir area) indicates that the current reservoir is still in the early stage of the secondary re-migration and has no impact on the whole paleo-reservoir (Fig. 13b). This caused the primitive and stable horizontal OWC in the northern paleo-reservoir and the highly tilted OWC in the southern secondary reservoir.

5 Development suggestion

The "unsteady reservoir" is a new type of reservoir that was proposed by Chinese scholars based on the exploration of an unconventional oil field in the Tarim Basin in the west of China and defined as a dynamically balancing oil entrapment which is still in the process of charging or

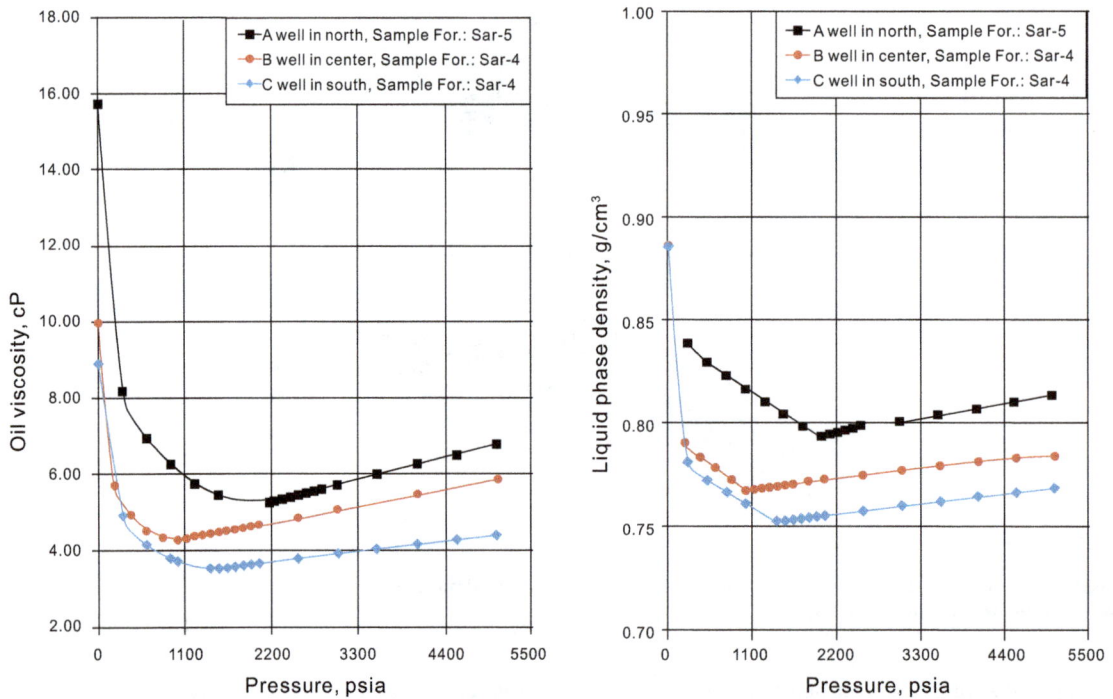

Fig. 14 Sketch map of PVT analysis in Sarvak (the well locations of A, B, C are shown in Fig. 4b)

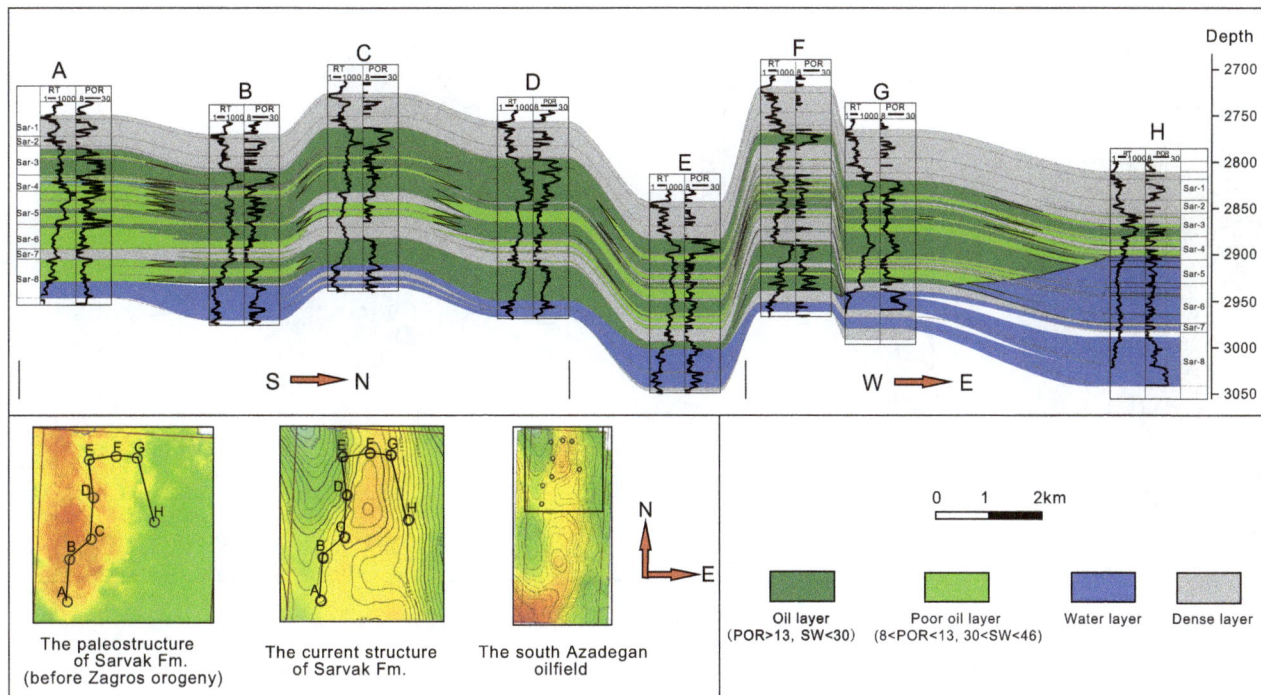

Fig. 15 The reservoir section of the northern high in the South Azadegan oil field

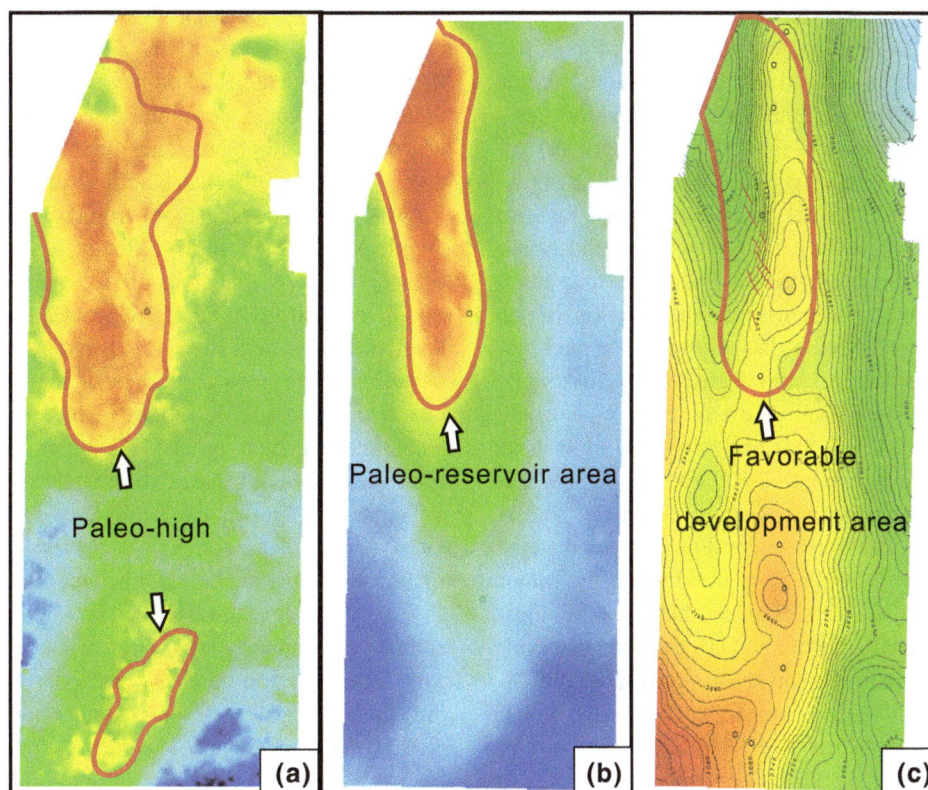

Fig. 16 Sketch map of the development suggestion for the Azadegan oil field. **a** The paleo-geomorphology of the depositional stage of the Sarvak; **b** The paleo-reservoir plane distribution of the Sarvak; **c** the favorable development area in the current structure

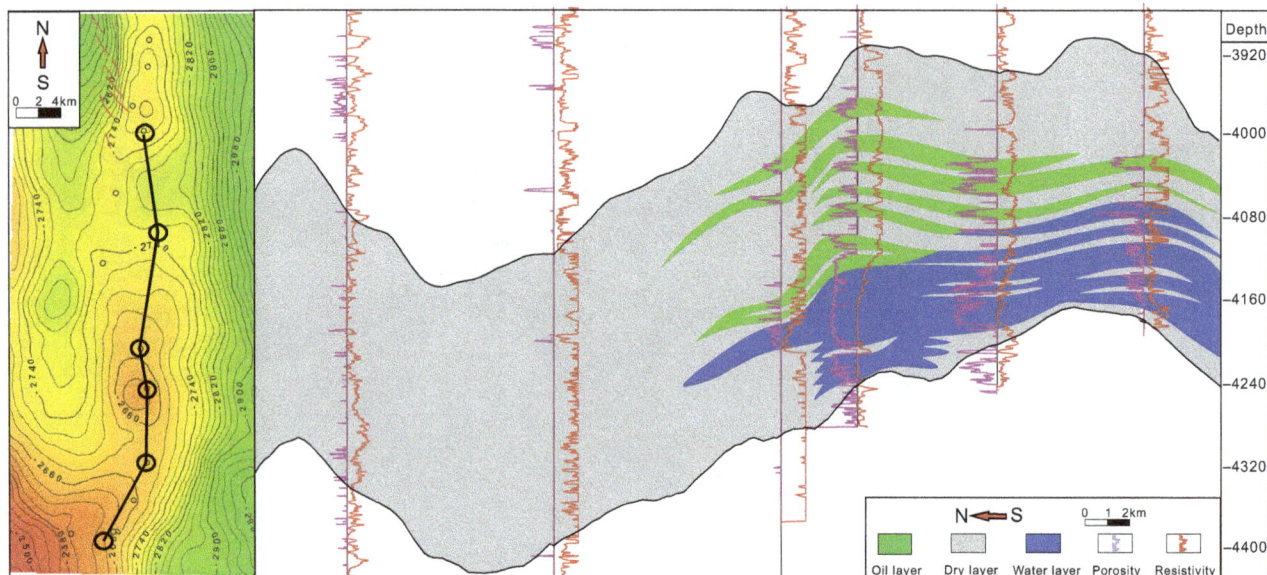

Fig. 17 Reservoir section of the lower Cretaceous Fahliyan Formation in the Azadegan oil field (N-S), Iran

adjusting by hydrocarbon migration as well as from structural activity and evolution, while the reservoir adjustment follows the trap adjustment (Sun et al. 2008, 2009; Jiang et al. 2008; Xu et al. 2008; Yang et al. 2012). According to this theory, the Sarvak reservoir of the Azadegan oil field belongs to the "charging type unsteady reservoir," and it can be divided into the retention reservoir which is a "paleo-reservoir" (the north high) and the "pre-secondary reservoir" which is still in the process of migration and accumulation in the secondary trap (the south high) (Yang et al. 2012). It has been noted that the Sarvak reservoir is still in the early readjustment phase, and the paleo-reservoir did not cause the integral re-migration. Therefore, the current northern high (the location of the paleo-reservoir) should be in its original state and not controlled by the current structure. Based on this, five wells were drilled in the west flank of the northern high in the south Azadegan field (Fig. 15). Results indicate all of these wells show a stable horizontal OWC and thicker oil columns relative to the structure position. Thereafter, the H well located in the east flank (paleo-low) proves that the tilted OWC also exists in the E-W trend along the minor axis. From the above, the northern high is the most favorable development zone due to both the good reservoir properties of the paleo-high and the paleo-reservoir situation (Fig. 16). Additionally, contrary to the conventional theory, the wells should not be drilled along the crest of structure but ought to be in the paleo-reservoir trap zone in the west flank of the current northern high (Fig. 16). Meanwhile, oil pay thickness which could be impacted by the tilted OWC should be considered as well in reserve calculations.

The southern high is the secondary trap formed by the Zagros activity in the very late stage, while the secondary reservoir is still in the process of re-migration. Regarding the southern reservoir, the pinching out of the reservoir thickness of the Sar-3 and the weaker reservoir quality of the Sar-4, 5, 6 is caused by the lower paleo-geography. Moreover, the decreasing oil pay thickness is caused by the higher OWC. Thus, the southern secondary reservoir is not the priority development zone. On the contrary, the authors have shown that the Kazhdumi (Burgan sandstone) and Gadwan (Zubair sandstone) reservoirs have formed a completed secondary reservoir in the new southern trap due to the good reservoir connectivity, reservoir quality, and low oil viscosity (Du et al. 2015c). Based on that, we can develop the southern lower sandstone reservoir first, while collecting more geological data of the upper Sarvak and then optimizing the development by using the static and dynamic data such as the 3D seismic and production dynamic analysis of the existing wells.

The lower Cretaceous Fahliyan reservoir is a matrix pore-type carbonate reservoir of which lithology is oolitic limestone and the average porosity/permeability is 19.6 %/ 12.7×10^{-3} μm^2, similar to the Sarvak. Based on the good tectonic inheritance of the Cretaceous formation and the same trap evolution trend in Azadegan field, it is assumed that the Fahliyan reservoir should possess the same unsteady characteristics. The newly drilled wells proved the prediction and showed a tilted OWC with around 120 m height difference north to south in the reservoir. Hence, it is an unsteady lithological-structural reservoir (Fig. 17). Therefore, the development method should follow the Sarvak example that confirms the favorable development zone by analyzing the reservoir distribution and paleo-reservoir location.

6 Conclusions

The genesis of the Sarvak unsteady reservoir in Azadegan oil field can be generalized as follows: the paleo-trap of the Sarvak formed at an early geological stage (after the upper Cretaceous) and the paleo-reservoir formed at a later geological stage (after the middle Miocene). The secondary reservoir adjustment caused by tectonic activity occurred at an ultra-late stage (after the Pliocene). The characteristics of the Sarvak reservoir are those of a massive carbonate reservoir with strong heterogeneity both vertically and laterally, and reservoir quality degrades along the secondary migration path, with a very short readjustment time span and high-viscosity crude oil. These factors together result in reservoir readjustment at a very slow rate and insufficient secondary readjustment. This has ultimately formed the unsteady reservoir with the irregular tilted OWC.

The "unsteady reservoir" is a new concept proposed by Chinese scholars in recent years. The relevant theory concerning this kind of reservoir has been successfully trialed in some Chinese oil fields, proving its objectivity and validity. It is believed that in the transition zone between the Arabian Platform and Zagros foreland basin, there should have been a greater number of oil fields similar to Azadegan oil field. Thorough analysis and identification of this kind of reservoir is highly recommended. It is also believed, based on this research, that the unsteady reservoir theory with relevant development methods has much scope for improvement and broad prospects of application in the Middle East.

References

Alsharhan AS, Nairn AEM. Sedimentary basins and petroleum geology of the middle east. Amsterdam: Elsevier; 1997. p. 813–43.

Alsharhan AS. Facies variations, diagenesis and exploration potential of the Cretaceous rudist-bearing carbonates of the Arabian Gulf. AAPG Bull. 1995;79(4):531–50.

Aqrawi AAM, Thehni GA, Sherwanni GH, et al. Mid-Cretaceous rudist-bearing carbonates of the Mishrif Formation: an important reservoir sequence in the Mesopotamian Basin, Iraq. J Pet Geol. 1998;21(1):57–82.

Alavi M. Chronology of trap formation and migration of hydrocarbons in Zagros sector of South West Iran. AAPG Bull. 1982;66(10):1535–41.

Alavi M. Tectonics of the Zagros orogenic belt of Iran, new data and interpretations. Tectonophysics. 1994;229(3–4):211–38.

Alavi M. Regional stratigraphy of the Zagros fold-thrust belt of Iran and its proforeland evolution. Am J Sci. 2004;304(1):1–20.

Berberian M, King GCP. Towards a paleogeography and tectonic evolution of Iran. Can J Earth Sci. 1981;18(2):210–65.

Beydoun ZR. Arabian plate hydrocarbon geology and potential—a plate tectonic approach, AAPG Stud Geol. 1991. p. 33–77.

Bordenave ML, Burwood R. Source rock distribution and maturation in the Zagros belt: provenance of the Asmari and Bangestan reservoir oil accumulations. Org Geochem. 1990;16(1):369–87.

Bordenave ML, Burwood R. The Albian Kazhdumi Formation of the Dezful Embayment, Iran: one of the most efficient petroleum generating systems. Petroleum Source Rocks. Heidelberg: Springer; 1995. p. 183–207.

Bordenave ML, Huc AY. The Cretaceous source rocks in the Zagros Foothills of Iran. Oil Gas Sci Technol. 1995;50(6):727–52.

Bordenave ML, Hegre JA. The influence of tectonics on the entrapment of oil in the Dezful embayment, Zagros fold belt. Iran. Pet Geol. 2005;28(4):339–68.

Colman SP. Fold development in Zagros simply folded belt, Southwest Iran. AAPG Bull. 1978;62(6):984–1003.

Du Y, Yi YJ, Xin J, et al. Genesis of large-amplitude tilting oil-water contact in Sarvak Formation in South Azadegan Oilfield, Iran. Pet Geol Exp. 2015a;37(2):187–93 (in Chinese).

Du Y, Xin J, Xu QC, et al. The rudist buildup depositional model based on reservoir architecture: a case from the Sarvak reservoir of the SA oilfield, Iran. Acta Sedimentol Sin. 2015b;33(6):1247–57 (in Chinese).

Du Y, Xin J, Chen J, et al. Review on the characteristic of Kazhdumi reservoir in SA Oilfield, Iran. J Southwest Pet Univ (Sci Technol Ed). 2015c;37(6):30–8 (in Chinese).

Gussow WC. Differential entrapment of oil and gas: a fundamental principle. AAPG Bull. 1954;38(5):816–53.

Ghabeishavi A, Vaziri MH, Taheri A, et al. Microfacies and depositional environment of the Cenomanian of the Bangestan anticline, SW Iran. Asian Earth Sci. 2010;37(3):275–85.

Glennie KW. Cretaceous tectonic evolution of Arabia's eastern plate margin: a tale of two oceans. Middle east models of Jurassic/Cretaceous Carbonate systems. SEPM Spec Publ. 2000;69:9–20.

Hessami K, Koyi HA, Talnot CJ, et al. Progressive unconformities within an evolving foreland fold-thrust belt, Zagros Mountains. J Geol Soc Lond. 2001;158(6):969–81.

Jiang TW, Xu HL, Lian ZG, et al. Origin of tilted oil-water contact and probe into the theory of unsteady hydrocarbon accumulation. J Southwest Pet Univ (Sci Technol Ed). 2008;30(5):1–6 (in Chinese).

Koop WJ, Kholief MM. Subsidence history of the Middle East Zagros Basin. Philos Trans R Soc Lond Ser A. 1982;305:149–67.

Li CL. Theoretical analysis of dipping water-oil contacts (II). Xinjiang Pet Geol. 2010;30(5):653–4 (in Chinese).

Li MC. Basic principles of migration and hydrocarbon exploration. Earth Sci—J China Univ Geosci. 2004;29(4):379–83 (in Chinese).

Liu H, Guo R, Dong JC, et al. Productivity evaluation and influential factor analysis for Sarvak reservoir in South Azadegan oil field, Iran. Pet Explor Dev. 2013a;40(5):585–90.

Liu Z, Tian CB, Zhang WM, et al. Causes of hydrodynamic pressure distribution: a case of the 4th Pay in Zubair Formation, Rumaila Oilfield, Iraq. Pet Explor Dev. 2013b;40(6):774–9.

McQuillan H. Small-scale fracture density in Asmari Formation of southwest Iran and its relation to bed thickness and structural setting. AAPG Bull. 1973;57(12):2367–85.

McQuillan H. Fracture patterns on Kuh-e Asmari anticline, southwest Iran. AAPG Bull. 1974;58(2):236–46.

Murris RJ. Middle East: stratigraphic evolution and oil habitat. AAPG Bull. 1980;64:587–618.

Sadooni FN, Aqrawi AAM. Cretaceous sequence stratigraphy and petroleum potential of the Mesopotamian basin Iraq. SEPM Spec Publ. 2000. p. 315–34.

Sadooni FN. The nature and origin of Upper Cretaceous basin margin rudist buildups of the Mesopotamian Basin, southern Iraq, with consideration of possible hydrocarbon stratigraphic entrapment. Cretac Res. 2005;26(2):213–24.

Sepehr M, Cosgrove JW. Structural framework of the Zagros fold-thrust belt, Iran. Mar Pet Geol. 2004;21(7):829–43.

Sherkati S, Letouzeh J. Variation of structural style and basin evolution in the Central Zagros (Izeh zone and Dezful Embayment) Iran. Mar Pet Geol. 2004;21(5):535–54.

Schowalter TT. Mechanics of secondary hydrocarbon migration and entrapment. AAPG Bull. 1979;63(5):723–60.

Stöcklin J. Structural history and tectonics of Iran. AAPG Bull. 1968;52(7):1229–58.

Soleimani M. Simulation of petroleum exploration based on a conceptual decision model: taking the Dezful Embayment in southwestern Iran as an example. Pet Explor Dev. 2013;40(4):476–80.

Sun LD, Jiang TW, Xu HL, et al. Exploration and practice for theory of unsteady-state hydrocarbon accumulation. Mar Origin Pet Geol. 2008;13(3):11–6 **(in Chinese)**.

Sun LD, Jiang TW, Xu HL, et al. Unsteady reservoir in Hadson Oilfield, Tarim Basin. Pet Explor Dev. 2009;36(1):62–7 **(in Chinese)**.

Xu DJ, Zhang WC, Du XJ, et al. Reservoir characteristics and development suggestion of chalky limestone in the Zagros Basin, Iran. Pet Geol Exp. 2010;32(1):15–8 **(in Chinese)**.

Xu HL, Jiang TW, Gu QY, et al. Probe into hydrocarbon accumulations in the Hadson oilfield, Tarim Basin. J Southwest Pet Univ (Sci Technol Ed). 2008;30(5):17–21 **(in Chinese)**.

Xu GQ, Zhang SN, Li ZD, et al. Carbonate sequence stratigraphy of a back-arc basin: a case study of the Qom Formation in the Kashan Area, Central Iran. Acta Geol Sin. 2007;81(3):488–500.

Wang Q, Cheng XB, Zhang WB. Approach to hydrocarbon accumulation in piedmont of Zagros foreland basin, Iran. Xinjiang Pet Geol. 2011;32(2):204–6 **(in Chinese)**.

Yang HJ, Sun LD, Zhu GY, et al. Characters and formation mechanism of unsteady reservoirs in Tarim Basin. Acta Pet Sin. 2012;33(6):1103–11 **(in Chinese)**.

Zhang Z, Li HW, Duan HZ, et al. Geological characteristics and hydrocarbon accumulation model of the Cenozoic Asmari Gachsran play, Zagros Basin. Oil Gas Geol. 2012;33(2):190–7 **(in Chinese)**.

PERMISSIONS

All chapters in this book were first published in PS, by Springer; hereby published with permission under the Creative Commons Attribution License or equivalent. Every chapter published in this book has been scrutinized by our experts. Their significance has been extensively debated. The topics covered herein carry significant findings which will fuel the growth of the discipline. They may even be implemented as practical applications or may be referred to as a beginning point for another development.

The contributors of this book come from diverse backgrounds, making this book a truly international effort. This book will bring forth new frontiers with its revolutionizing research information and detailed analysis of the nascent developments around the world.

We would like to thank all the contributing authors for lending their expertise to make the book truly unique. They have played a crucial role in the development of this book. Without their invaluable contributions this book wouldn't have been possible. They have made vital efforts to compile up to date information on the varied aspects of this subject to make this book a valuable addition to the collection of many professionals and students.

This book was conceptualized with the vision of imparting up-to-date information and advanced data in this field. To ensure the same, a matchless editorial board was set up. Every individual on the board went through rigorous rounds of assessment to prove their worth. After which they invested a large part of their time researching and compiling the most relevant data for our readers.

The editorial board has been involved in producing this book since its inception. They have spent rigorous hours researching and exploring the diverse topics which have resulted in the successful publishing of this book. They have passed on their knowledge of decades through this book. To expedite this challenging task, the publisher supported the team at every step. A small team of assistant editors was also appointed to further simplify the editing procedure and attain best results for the readers.

Apart from the editorial board, the designing team has also invested a significant amount of their time in understanding the subject and creating the most relevant covers. They scrutinized every image to scout for the most suitable representation of the subject and create an appropriate cover for the book.

The publishing team has been an ardent support to the editorial, designing and production team. Their endless efforts to recruit the best for this project, has resulted in the accomplishment of this book. They are a veteran in the field of academics and their pool of knowledge is as vast as their experience in printing. Their expertise and guidance has proved useful at every step. Their uncompromising quality standards have made this book an exceptional effort. Their encouragement from time to time has been an inspiration for everyone.

The publisher and the editorial board hope that this book will prove to be a valuable piece of knowledge for researchers, students, practitioners and scholars across the globe.

LIST OF CONTRIBUTORS

N. Farzin Nejad and A. A. Miran Beigi
Petroleum Refining Technology Development Division, Research Institute of Petroleum Industry, 14857-33111 Tehran, Iran

Hamed Firoozinia, Kazem Fouladi Hossein Abad and Akbar Varamesh
Research Institute of Petroleum Industry, Tehran 1485733111, Iran

Mohammad O. Eshkalak, Umut Aybar and Kamy Sepehrnoori
The University of Texas at Austin – PGE, 200 E Dean Keeton, Austin, TX 78712, USA

Wei Wang, Jun Yao, Hai Sun and Wen-Hui Song
School of Petroleum Engineering, China University of Petroleum, Qingdao 266580, Shandong, China

Xin-Shun Zhang, Hong-Jun Wang, Feng Ma, Yan Zhang and Zhi-Hui Song
Research Institute of Petroleum Exploration & Development, Petro China, Beijing 100083, China

Xiang-Can Sun
Oil & Gas Survey, China Geological Survey, Beijing 100029, China

Yan-Yu Zhang and Xue-Wei Duan
College of Petroleum Engineering, China University of Petroleum, Qingdao 266580, Shandong, China

Xiao-Fei Sun
College of Petroleum Engineering, China University of Petroleum, Qingdao 266580, Shandong, China
School of Geosciences, China University of Petroleum, Qingdao 266580, Shandong, China
The Department of Chemical and Petroleum Engineering, University of Calgary, Calgary, AB T2N 1N4, Canada

Xing-Min Li
Research Institute of Petroleum Exploration and Development, Petro China, Beijing 100083, China

Ji-Rui Hou, Ze-Yu Zheng, Zhao-Jie Song, Min Luo, Hai-Bo Li, Li Zhang and Deng-Yu Yuan
Research Institute of Enhanced Oil Recovery, China University of Petroleum, Beijing 102249, China
Key Laboratory of Marine Facies, Sinopec, Beijing 102249, China
Key Laboratory of Petroleum Engineering, Ministry of Education, Beijing 102249, China

Hong-Kui Ge, Liu Yang, Ying-Hao Shen, Kai Ren, Fan-Bao Meng, Wen-Ming Ji and Shan Wu
State Key Laboratory of Petroleum Resources and Prospecting, China University of Petroleum, Beijing 102249, China

Farhad Salimi and Mozafar Abdollahifar
Department of Chemical Engineering, Faculty of Basic Sciences, Kermanshah Branch, Islamic Azad University, Kermanshah, Iran

Javad Salimi
Young Research and Elite Club, Kermanshah Branch, Islamic Azad University, Kermanshah, Iran

Ying-Hui Bian, Shao-Tang Xu, Le-Chun Song, Yu-Lu Zhou, Li-Jun Zhu, Yu-Zhi Xiang and Dao-Hong Xia
State Key Laboratory of Heavy Oil Processing, College of Chemical Engineering, China University of Petroleum (East China), Qingdao 266580, China

Yan-Bao Guo, Cheng Liu, De-Guo Wang and Shu-Hai Liu
College of Mechanical and Transportation Engineering, China University of Petroleum, Beijing 102249, China

Ajay Mandal
Department of Petroleum Engineering, Indian School of Mines, Dhanbad 826004, India

Achinta Bera
Department of Civil & Environmental Engineering, School of Mining and Petroleum Engineering, University of Alberta, Edmonton, AB T6G 2W2, Canada

Saeed Mojeddifar, Gholamreza Kamali and Hojjatolah Ranjbar
Department of Mining Engineering, Shahid Bahonar University of Kerman, Kerman, Iran

Shao-Gui Deng, Li Li and Yi-Ren Fan
School of Geosciences, China University of Petroleum, Qingdao 266580, Shandong, China

Zhi-Qiang Li
China Research Institute of Radiowave Propagation, Xinxiang 453003, Henan, China

Xu-Quan He
China Petroleum Southwest Oil and Gas Field Branch, Chengdu 610051, Sichuan, China

Yan-Zhong Wang, Ying-Chang Cao, Shao-Min Zhang, Fu-Lai Li and Fan-Chao Meng
School of Geosciences, China University of Petroleum, Qingdao 266580, Shandong, China

Saeed Majidaie
Leap Energy- Subsurface Consulting Services, Kuala Lumpur, Malaysia

Mustafa Onur
Department of Petroleum and Natural Gas Engineering, Istanbul Technical University, Istanbul, Turkey

Isa M. Tan
Applied Science Department, Universiti Teknologi Petronas, Tronoh, Perak, Malaysia

Ke-Lai Xi
School of Geosciences, China University of Petroleum, Qingdao, Shandong 266580, China
Department of Geosciences, University of Oslo, Oslo 0316, Norway

Ying-Chang Cao, Yan-Zhong Wang, Qing-Qing Zhang, Jie-Hua Jin, Shao-Min Zhang, Jian Wang, Tian Yang and Liang-Hui Du
School of Geosciences, China University of Petroleum, Qingdao, Shandong 266580, China

Ru-Kai Zhu
Research Institute of Petroleum Exploration and Development, CNPC, Beijing 100083, China

Yang Du
School of Geoscience and Technology, Southwest Petroleum University, Chengdu 610500, Sichuan, China
Geology and Exploration Research Institute, CNPC Chuanqing Drilling Engineering Company Limited, Chengdu 610051, Sichuan, China

Jie Chen, Jun Xin, Juan Wang, Yi-Zhen Li and Xiao Fu
Geology and Exploration Research Institute, CNPC Chuanqing Drilling Engineering Company Limited, Chengdu 610051, Sichuan, China

Yi Cui
Iraq Branch Company of CNODC, CNPC, Dubai 500486, United Arab Emirates

Index